About Island Press

Island Press is the only nonprofit organization in the United States whose principal purpose is the publication of books on environmental issues and natural resource management. We provide solutions-oriented information to professionals, public officials, business and community leaders, and concerned citizens who are shaping responses to environmental problems.

In 2006, Island Press celebrates its twenty-second anniversary as the leading provider of timely and practical books that take a multidisciplinary approach to critical environmental concerns. Our growing list of titles reflects our commitment to bringing the best of an expanding body of literature to the environmental community throughout North America and the world.

Support for Island Press is provided by the Agua Fund, The Geraldine R. Dodge Foundation, Doris Duke Charitable Foundation, The William and Flora Hewlett Foundation, Kendeda Sustainability Fund of the Tides Foundation, Forrest C. Lattner Foundation, The Henry Luce Foundation, The John D. and Catherine T. MacArthur Foundation, The Marisla Foundation, The Andrew W. Mellon Foundation, Gordon and Betty Moore Foundation, The Curtis and Edith Munson Foundation, Oak Foundation, The Overbrook Foundation, The David and Lucile Packard Foundation, The Winslow Foundation, and other generous donors.

Ecological Consequences *of* Artificial Night Lighting

Ecological Consequences *of* Artificial Night Lighting

Edited by

Catherine Rich • Travis Longcore

ISLANDPRESS / Washington • Covelo • London

Library of Congress Cataloging-in-Publication Data

Ecological consequences of artificial night lighting / edited by Catherine
Rich and Travis Longcore
p. cm.
Includes bibliographical references.
ISBN 1-55963-128-7 (cloth : alk. paper) — ISBN 1-55963-129-5 (pbk. : alk. paper)
1. Exterior lighting—Environmental aspects. I. Rich, Catherine.
II. Longcore, Travis.
QH545.E98E26 2005
577.27'2—dc22

2005020202

British Cataloguing-in-Publication Data available.

Book design: Brighid Willson

Printed on recycled, acid-free paper

Manufactured in the United States of America
10 9 8 7 6 5 4 3 2

For our parents,
and in memory of Cherryl Wilson

Night is certainly more novel and less profane than day.

—Henry David Thoreau (1817–1862),
"Night and Moonlight," 1863

Contents

PART V. INVERTEBRATES

PART VI. PLANTS

PART V. INVERTEBRATES

PART VI. PLANTS

Preface

For people who care about the natural world, the unrelentingly destructive activities of humanity are all too familiar. Most of these are painfully obvious to anyone watching the transformation of long-appreciated landscapes. We know death by bulldozer when we see it.

Seeing may be the key. We are for the most part diurnal, and visual, and as we consider the transformation of the planet, mostly we envision changes that we are able to see by day. It is an inherent bias in our collective thinking. But as we light up the night, with ever more powerful lights, is it not obvious, giving the matter some thought, that we are wreaking havoc on creatures with physiologies far more delicate than ours, interfering with the lifeways of suites of organisms that have evolved over the millennia with a dependable pattern of light and dark? All this happens at times when most of us are sleeping. As the night is ever more brightly lit, at least we can close the shutters. But what of the animals and plants?

I imagine myself in the position of other creatures, unable to control or escape their environments, which we have controlled for them—whales subject to sonar far beyond any sound their ears were meant to accommodate, birds migrating thousands of miles, thrown off course by lights with no limits. We are ultimately deprived of their beauty as they die off by the millions, from our careless expansion into all realms of the Earth.

My own love for night started simply as an aesthetic appreciation, and over the last twenty-five years or so I have been every year more profoundly saddened by the glare foisted on us by brighter and brighter lights. Long ago, as an undergraduate student, I loved to walk the neighborhoods of Berkeley on windy autumn nights, catching glimpses of softly lit walls of books in beautiful libraries in beautiful old homes, on winding streets lined with old, shady trees. The wind, the trees, the books, the soft lights—they were solace and fodder for dreams of my own future. I cannot say that I considered back then that the streets were probably dark enough, the houses lit softly enough, for there still to be myriad creatures living in lush yards and nearby unbuilt hillsides. I was then studying chemistry and physics, later psychology and anthropology, and never thought about studying ecology, despite being deeply committed to protecting "the environment." My lifelong love of trees and wild creatures—birds, caterpillars, lobsters, lizards—had not been nurtured into a career. (Even by then, the kind of success envisioned for the children of Beverly Hills did not require nature study. Indeed, it was not even an option.)

I went straight to law school after college. I had not yet found a discipline where I felt to be at home. After ten years of doing everything I could to avoid using my law degree except when it involved environmental protection, I reentered the academic world, looking to reclaim the part of myself that loved nature and to supplement that love with deeper knowledge. When a consortium of Russian business interests sought Western corporate support to perpetually light up the tundra with a giant space mirror, I had a visceral reaction, shared by millions I'm sure, to the prospect of farming and industrializing the Arctic—where would so many of the world's birds gather to breed? As a student of biogeography I also wondered what such an assault would mean to the ecology of the tundra itself. My old, deep appreciation for night, and the feelings engendered by catching glimpses of other people's lives in softly lit homes filled with books, had become an ecologically informed awareness of the importance of night to nonhuman creatures, which allowed for a deeper concern about their welfare as night is transformed into something akin to day.

Some people need science to be convinced. Others are moved by something different, something better elicited by words, art, or music. Only 150 years ago, Henry David Thoreau walked his woods and fields at ten o'clock in the evening—about the time for the early news—and discovered a world apart from that which he knew so well by day. The wholesale transformation of the Earth in this almost unfathomably short time is perhaps nowhere more evident than in considering night landscapes. Thank goodness for the writers who know nature, who can help us see, who can help us remember. The vignettes interspersed in this book are meant to remind the reader that what we take for granted as night is nothing like what a natural night should be. The descriptions share a common thread, an unspoken recognition that night is a place all its own.

This is a book of science. But if it also infuses a little bit of reverence back into the tasks associated with planning for the nighttime environment, then we will have done our job.

—CATHERINE RICH

Ten years ago, Catherine started asking a seemingly simple question—"What happens to animals and plants when subjected to artificial night lighting?" We were both students of geography at UCLA; she was finishing her M.A. and thinking about a dissertation topic. As logical a question as it seemed, information was not widely available, save for the well-known examples of sea turtles and birds.

We started collecting references, assembling bits of information from reports and articles, often incidental observations in writings on often unrelated topics. Other people were investigating questions about the effects of artificial light on plants and animals, but only a few had started to integrate findings across taxa.

Although the science supported the intuitive observation that lighting ecosystems would affect species, policymakers for the most part never considered the effects of artificial light on nature. In 1999, Catherine received a call from a distressed Fish and Game biologist concerned about a decorative lighting project proposed for the Vincent Thomas Bridge at the Port of Los Angeles. The proposal included high-powered spotlights aimed directly into the sky, and project proponents had given little or no thought to the environmental consequences. Through the nonprofit that we founded together, The Urban Wildlands Group, Catherine organized a group of scientists, both biologists and astronomers, to testify before the California Coastal Commission (which had asserted jurisdiction over the proposal after we brought it to their attention) about the potential adverse effects of lights on migratory birds and other wildlife. The commissioners were receptive to this information and denied the proposal. (Later a much more environmentally sensitive design was approved with our support.) It was evident that absent the sort of concerted effort to oppose this project, decisionmakers such as the appointed members of the California Coastal Commission and their staff lacked the scientific information necessary to evaluate the effects of artificial night lighting on ecosystems in their review of projects. Catherine and I have found this lack of information time and again as we consult to those trying to protect natural places—environmental assessment documents either do not consider the effects of artificial night lighting on biological resources or do so poorly.

After working with us in opposition to the original Vincent Thomas Bridge proposal, Bob Gent of the International Dark-Sky Association asked that we "write a paper" reviewing the effects of artificial lighting on nature, to help in his advocacy and outreach efforts and to counteract the resistance he met when raising the topic. A few such articles had already been published, and we did not think we could improve on them without the input of scientists with special expertise in this topic. So we decided instead to convene an international conference. With the support of the UCLA Institute of the Environment, its then-director Rich Turco, and several key funders, the Ecological Consequences of Artificial Night Lighting conference was held in February 2002 on the UCLA campus.

The conference was an effort made possible, or at least made easier, by the Internet. With our collection of scientific articles as a starting point, extensive Web searches allowed us to find scientists who were investigating aspects of this topic. Their response when contacted about a possible meeting was enthusiastic; all were eager to meet others with similar interests.

This book contains chapters by many of the presenters at the conference. Some chapters are written by experts who were not presenters.

If we were building a house instead of editing a book, Catherine would be the architect and finish carpenter, and I would be the general contractor. Like all teams, we have specialized to capitalize on our different strengths. She is the visionary, and I am thankful to have had the opportunity to be in on the project.

We owe a debt of gratitude to those who have made this book possible and who inspired and encouraged its development and publication. The authors of the chapters deserve special recognition for their unique contributions, for accommodating our editorial requests, and for their patience. We appreciate and acknowledge the reviewers of each of the chapters, whose comments and insights improved the book as a whole. Our heartfelt thanks go to Bernd Heinrich, Carl Safina, and Phil DeVries for embracing the topic and generously agreeing to write about their experiences of the night.

We confirmed nearly all of the citations in the book to the original source, a task that would have been insurmountable without UCLA's amazing collections and librarians. We are especially grateful to the interlibrary loan staff for processing scores of requests for obscure books, unpublished reports, and journal articles in several languages. We are indebted to the UCLA Department of Geography for providing uninterrupted access to library and academic Internet resources. Our research assistant, Sarah Casia, tracked down and copied references quickly and accurately. Attorney and friend Jonathan Kirsch graciously provided advice about publishing a book.

At Island Press we found editors, Barbara Dean and Laura Carrithers, who shared our enthusiasm for the subject and have helped us to shape a book to best convey this information.

We deeply appreciate the efforts of those who are working, or have worked, in the trenches on this issue: Virginia Brubeck and the many other agency staff members who have fought to protect species and habitats from the adverse effects of artificial night lighting, sometimes to their own detriment; Bob Gent, Dave Crawford, Jack Sales, and others involved with the International Dark-Sky Association who have eagerly incorporated these issues into their important work to protect the night sky; Michael Mesure and the Fatal Light Awareness Program for having the fortitude to document the incredible losses of migratory birds from collisions with buildings, both day and night; and countless others working to protect species and habitats from the deleterious effects of artificial night lighting.

The Conservation and Research Foundation provided a grant for the preparation of the book, a vote of confidence that we appreciate greatly. The Biological Resources Division of the U.S. Geological Survey provided a generous grant supporting publication.

A few individuals deserve special recognition. Larry and Sara Wan have supported us across many realms, for years. They are a continuing source of inspiration. The late Cherryl Wilson—our neighbor and Catherine's dear friend—for nearly twenty years recognized and nurtured Catherine's gift for protecting what is good in the world. Cherryl's interest in protecting the beauty of the night, shared and enhanced by her life partner George Eslinger, seamlessly blended with our interest in saving the night for nature. We thank George, too, for years ago offering a research site to Catherine for a dissertation that was not to be, for continuing to provide technical assistance by phone at any hour, and for his friendship.

Our parents have provided support in many ways, from generous financial contributions, to helping edit chapters while visiting us from

Maine, to caring about each deadline and encouraging our progress. My parents are both research biologists; Catherine's mother is a dancer and choreographer and her father is a television director. As we look over the manuscript, we see their influence.

—Travis Longcore

Chapter 1

Introduction

Catherine Rich and Travis Longcore

What if we woke up one morning only to realize that all of the conservation planning of the last thirty years told only half the story—the daytime story? Our diurnal bias has allowed us to ignore the obvious, that the world is different at night and that natural patterns of darkness are as important as the light of day to the functioning of ecosystems.

There have always been naturalists with a preference for night, those who study bats and badgers, moths and owls, who awaken when the sun goes down (e.g., Ferris 1986, Ryden 1989). But as a whole, professional conservationists have yet to recognize the implications of the dramatic transformation of the nighttime environment by ever-increasing artificial lights, except for the few well-known situations that leave dead bodies on the ground.

Lighted towers and tall buildings so confuse migrating birds that they circle and die of exhaustion or of collisions with each other or the structures themselves. Sea turtle hatchlings attracted to coastal streetlights end up desiccated, crushed under foot and wheel, or killed by predators. Yet beyond these high-profile examples, the magnitude of the ecological consequences of artificial night lighting is only beginning to be known. But all indications are that unless we consider protection of the night, our best-laid conservation plans will be inadequate.

This book provides a scientific basis to begin addressing the challenge of conserving the nighttime environment, but it remains critically necessary to expand basic research into the effects of altered light regimes on species and ecosystems.

The chapters here are meant to complement ongoing efforts to reduce, for other reasons, unnecessary and wasteful lighting. Loss of the view of the night sky across the developed world saddens poets and frustrates backyard astronomers (Riegel 1973). Excessive and improperly shielded lighting burdens society with the economic and environmental costs of wasted energy. These important issues are not addressed in detail here; rather, this book concentrates on the effects of artificial night lighting on nonhuman species and ecosystems.

A History of Artificial Light Ecology

Humans have long manipulated nighttime lighting levels, often with the intention of affecting wildlife behavior. Stoking the campfire at night has kept predators at bay since prehistoric times. As with many destructive human activities, the awareness that nighttime illumination might harm the natural world has developed relatively recently as technological innovations have facilitated a nearly unlimited ability to light the night. For birds, concern about needless deaths at lighthouses and other lights was expressed in the late 1800s and increased through the early 1900s (see Chapter 4, this volume). For other taxa, only the recent rapid urbanization of the developed world has resulted in sufficient effects to stimulate investigation.

The attraction of many groups of animals to light has been well known and documented since Aristotle (*The History of Animals*). Verheijen produced a monograph in 1958 that reviewed the mechanisms by which animals were attracted to lights, drawing on an extensive, predominantly European and Japanese literature dating from the late 1800s and early 1900s. Verheijen's (1958) review documents the adverse effects of lights on wildlife, and in 1985 he proposed the term *photopollution* to mean "artificial light having adverse effects on wildlife" (Verheijen 1985:1). Also in the 1980s, Raymond (1984) raised concerns about the increasing problem of sea turtle disorientation from lights at beaches, which had been described earlier by McFarlane (1963; see Chapter 7, this volume). In 1988, Frank published a thorough review of the influence of artificial night lighting on moths. With the exception of Verheijen's (1985) article, studies of the effects of artificial night lighting remained focused on single taxa. An approach integrating findings across different taxonomic groups that might be called artificial light ecology did not emerge.

Synthesis of the kind likely envisioned by Verheijen (1985) began in the 1990s. Alan Outen of the Hertfordshire Biological Records Centre produced a white paper, "The Possible Ecological Implications of Artificial Night Lighting," in 1994, which he revised in 1997 and 1998 and published as a book chapter in 2002 (Outen 2002). The view that light pollution posed a broad problem for whole ecosystems remained largely in the gray literature, notably astronomer Arthur Upgren's (1996) review published in the Natural Resources Defense Council's magazine *The Amicus Journal*, and Wilson's (1998) report for *Environmental Building News*. Witherington (1997) reviewed the deleterious effects of photopollution on sea turtles and other nocturnal animals and suggested that animal behaviorists could make an important contribution to conservation biology by studying "biological photopollution." In Europe, public awareness of light pollution led to a series of reports and studies in the Netherlands in the late 1990s (Health Council of the Netherlands 2000, de Molenaar et al. 1997, 2000, 2003), several studies were completed and an academic conference was held in Germany (Scheibe 1999, Eisenbeis and Hassel 2000, Schmiedel 2001, Kolligs 2000), and a conference review was produced in France (Raevel and Lamiot 1998). In 2002 we convened the first North American conference on this topic, which provided the basis for this book.

Purpose and Scope

This book reviews the state of knowledge about the ecological consequences of artificial night lighting. The phrase "ecological consequences of artificial night lighting" communicates the essential elements that distinguish this field of inquiry from others. The term *ecological consequences* highlights that we are concerned with ecology. Because the term *light pollution* has come to be understood as referring to the degradation of human views of the night sky, we have largely avoided its use. We have found it helpful to distinguish between "astronomical light pollution," in which stars and other celestial bodies are washed out by light that is either directed or reflected upward, and "ecological light pollution," which disrupts ecosystems (Longcore and Rich 2004; Figure 1.1). The term *artificial night lighting* is meant to communicate our focus on light generated by human activity rather than on the effects of natural patterns of light and dark, although understanding natural conditions is central to describing disruptions.

Ecological light pollution includes direct glare, chronically increased

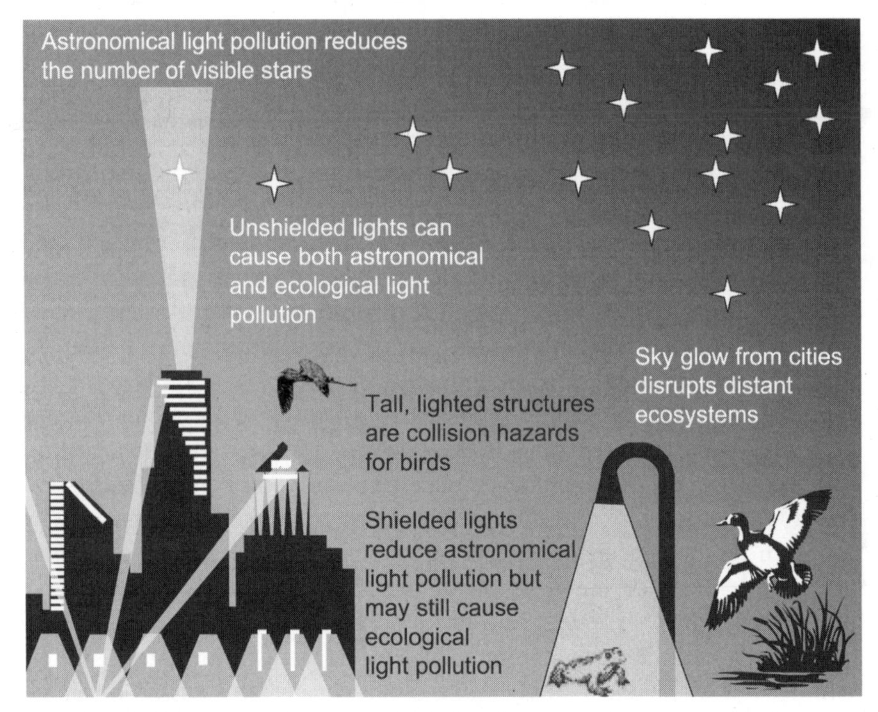

Figure 1.1. Diagram depicting ecological and astronomical light pollution. From Longcore and Rich (2004).

illumination, and temporary, unexpected fluctuations in lighting. Sources of ecological light pollution include sky glow, lighted structures (e.g., office buildings, communication towers, bridges), streetlights, security lights, lights on vehicles, fishing boats, flares on offshore hydrocarbon platforms, and even lights on undersea research vessels (see Kochevar 1998). The phenomenon therefore involves potential effects across a range of spatial and temporal scales.

The extent of ecological light pollution is global (Figure 1.2; Elvidge et al. 1997). The first atlas of "artificial night sky brightness" illustrates that astronomical light pollution extends to every inhabited continent (Cinzano et al. 2001). Cinzano et al. (2001) calculated that only 56% of Americans live where it becomes sufficiently dark at night for the human eye to make a complete transition from cone to rod vision and that fully 18.7% of the terrestrial surface of the Earth experiences night sky brightness that is polluted by astronomical standards. As discussed in the chapters that follow, species and ecosystems may be affected by sky glow from distant sources. Furthermore, even shielded lights that are pointed down-

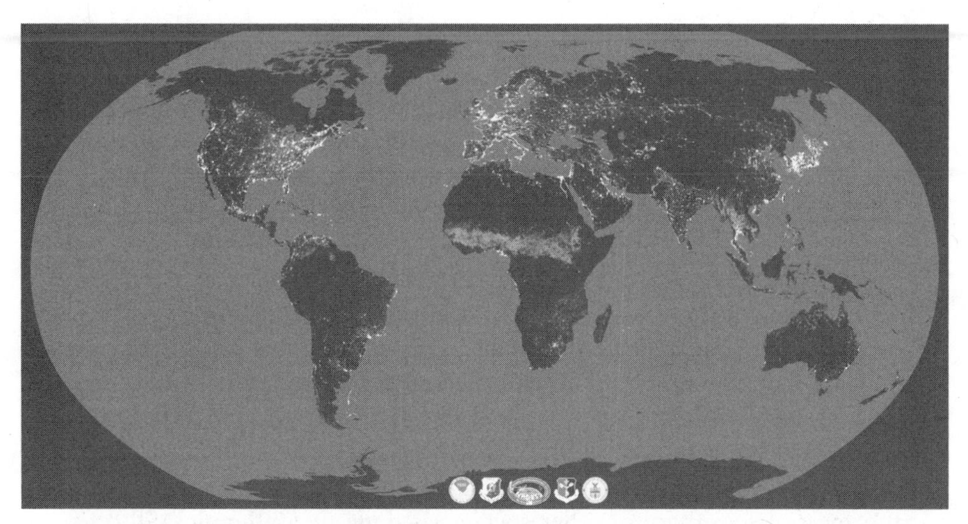

Figure 1.2. Distribution of artificial lights visible from space. Produced using cloud-free portions of low-light imaging data acquired by the U.S. Air Force Defense Meteorological Satellite Program (DMSP) Operational Linescan System (OLS). Four types of lights are included: human settlements (cities, towns, and villages), fires (defined as ephemeral lights on land), gas flares, and heavily lit fishing boats. See Elvidge et al. (2001) for details. Image, data processing, and descriptive text by the National Oceanic and Atmospheric Administration's National Geophysical Data Center.

ward, and thereby not contributing to sky glow, may have ecological consequences.

As is evident in Figure 1.2, excessive lighting is associated with the wealthy countries of the world, places where people can afford to consume energy to illuminate the environment all night. The developing world, although supporting much higher population densities, is shown to be much darker at night, with fires used as lights rather than electric fixtures. The near absence of outdoor electric lighting across heavily populated regions of Africa illustrates this point.

Even in the developing world, however, industrial resource extraction is associated with artificial lighting. Flares from oil wells are visible off the coast of Nigeria, in otherwise dark regions of North Africa, and across the sparsely populated regions of Siberia. Lights from fishing vessels virtually eliminate night in the Sea of Japan and are visible off portions of the coasts of Southeast Asia and South America. This wasted, ecologically disruptive light is itself the end product of extractive and consumptive processes that are themselves environmentally damaging.

Units and Measurement

Illumination, or illuminance, is the amount of light incident per unit area; it is not the only measurement relevant to ecological light pollution, but it is the most commonly used. Light varies in its intensity (the number of photons per unit area) and in its spectral content (expressed by wavelength). Ideally, ecologists should measure illumination in photons per square meter per second, with associated measurements of the wavelengths of light present. More often, illumination is measured in lux (or footcandles, the non-SI unit), which expresses the intensity of light incident on a surface weighted for the spectral sensitivity of the human eye. The lux measurement places more emphasis on wavelengths of light that the human eye detects best and less on wavelengths that humans do not perceive as well. It is possible to avoid this human bias and adjust lux for the spectral sensitivity of other species, as done by Gal et al. (1999) for mysid shrimp. But because most engineering and planning professionals use lux, we use it as the measure of illuminance in this book. Table 1.1 illustrates familiar situations and their associated illumination. A sudden change in illumination is disruptive for some species (Buchanan 1993; see Chapters 2 and 9, this volume), so the percentage change in illumination, rate of change in illumination, or similar measures may be relevant.

Ecologists may measure luminance of light sources that are visible to organisms. Luminance is measured as the intensity of light per unit area of the source (e.g., candela/m^2). How bright these sources appear to organisms depends on ambient conditions; in dark conditions a dim light appears very bright, whereas it would be practically invisible in daylight.

Table 1.1. Illumination from common sources.

Source	*Illumination (lux)*
Full sunlight	103,000
Partly sunny	50,000
Operating table	18,000
Cloudy day	1,000–10,000
Bright office	400–600
Most homes	100–300
Lighted parking lot	10
Full moon under clear conditions	0.1–0.3
Quarter moon	0.01–0.03
Clear starry sky	0.001
Overcast night sky	0.00003–0.0001

Organization of the Book

We have divided the book into six parts, each addressing the effects of artificial night lighting on a taxonomic group. These divisions—mammals, birds, reptiles and amphibians, fishes, invertebrates, and plants—follow the subdisciplinary boundaries of modern zoology and botany defined by evolutionary relationships. They also follow the divisions of life described by Aristotle and Linnaeus without the benefit of modern evolutionary thought. In this division, some parts have more chapters than others, which reflects the unequal attention received by different groups. The taxonomic coverage is entirely disproportionate to the number of species in each group and does not reflect their importance in ecosystems. Little information is available about the effects of artificial light on marine mammals, for example, except for accounts of increased foraging on salmon by seals under artificial lights (Yurk and Trites 2000). Much work remains to be done to investigate the effects of artificial night lighting across the diversity of species on Earth.

Each section begins with a vignette about nature at night, either written specially for this book or excerpted from another source. The vignettes serve several purposes. They offer anecdotal observations of the ecology of organisms at night. From Henry David Thoreau's moonlit walks to Bernd Heinrich's night in the Maine woods, they illustrate that things are indeed different in the dark and that naturalists and scientists have recorded these differences for a long time. Anecdotal natural history observations such as these are often the source of scientific hypotheses. The vignettes are also meant to be evocative. We hope that an appreciation for the nature of night will remind lay and scientific readers alike why this topic is important.

Part I, on mammals, opens with Alexander von Humboldt's account of the clamor of animals at night in the tropical rainforests of South America. He describes tumultuous activity during the full moon, especially by larger mammals.

Paul Beier's chapter on terrestrial mammals provides insight into this phenomenon, reviewing many examples of the influence of lighting levels on predation risk and activity in mammals. He discusses the potential disruption of circadian, circalunar, and circannual cycles by artificial lighting and identifies situations in which artificial night lighting would be particularly hazardous to mammals.

Jens Rydell reports on the interaction between bats and insects at streetlights in Chapter 3. Although bats exploit the aggregations of insects attracted to streetlights, Rydell reports evidence that such lights are not necessarily beneficial to all bats. As with studies of small terrestrial mammals, competition and predation risk emerge as important factors restructuring and potentially reducing diversity in animal communities affected by outdoor lighting.

Bernd Heinrich's account of nights outside growing up in Maine begins Part II on birds. He describes the transformation of the woods in the dark and the nocturnal flight song of the ovenbird. This species happens to be particularly vulnerable to death by collision with tall lighted structures during its nocturnal migration. Sidney A. Gauthreaux Jr. and Carroll G. Belser (Chapter 4) document this hazard to migratory birds through time, from lighthouses and lightships to today's proliferating communication towers. They present the mechanisms of bird attraction to lights at night and report original research on the behavior of migratory birds around tall towers with different lighting types.

In Chapter 5, William A. Montevecchi addresses the risks of artificial night lighting to seabirds, including the uniquely dangerous flares of hydrocarbon platforms that both attract and incinerate birds. He considers direct, indirect, and cumulative effects of attraction to artificial light and provides detailed recommendations to reduce these effects, especially emphasizing the important role of independent observers in gathering useful data and enforcing compliance of regulations to protect birds.

Johannes G. de Molenaar, Maria E. Sanders, and Dick A. Jonkers contributed Chapter 6, which considers the effects of artificial night lighting on the nest choice and success of meadow birds during their breeding season. Their experiment in the Netherlands investigating the effect of roadway lighting on breeding black-tailed godwits has a before–after–control–impact design that is best suited to the investigation of this type of question but too rarely implemented. As they report, the small but statistically significant effect of lighting on breeding behavior was sufficient basis for the Dutch government to change the lighting system to reduce roadway illumination after peak traffic hours.

Reptiles and amphibians are the subject of Part III. David Ehrenfeld sets the tone in his reprinted essay on night and place from his research on sea turtles in Costa Rica. The darkness he describes, which is fundamental to the female turtle's choice of nest site, has been eliminated by artificial light in many other places. Michael Salmon and his students and colleagues have long researched the effects of artificial lighting on sea turtles and their hatchlings. In Chapter 7, Salmon describes the lessons learned on Florida's beaches—the interference with female nest site location and hatchling seafinding—and elaborates on the various solutions to reduce these effects, ranging from partial measures such as moving nests, to local controls on artificial night lighting, to comprehensive regional plans to restore darkness at nesting beaches.

Gad Perry and Robert N. Fisher discuss the effects of night lights on all other reptile groups (Chapter 8). They document how lights allow diurnal reptiles to extend activity into the nighttime and how nocturnal species may exploit aggregations of prey at lights. They also describe activity patterns that vary with moonlight and avoidance of moonlight in some species.

In Chapter 9, Bryant W. Buchanan describes the observed and potential effects of artificial night lighting on anuran amphibians. He considers the effects of chronic or dynamic shifts in illumination on behavior, physiology, and development. His own research, including a study to determine whether headlamps worn by researchers affect frog behavior and observations of the interruption of breeding choruses by artificial night lighting, provides important evidence of these effects.

Sharon E. Wise and Bryant W. Buchanan present the effects of artificial lighting on salamanders in Chapter 10. They draw on the extensive literature on salamanders and light from laboratory and field studies to describe changes in behavior and physiology resulting from artificial lighting, with consideration of its duration and spectral content. They include the results of several of their own unpublished experiments, including observations of salamanders in the field delayed in their foraging activity by an experimental treatment of dim light (ingeniously provided by a string of holiday lights).

Part IV concerns fishes. As the opening vignette by Carl Safina describes, changes in ambient illumination affect fish behavior. He reports the lore of old fishermen that the big fish are active during the full moon but avoid the new moon. The single chapter in this section (Chapter 11), by Barbara Nightingale, Travis Longcore, and Charles A. Simenstad, addresses the effects of artificial lighting on the bony fishes, with salmon species as exemplars. Nightingale et al. discuss the mechanism of fish vision and the response to light as influenced by age, species, ambient conditions, and lighting type. They describe the observed and potential effects of increased illumination on foraging and schooling, predator–prey relations, migration, reproduction, and harvest. More information about fishes is found in Chapter 15, which is primarily about aquatic invertebrates. The effects of artificial light on sharks, rays, and other "lower" vertebrates await future investigation.

Part V contains four chapters on invertebrates. Ecologist Philip J. DeVries wrote the opening essay on night and light in the tropics. He reminds us that many butterfly caterpillars feed at night and that the number of insects at lights decreases over time as they are consumed by predators and as the surrounding environment is destroyed.

In Chapter 12, Gerhard Eisenbeis discusses the attraction of flying insects in many taxonomic groups to streetlights. He classifies the effects that lights may have on insect behavior and considers the potential reduction in insect diversity around lights. He draws on his work with German colleagues to document and explain the patterns of insect death at streetlights in rural Germany, estimates total insect mortality from streetlights in Germany, and recommends lighting types to decrease such mortality.

Chapter 13 continues on this theme, as Kenneth D. Frank reports in detail the effects of outdoor lights on moths and moth populations. He explains the influence of artificial light on individual moths and works through the apparent contradiction that many moth species survive in heavily lit areas.

The effects of stray light on fireflies have the potential to be very different from those on other taxa. These bioluminescent beetles, which are the subject of Chapter 14 by James E. Lloyd, cue their behavior on ambient light intensity. In this chapter, Lloyd reviews the mechanisms by which light may interfere with the intraspecific and interspecific visual communication of firefly species. He also identifies needed research on this subject, listing useful projects that could be conducted by students.

Chapter 15 documents light pollution on freshwater lakes and its recorded and potential effects on invertebrates and their vertebrate predators. Marianne V. Moore, Susan J. Kohler, and Melani S. Cheers developed instrumentation to record nighttime illumination levels at lakes across an urban-to-rural gradient in New England. With these illumination levels, and incorporating previous research on the response of invertebrates and fishes to light, they predict the biological effects of light pollution on animals in the water column.

An excerpt from Henry David Thoreau's essay "Night and Moonlight" begins Part VI, about plants. He describes how very differently he experiences the landscape and its plants and animals at night. Plants perceive light in the environment and respond to these cues physiologically. In Chapter 16, Winslow R. Briggs describes how plants detect light in the environment and some of the physiological responses of plants to light. Few published studies address the direct effects of artificial night lighting on plants, but Briggs reviews the mechanisms by which these effects would occur.

Our final chapter, Chapter 17, situates the examples of the book in the framework of ecology and identifies general mechanisms by which artificial night lighting influences species.

The range of ecological consequences of artificial night lighting is broad, including desynchronization of the mating flight of ants, disruption of the daily movement of zooplankton, altered nest site choice in breeding birds, interference with dispersal patterns of mammals, delay of the downstream migration of salmon, and disorientation and death of migratory birds. Many examples are found in the pages that follow, but much more remains to be learned. Only the imagination and creativity of current and future readers and researchers limits the questions that might be productively investigated.

Literature Cited

Buchanan, B. W. 1993. Effects of enhanced lighting on the behaviour of nocturnal frogs. *Animal Behaviour* 45:893–899.

Cinzano, P., F. Falchi, and C. D. Elvidge. 2001. The first world atlas of the artificial night sky brightness. *Monthly Notices of the Royal Astronomical Society* 328:689–707.

Eisenbeis, G., and F. Hassel. 2000. Zur Anziehung nachtaktiver Insekten durch Straßenlaternen – eine Studie kommunaler Beleuchtungseinrichtungen in der Agrarlandschaft Rheinhessens [Attraction of nocturnal insects to streetlights: a study of municipal lighting systems in a rural area of Rheinhessen]. *Natur und Landschaft* 75(4):145–156.

Elvidge, C. D., K. E. Baugh, E. A. Kihn, H. W. Kroehl, and E. R. Davis. 1997. Mapping city lights with nighttime data from the DMSP Operational Linescan System. *Photogrammetric Engineering & Remote Sensing* 63:727–734.

Elvidge, C. D., M. L. Imhoff, K. E. Baugh, V. R. Hobson, I. Nelson, J. Safran, J. B. Dietz, and B. T. Tuttle. 2001. Night-time lights of the world: 1994–1995. *ISPRS Journal of Photogrammetry & Remote Sensing* 56:81–99.

Ferris, C. 1986. *The darkness is light enough: the field journal of a night naturalist.* Ecco, New York.

Frank, K. D. 1988. Impact of outdoor lighting on moths: an assessment. *Journal of the Lepidopterists' Society* 42:63–93.

Gal, G., E. R. Loew, L. G. Rudstam, and A. M. Mohammadian. 1999. Light and diel vertical migration: spectral sensitivity and light avoidance by *Mysis relicta*. *Canadian Journal of Fisheries and Aquatic Sciences* 56:311–322.

Health Council of the Netherlands. 2000. *Impact of outdoor lighting on man and nature*. Publication No. 2000/25E. Health Council of the Netherlands, The Hague.

Kochevar, R. E. 1998. *Effects of artificial light on deep sea organisms: recommendations for ongoing use of artificial lights on deep sea submersibles*. Technical Report to the Monterey Bay National Marine Sanctuary Research Activity Panel, Monterey, California.

Kolligs, D. 2000. Ökologische Auswirkungen künstlicher Lichtquellen auf nachtaktive Insekten, insbesondere Schmetterlinge (Lepidoptera) [Ecological effects of artificial light sources on nocturnally active insects, in particular on moths (Lepidoptera)]. *Faunistisch-Ökologische Mitteilungen* Supplement 28:1–136.

Longcore, T., and C. Rich. 2004. Ecological light pollution. *Frontiers in Ecology and the Environment* 2:191–198.

McFarlane, R. W. 1963. Disorientation of loggerhead hatchlings by artificial road lighting. *Copeia* 1963:153.

Molenaar, J. G. de, R. J. H. G. Henkens, C. ter Braak, C. van Duyne, G. Hoefsloot, and D. A. Jonkers. 2003. *Road illumination and nature, IV. Effects of road lights on the spatial behaviour of mammals*. Alterra, Green World Research, Wageningen, The Netherlands.

Molenaar, J. G. de, D. A. Jonkers, and R. J. H. G. Henkens. 1997. *Wegverlichting en natuur. I. Een literatuurstudie naar de werking en effecten van licht en verlichting op de natuur* [Road illumination and nature. I. A literature review on the function and effects of light and lighting on nature]. DWW Ontsnipperingsreeks deel 34, Delft.

Molenaar, J. G. de, D. A. Jonkers, and M. E. Sanders. 2000. *Road illumination and nature. III. Local influence of road lights on a black-tailed godwit (*Limosa l. limosa*) population*. DWW Ontsnipperingsreeks deel 38A, Delft.

Outen, A. R. 2002. The ecological effects of road lighting. Pages 133–155 in B. Sherwood, D. Cutler, and J. Burton (eds.), *Wildlife and roads: the ecological impact*. Imperial College Press, London.

Raevel, P., and F. Lamiot. 1998. *Impacts écologiques de l'éclairage nocturne* [Ecological impacts of night lighting]. Premier Congrès Européen sur la Protection du Ciel Nocturne, June 30–May 1, Cité des Sciences, La Villette, Paris.

Raymond, P. W. 1984. *Sea turtle hatchling disorientation and artificial beachfront lighting: a review of the problem and potential solutions*. Center for Environmental Education, Washington, D.C.

Riegel, K. W. 1973. Light pollution: outdoor lighting is a growing threat to astronomy. *Science* 179:1285–1291.

Ryden, H. 1989. *Lily pond: four years with a family of beavers*. William Morrow & Company, New York.

Scheibe, M. A. 1999. Über die Attraktivität von Straßenbeleuchtungen auf Insekten aus nahegelegenen Gewässern unter Berücksichtigung unterschiedlicher UV-Emission der Lampen [On the attractiveness of roadway lighting to insects from nearby waters with consideration of the different UV emission of the lamps]. *Natur und Landschaft* 74:144–146.

Schmiedel, J. 2001. Auswirkungen künstlicher Beleuchtung auf die Tierwelt – ein Überblick [Effects of artificial lighting on the animal world: an overview]. *Schriftenreihe für Landschaftspflege und Naturschutz* 67:19–51.

Upgren, A. R. 1996. Night blindness: light pollution is changing astronomy, the environment, and our experience of nature. *Amicus Journal* 17(4):22–25.

Verheijen, F. J. 1958. The mechanisms of the trapping effect of artificial light sources upon animals. *Archives Néerlandaises de Zoologie* 13:1–107.

Verheijen, F. J. 1985. Photopollution: artificial light optic spatial control systems fail to cope with. Incidents, causations, remedies. *Experimental Biology* 44:1–18.

Wilson, A. 1998. Light pollution: efforts to bring back the night sky. *Environmental Building News* 7(8):1, 8–14.

Witherington, B. E. 1997. The problem of photopollution for sea turtles and other nocturnal animals. Pages 303–328 in J. R. Clemmons and R. Buchholz (eds.), *Behavioral approaches to conservation in the wild*. Cambridge University Press, Cambridge.

Yurk, H., and A. W. Trites. 2000. Experimental attempts to reduce predation by harbor seals on out-migrating juvenile salmonids. *Transactions of the American Fisheries Society* 129:1360–1366.

Part I

Mammals

Night, Venezuela

After eleven o'clock, such a noise began in the contiguous forest, that for the remainder of the night all sleep was impossible. The wild cries of animals rung through the woods. Among the many voices which resounded together, the Indians could only recognize those which, after short pauses, were heard singly. There was the monotonous, plaintive cry of the Aluates (howling monkeys), the whining, flute-like notes of the small sapajous, the grunting murmur of the striped nocturnal ape (*Nyctipithecus trivirgatus*, which I was the first to describe), the fitful roar of the great tiger, the Cuguar or maneless American lion, the peccary, the sloth, and a host of parrots, parraquas (*Ortalides*), and other pheasant-like birds. Whenever the tigers approached the edge of the forest, our dog, who before had barked incessantly, came howling to seek protection under the hammocks. Sometimes the cry of the tiger resounded from the branches of a tree, and was then always accompanied by the plaintive piping tones of the apes, who were endeavouring to escape from the unwonted pursuit.

If one asks the Indians why such a continuous noise is heard on certain nights, they answer, with a smile, that "the animals are rejoicing in the beautiful moonlight, and celebrating the return of the full moon." To me the scene appeared rather to be owing to an accidental, long-continued, and gradually increasing conflict among the animals. Thus, for instance, the jaguar will pursue the peccaries and the tapirs, which, densely crowded together, burst through the barrier of tree-like shrubs which opposes their flight. Terrified at the confusion, the monkeys on the tops of the trees join their cries with those of the larger animals. This arouses the tribes of birds who build their nests in communities, and suddenly the whole animal world is in a state of commotion. Further experience taught us, that it was by no means always the festival of moonlight that disturbed the stillness of the forest; for we observed that the voices were loudest during violent storms of rain, or when the thunder echoed and the lightning flashed through the depths of the woods. The good-natured Franciscan monk who (notwithstanding the fever from which he had been suffering for many months) accompanied us through the cataracts of Atures and Maypures to San Carlos, on the Rio Negro, and to the Brazilian coast, used to say, when apprehensive of a storm at night, "May Heaven grant a quiet night both to us and to the wild beasts of the forest!"

Alexander von Humboldt

From "The Nocturnal Life of Animals in the Primeval Forest," *Views of Nature*, 1850.

Chapter 2

Effects of Artificial Night Lighting on Terrestrial Mammals

Paul Beier

All 986 species of bats, badgers and most smaller carnivores, most rodents (with the notable exception of squirrels), 20% of primates, and 80% of marsupials are nocturnal, and many more are active both night and day (Walls 1942). Thus it would be surprising if night lighting did not have significant effects on mammals. Compared with investigations on birds, lepidopterans, other insects, and turtles, however, few studies, or even anecdotal reports, document the effects of artificial night lighting on mammals in the wild. Because of the dearth of empirical evidence, this chapter begins with a review of the biology of mammalian vision, including the extensive literature on how moonlight affects nocturnal behavior of mammals and how light influences mammalian biological clocks. I then discuss several classes of likely effects of artificial night lighting on mammals, namely disruption of foraging patterns, increased predation risk, disruption of biological clocks, increased mortality on roads, and disruption of dispersal movements through artificially lighted landscapes. I include recommendations for experiments or observations that could advance our understanding of the most likely and significant effects.

Light and the Ecology and Physiology of Mammals

Insight into the potential consequences of artificial night lighting on mammals can be gained from an understanding of the activity patterns, visual ability, and physiological cycles of species under normal patterns of light and dark. Artificial light at night may disrupt the various daily, monthly, and annual cycles described in this section.

Mammals vary in their activity periods, with corresponding adaptations in their visual systems (Walls 1942). Activity patterns can be classified into five types (Halle and Stenseth 2000). Mammals with a nocturnal pattern obviously are most likely to be affected by artificial night lighting. I will treat the crepuscular pattern, defined as nocturnal with activity peaks at dawn and dusk, as a variant on the nocturnal theme; this group includes most lagomorphs. Diurnal mammals include all squirrels, except the flying squirrels, and most primates, including humans. Indeed, if human vision were not so anatomically diurnal, artificial lighting would not be necessary. Mammals with the 24-hour pattern include ungulates and larger carnivores, plus some smaller carnivores. These species have excellent night vision and usually are most active at night but have regular daytime activity periods as well. I ignore the ultradian pattern—periodicity less than 24 hours, typically 3- to 5-hour cycles—because it has been documented only in voles and is light-independent (Gerkema et al. 1990).

Anatomy and Physiology of Vision in Mammals

How various mammals respond to light depends, among other things, on the architecture of the eye, including its pupil, type of lens, and especially whether the photosensitive cells in the retina are dominated by rods or cones. Nocturnal mammals have large pupils to admit more light, huge lenses to minimize spherical aberration, and rod-rich retinas (Walls 1942). The rod system has high sensitivity but low acuity; that is, it can be stimulated by relatively few photons, but ability to see detail is poor because many rod cells connect to a single neuron. This means that small stimuli from several rods can act in concert to stimulate a neuron and thus deliver a signal to the brain. Because the brain is unable to determine exactly which rods were stimulated, however, it cannot discern the exact size and shape of the perceived object. In contrast, there is little summation among neurons where cones and neurons approach a 1:1 ratio in parts of some mammalian retinas.

Most nocturnal mammals have few cones; bats and armadillos have nearly cone-free retinas (Walls 1942). Nocturnal mammals with few cones are temporarily blinded by bright light because the rods become unresponsive (i.e., saturated) above 120 candela/m^2, approximately the light level at twilight. Narrowing the pupil is the primary short-term defense of cone-poor mammals against rod saturation in bright light but is only marginally effective at reducing the blinding effect of light (Perlman and Normann 1998).

Because they lack high-resolution cones, few nocturnal mammals eat seeds, small fruits, or small mobile insects unless such foods are clumped into large, visually detectable aggregations such as inflorescences or anthills or are detectable by other means such as echolocation or scent. Nocturnal animals can partially overcome the poor resolving power of the rod-dominated retina by having large eyes that permit large retinal images. Because the size of rods does not decrease with body size, what matters here is the absolute, not relative, size of the retinal image (Walls 1942). Thus the limited skull size of small nocturnal mammals limits their evolutionary ability to improve visual resolution.

The retina of diurnal mammals is rich in cones, which provide clear images at close range or in good light. A large number of photons is needed to stimulate a cone, however, which makes cones useless in dim light. Most, perhaps all, diurnal squirrels are similar to diurnal birds in having retinas so poor in rods that they are nearly blind at night. Although most diurnal mammals, including humans, have fewer cones than rods, most of these mammals are large, and their large retinal image ensures high visual acuity in daylight. The lenses of diurnal mammals resemble those of 24-hour mammals.

Like some nocturnal and crepuscular mammals, most mammals capable of 24-hour activity have a retina composed mostly of rods, but they have enough cones for a second image-forming system useful in bright light (Perlman and Normann 1998). Changes in pupil size are less important than photon saturation of the rods in switching between systems (Perlman and Normann 1998). When a mammal with a 24-hour eye comes from darkness into light, the rods saturate, thereby becoming incapable of stimulation, and the shift to the cone system occurs within about two seconds. The shift from bright to low light takes much longer (Lythgoe 1979) and involves more complex chemical reactions for the rods to fully resensitize (Perlman and Normann 1998). Although the rod system may gain a 100-fold increase in sensitivity within 10 minutes after the transition to darkness, another 10-fold gain in sensitivity can occur between 10 and 40 minutes (Lythgoe 1979). The presence of a bright light in an otherwise dark environment may suppress the rod system in part or all of the retina, leaving the animal not fully adjusted to the dark.

Many 24-hour mammals, and some nocturnal and crepuscular mammals, have a highly reflective layer behind the photoreceptive cells, the tapetum lucidum, that amplifies the light reaching those cells. The tapetum is found in most carnivores and ungulates but rarely in rodents, lagomorphs, or higher primates.

In mammals with both rod and cone systems, the shift between systems is accompanied by a change in spectral sensitivity called the Purkinje shift. Cone cells have a variety of photoreactive pigments, and this variety creates a capacity for color vision in the cone system. Because rods rely on only one photoreactive pigment, rhodopsin, with maximum absorption around 496 nm, the color-blind rod system discriminates only on the basis of brightness.

Influence of Moonlight on Behavior of Nocturnal Mammals

Most nocturnal mammals react to increasing moonlight by reducing their use of open areas, restricting foraging activity and movements, reducing total duration of activity, or concentrating foraging and longer movements during the darkest periods of night. Such behaviors have been recorded in studies of desert rodents (Lockard and Owings 1974, Price et al. 1984, Bowers 1988, Alkon and Saltz 1988), temperate-zone rodents (Kaufman and Kaufman 1982, Travers et al. 1988, Vickery and Bider 1981, Wolfe and Summerlin 1989, Topping et al. 1999), desert lagomorphs (Butynski 1984, Rogowitz 1997), arctic lagomorphs (Gilbert and Boutin 1991), fruit bats (Morrison 1978, Law 1997, Elangovan and Marimuthu 2001), a predatory bat (Subbaraj and Balasingh 1996), some primates (Wright 1981), male woolly opossums (Julien-Laferrière 1997), and European badgers (Cresswell and Harris 1988).

Most authors attributed these changes to increased predation risk in open habitats under bright moonlight. Although no field study conclusively confirms or refutes this explanation, circumstantial evidence supports it. Increased coyote howling during the new moon is consistent with the unprofitability of hunting rodents under these conditions (Bender et al. 1996). In laboratory studies (Clarke 1983, Dice 1945), owls were better able to catch deer mice in brighter light. However, as Clarke (1983) explained, these laboratory results may not reveal much about the effect on predation rate under natural conditions. On bright nights, most prey remain in secure places, but the few that are in bright conditions may be readily killed. On dark nights, owl efficiency per prey may be reduced, but with many active prey available, the total prey consumption and the prey's mortality rate from the owl may be unchanged (Daly et al. 1992). Similarly, ocelot behavior is consistent with the hypothesis that fewer but more successful prey encounters occur under bright light (Emmons et al. 1989).

Some nocturnal species neither decrease activity nor seek habitats with canopy cover during bright moonlight. Many insectivorous bats do not decrease activity during bright moonlight (Negraeff and Brigham 1995, Hecker and Brigham 1999), although some species do, at least in captivity (Erkert 2000). Some insectivorous bats prefer to forage in upper canopy under bright moonlight (Hecker and Brigham 1999) or under artificial night lighting (Rydell and Baagøe 1996), in both cases because insect prey are more abundant in the brighter areas (for further discussion of bats see Chapter 3, this volume). Moonlight is associated with increased activity in woodland rodents such as *Peromyscus leucopus* (Barry and Francq 1982), the nocturnal monkey *Aotus trivirgatus* (Wright 1981), and the galagos (Galagonidae; Nash 1986). In most instances, these studies provided adaptive reasons for increased activity in moonlight. For example, the galagos, although nocturnal, visually detect their insect prey, and they avoid predation not by concealment but by visual detection, mobbing, and flight. Moonlight does not change the activity pattern of ocelots (Emmons et al. 1989) or white-tailed deer (Beier and McCullough 1990; but see Kie 1996).

The Circadian Clock in Mammals

The freerunning period of activity, the activity cycle for an animal under constant light or darkness, ranges from 23 to 25 hours for most vertebrates, with extremes of 21 to 27 hours (Foster and Provencio 1999). Because the freerunning clock is not exactly 24 hours, the internal circadian system must be synchronized to local time by a cue in the animal's environment. This process is called entrainment, and the cue used to synchronize the internal clock is called a *Zeitgeber*. For all vertebrates, the primary *Zeitgeber* is change in the quantity, and perhaps the spectral quality, of light at dawn and dusk (Foster and Provencio 1999). In vertebrates, the two image-forming visual systems (i.e., the rod and cone systems) do not entrain the biological clock, which is governed by a special photoreceptor system separate from them. In mammals, this photoreceptor system lies in the retina and communicates to a different part of the brain, the suprachiasmatic nuclei (SCN), via a different neural system, comprising less than 0.01% of retinal ganglion cells (Foster and Provencio 1999). Loss of the eyes or SCN blocks entrainment of the circadian clock in all mammals studied. Shifting circadian rhythm requires more light than that needed to form a visual image, and the stimulus must be of longer duration, 30 seconds to 100 minutes (Figure 2.1; Foster and Provencio

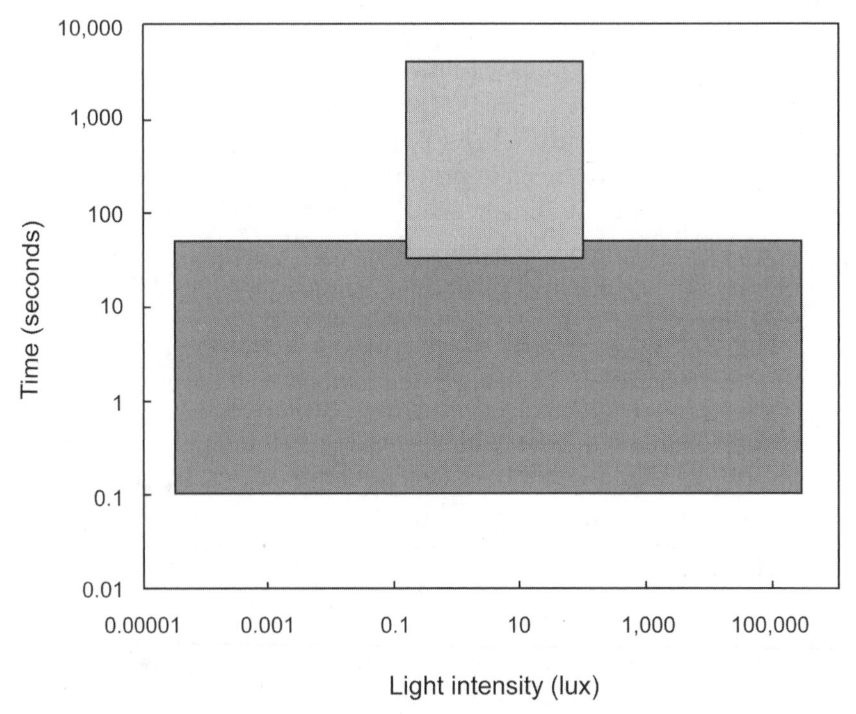

Figure 2.1. The response range of the visual-imaging system (large box) has minimal overlap with the response range of the circadian system in vertebrates (small box). Influencing the biological clock requires both more light (x-axis) and longer duration (y-axis) than forming a visual image. This protects the circadian system from many photic stimuli that do not provide reliable time cues. The upper threshold in light intensity makes the circadian clock more sensitive to twilight intensities than to full sunlight. Artificial lights within the range of duration and intensity described by the small box disrupt the mammalian biological clock. Figure adapted from Foster and Provencio (1999: Figure 3), with the x-axis converted from photons per unit area. Although there is no exact conversion to lux, this approximation allows the reader to compare these light intensities with those illustrated in Figure 2.2.

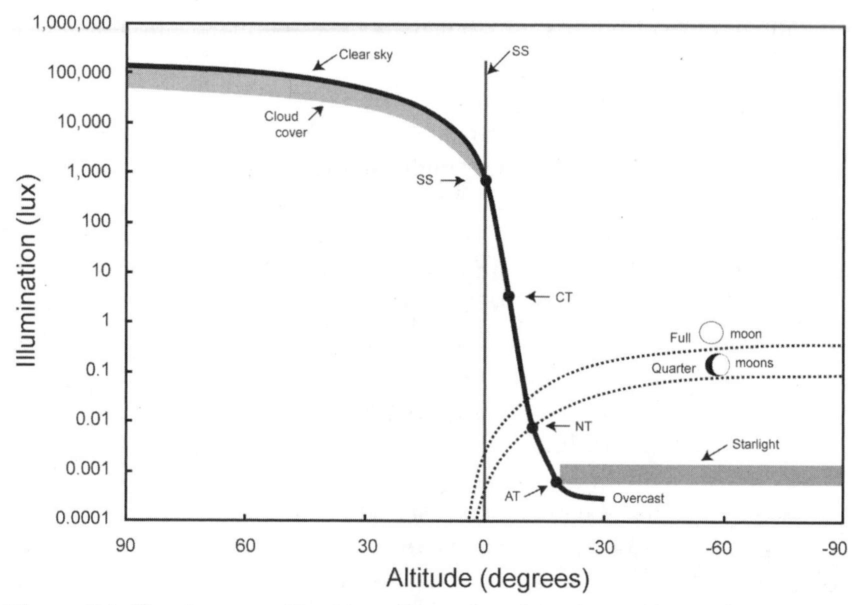

Figure 2.2. Illumination at Earth's surface varies with solar and lunar altitude above the horizon. For comparative purposes, the Illuminating Engineering Society of North America recommends 3–16 lux illumination for U.S. highways or as a maximum for off-site spill from recreational sports facilities. In practice, these recommendations often are exceeded by an order of magnitude. Note log scale on *y*-axis. The altitude of the moon above the horizon is deliberately displayed on the negative (below horizon) half of the *x*-axis so that the *x*-axis can be interpreted as time relative to sunset. SS, sunset; CT, civil twilight with sun 6° below horizon; NT, nautical twilight with sun 12° below horizon; AT, astronomical twilight with sun 18° below horizon. Figure adapted from McFarland et al. (1999: Figure 1).

1999). Light level at twilight falls at the lower end of this range (Figure 2.2; McFarland et al. 1999). These thresholds—as well as the upper limits—are useful in preventing photic noise from resetting the circadian clock. For instance, lightning, which can be fifty times brighter than direct sunlight, would confuse circadian rhythm if it were of sufficient duration. It has long been thought that the irradiance of starlight and the full moon both fall below the threshold for entrainment and cannot reset the circadian clock, although entrainment of circadian rhythm recently has been recorded at illuminances as low as 10^{-5} lux in bats (Erkert 2004). Low-intensity stimuli of sufficient duration can suppress melatonin production in rats (Dauchy et al. 1997) and humans (Brainard et al. 1997), suggesting that such stimuli also affect the circadian clock, at least in humans (Shanahan et al. 1997) in addition to bats.

The light regime and the circadian clock also influence production of some hormones, notably melatonin, which mediates not only the activity patterns discussed earlier but also almost every physiological or behavioral rhythm in mammals (Bartness and Goldman 1989). In all species, melatonin production is high at night and suppressed during daytime, although reaction to melatonin often differs between diurnal and nocturnal species. Among its many roles, melatonin suppresses tumor growth by regulating production and tumor use of linoleic acid. In a laboratory experiment, Dauchy et al. (1997) determined that minimal light contamination of 0.2 lux, simulating a light leak around a laboratory door during an otherwise normal dark phase, disrupted normal circadian production of melatonin and promoted tumor growth in rats. Compared with rats experiencing a cycle of 12 hours light and 12 hours total darkness per day, rats experiencing light contamination produced 87% less melatonin, similar to the 94% decline observed in rats held in full light 24 hours per day. There were corresponding dramatic increases in tumor growth. Remarkably, low-intensity light exposure during the subjective dark phase had virtually the same effect as constant light in blocking melatonin production and stimulating tumor growth.

The Circannual Clock and Lunar Clock in Mammals

Mammals also have an endogenous rhythm with a freerunning period of about one year. The circannual clock influences annual changes in body mass, hormones, reproductive status, hibernation, and the circadian activity pattern over the course of the year. By controlling breeding season, delayed fertilization of the ovum, and delayed implantation of the blastocyst, the circannual clock causes parturition of most species of mammals to occur in a highly compressed period. This reduces the neonatal mortality rate by predator swamping and synchronizes parturition with favorable foraging conditions (Vaughan 1978, Gwinner 1986).

Because experiments on the circannual clock take years to complete, our understanding of it remains poor, and only three mammal species have been studied in any detail, namely the golden hamster (*Mesocricetus auratus*; Bronson 1989), domestic sheep (Bronson 1989), and golden-mantled ground squirrel (*Spermophilus lateralis*; Dark et al. 1990, Zucker et al. 1983, Pengelley and Fisher 1963, Lee et al. 1986). Light appears to be the most important—perhaps the only—*Zeitgeber* for the circannual clock of hamster and sheep (Bronson 1989). Both of these species, however, are highly domesticated and all laboratory stocks of the hamster are highly inbred, having descended from a single mother and her litter captured in 1930. These factors may limit the extent to which we can extrapolate to wild mammals.

Light may be of equal or lesser importance than temperature in setting the circannual clock of the golden-mantled ground squirrel and especially in governing the hibernation cycle of the species. Zucker et al. (1983) demonstrated that light was involved in entraining the circannual clock in golden-mantled ground squirrels. Loss of the SCN, however, disrupted the annual reproductive cycle and the annual cycle of body mass in only 8 of 19 squirrels, indicating existence of a circannual oscillator that is anatomically separate from the SCN. Although the neural structure that functions as the circannual oscillator has not been identified, it is influenced by the retinal system that terminates in the SCN (Dark et al. 1990). Although Hock (1955) reported a strong role for light in initiating hibernation of the Arctic ground squirrel (*Spermophilus undulatus*), Pengelley and Fisher (1963) reported that although an artificially reversed thermal regime caused golden-mantled ground squirrels to hibernate in summer, it was impossible to produce a similar reversal in the phase of the hibernation cycle by changing light conditions. Emergence from hibernation in spring cannot possibly be influenced by photoperiod because these squirrels hibernate in dark burrows.

In summary, studies of circannual cycles of a few mammalian species suggest that light is an important *Zeitgeber* but perhaps not the only one. The importance of light as a circannual regulator is also a logical necessity, given the crucial role of light in production of melatonin and the well-documented importance of melatonin in governing reproductive activity (Bartness and Goldman 1989). Bronson (1989) and Gwinner (1986) provide excellent overviews of this complex topic.

Lunar cycles also may play an important role in timing of mammalian reproductive behaviors. Murray (1982) and Skinner and van Jaarsveld (1987) suggested that moonlight may synchronize estrus in some ungulates. Both of these were observational studies, and there appears to be no experimental work on how lunar cycles affect mammalian reproduction or whether the mammalian brain has a neural circalunar oscillator that is entrained by moonlight. The absence of such evidence is a result of a lack of effort and cannot be construed as refuting the existence or importance of a circalunar clock.

Plausible Effects of Artificial Night Lighting on Mammals

In the rest of this chapter, I make inferences about plausible effects of artificial night lighting by considering the foregoing information in relation to the properties of artificial night light and evaluating the handful of studies on how artificial lighting influences mammal behavior in the wild. Potential influences of artificial lights at night on mammals include disruption of foraging behavior, increased risk of predation, disruption of biological clocks, increased deaths in collisions on roads, and disruption of dispersal movements and corridor use.

Disruption of Foraging Behavior and Increased Risk of Predation

Many studies cited in this chapter have shown that bats, nocturnal rodents, and other nocturnal mammals respond to moonlight by shifting their activity periods, reducing their activity, traveling shorter distances, and consuming less food. Artificial light of similar intensity to moonlight caused rodents in experimental arenas to reduce their activity, movement, and food consumption (Vasquez 1994, Kramer and Birney 2001, Brillhart and Kaufman 1991, Clarke 1983, Falkenberg and Clarke 1998). These experiments used both fluorescent and incandescent lights to simulate moonlight, with rodents responding to stimuli equivalent to that of a half moon (0.1 lux) as well as a full moon (0.3 lux). Thus, artificial night lighting of similar intensity to moonlight reduces activity and movement of many nocturnal animals, particularly those that rely on concealment to reduce predation risk during nocturnal foraging. Because roadway lighting in the United States is designed to illuminate the road surface at a minimum of 3 lux (the lowest acceptable value midway between light standards) and an average of 4–17 lux, depending on type of pavement and roadway, with maximum values two or three times the average directly under lampposts (IESNA 2000), all artificial night lighting can be expected to have such effects along road edges.

Although small mammals can respond to bright moonlight by shifting foraging and ranging activities to darker conditions, this option is not available to animals experiencing artificially increased illumination throughout the night. Under these circumstances, unless they abandon the lighted area, nocturnal animals have only two unfortunate choices. One is to accept the risk of predation by foraging under bright light, as Alkon and Saltz (1988) observed when food shortages forced crested porcupines (*Hystrix indica*) to abandon their light-phobic behaviors. The other option is to continue to minimize predation risk even at the cost of loss of body mass, as observed in an experiment on the cricetid rodent *Phyllotis darwini* (Vasquez 1994). The rodents responded to simulated moonlight by carrying 40% of their food to the refuge site in the arena and consuming it there, compared with less than 4% of food consumption under dark conditions. On bright nights, the rodents consumed 15% less food and lost 4.4 g, compared with a 1.1-g weight loss on dark nights. Despite difficulties in translating these experimental results to field conditions, artificial night lighting undoubtedly reduces food consumption and probably increases predation risk for nocturnal rodents in the wild.

Few studies have investigated the effects of artificial light on feeding behavior of mammals in natural populations. In one study Kotler (1984) strongly confirmed that artificial night lighting affects nocturnal rodents. During the new moon, Kotler observed that seed harvest by the desert rodent community (four species of *Dipodomys*, *Peromyscus maniculatus*, and possibly *Perognathus longimembris* and *Microdipodops pallidus*) decreased an average of 21% in response to a single fluorescent or gasoline camping lantern placed to cast light equivalent to 160% (8 m [26 ft] from lantern) to 25% (35 m [115 ft] from lantern) of the light of a full moon. He also reported that, within trials, harvesting rate was lower at feeding sites that were most brightly illuminated, but he did not quantitatively describe that relationship. To help planners estimate the magnitude of this effect, future research should determine the functional relationship between food harvest (or other variables related to fitness) and illumination and determine whether there is a threshold illumination below which no effect occurs. Although lighting at sports stadiums, gas stations, and some commercial operations is brighter than highway lighting, the latter probably is the brightest lighting that affects large areas of wildlands. Thus, research focusing on the intensities and heights of lighting that are prescribed or implemented along highways, and their effects in a landscape context, would be most helpful.

Bird et al. (2004) also investigated the effects of artificial lighting on rodent foraging. In coastal Florida, they measured foraging of Santa Rosa beach mouse (*Peromyscus polionotus leucocephalus*) as a proxy for the other threatened and endangered subspecies of *Peromyscus polionotus*. Resource patches of food were placed along transects with arrays of low pressure sodium lights, "bug" lights, and no lights. The percentage of resource patches foraged by mice was significantly higher in dark arrays than light arrays and higher at arrays with bug lights than low pressure sodium lights. Effects of actual beachfront lighting were presumed to be greater than those observed in the experiment because taller and more intense light sources are commonly used in coastal development.

De Molenaar et al. (2003) studied mammal response to streetlamps experimentally installed on small earthen dams that crossed flooded drainage ditches in the Netherlands. Aquatic mammals such as muskrats (*Ondatra zibethicus*) had to cross these dams to move along the ditch, and other mammals used the dams to pass between patches of upland habitat without swimming. The four predators—polecat (*Mustela putorius*), stoat (*Mustela erminea*), weasel (*Mustela nivalis*), and fox (*Vulpes vulpes*)—were more likely to walk on or near illuminated dams than unlit ones, and the brown rat (*Rattus norvegicus*) seemed to avoid lighted dams. The four other species studied (muskrat, hedgehog [*Erinaceus europaeus*], hare [*Lepus europaeus*], and roe deer [*Capreolus capreolus*]) showed no marked response.

With their cone-rich retinas, most sciurids probably are nearly blind at night, even under moonlight or artificial night lighting. To conceal themselves from visual predators, most tree squirrels spend the night in nests in trees, and ground squirrels sleep underground. To the extent that artificial night lighting assists visual predators at night, it could decrease squirrel survival rates.

Does artificial night lighting benefit owls, bats, or other predators? If desert rodents are more vulnerable to owls and other nocturnal predators under moonlight or its equivalent, it is tempting to think of artificial night lighting as enhancing habitat for these predators. Many species of insectivorous bats aggregate at streetlamps to exploit aggregations of moths and other insects that are attracted to the light (Blake et al. 1994, Rydell and Baagøe 1996). Some reports have implied that this is good for bats, but this makes sense only under the nonecological valuation that more is better. Certainly such aggregations are not natural, nor are they beneficial to insect prey of the bats. Such lighting should not be justified in terms of benefits to bats unless the feeding stations are explicitly intended to compensate for human-caused loss of other food sources or human-caused excess of the insect populations attracted to the lights.

Disruption of Biological Clocks

Assuming that the circadian clock evolved to maximize foraging efficiency, to reduce risk of predation, to enhance parental care, or for similarly important reasons, artificial night lighting can adversely affect animals by disrupting that clock. These individuals also would be out of phase with their neighbors living in a natural light–dark cycle; in more social mammals this could affect mating success, group-mediated antipredator vigilance, and other processes.

Almost all studies of how light pulses can shift the biological clock used artificial light, either fluorescent or incandescent, as the stimulus. All of these studies demonstrate that brief (10- to 15-minute duration) and moderately bright (about 1,000 lux, equivalent to bright twilight) stimuli can shift the circadian clock by 1–2 hours (Halle and Stenseth 2000). This finding suggests that artificial night lighting can disrupt circadian patterns in the wild. These experiments, however, were conducted only on captive animals held in 24-hour darkness except for the experimental stimuli. One experiment on the nocturnal flying squirrel *Glaucomys volans* came much closer to natural conditions in that the experimental animals had free access to a completely dark nest box and could choose when to emerge to a larger chamber where they might encounter artificial light (DeCoursey 1986). If the squirrel encountered light at arousal time, when it expected to enter a dark world, it would return to its nest box to sleep, delaying its circadian clock by 40 minutes. Because most nocturnal animals spend the day in burrows or cavities with unmeasured but presumably very low light levels, these experimental results probably are ecologically relevant to all nocturnal mammals.

Only two studies compared artificial light with daylight in terms of their effects on the circadian clock. In one study, wild-caught nocturnal mice were subjected to pulses of daylight, incandescent light, and fluorescent light, each 1,000 lux and 15 minutes in duration, at various points in the circadian cycle (Sharma et al. 1997). The phase shift response was strongest 2–3 hours after the transition from subjective day to subjective night, at which time the daylight stimulus produced a greater delay in activity (about 2.5 hours) than the two types of artificial light (each about 1.5 hours). The other study (Joshi and Chandrashekaran 1985) applied the same experimental protocol on a bat and found that incandescent lights produced large phase shifts in the opposite direction as the shifts elicited by daylight and fluorescent light. Artificial night lighting is about as effective as natural light in setting—or disrupting—the circadian clock.

The effect of the circadian clock on production of melatonin may have serious ecological consequences. Dauchy et al. (1997) documented that modest levels of nocturnal light suppressed melatonin production with dramatic effects on tumor growth in rats. Although these results cannot be directly translated to wild mammals, this study suggests that disruption of biological clocks by artificial night lighting could have profound effects on individual animals. If a significant fraction of individuals in a population is affected, population and ecosystem effects are also possible. In the golden hamster, the visual system that regulates the circadian clock is responsive to stimuli between 300 and 500 nm but insensitive towavelengths of 640 nm or longer and 290 nm or shorter (Brainard et al. 1994). Further research on the spectral sensitivity of additional mammals may provide guidance that would allow the selection of outdoor lighting to avoid or minimize this potential effect, perhaps in the red-yellow spectrum.

Despite ample evidence that artificial lighting can disrupt circadian and circannual clocks in the laboratory setting—where all existing research has been conducted—there is no confirmation of these effects in wild populations. In part this is an intractable problem because phase shifts have been defined in a way (Gwinner 1986) that can be measured only in a laboratory. Melatonin levels in wild populations subject to artificial night lighting, however, could be compared with levels in undisturbed populations, controlling for time of day, to yield a biologically meaningful estimate of the magnitude of this problem in nature. In addition, population-level studies can demonstrate the overall effect of artificial night lighting on mammal populations, although it may be difficult or impossible to disentangle the effects of disrupted biological clocks from those of other mechanisms, such as reduced foraging or increased predation risk.

Effect of Street Lighting on Roadkill of Mammals

Intensity and type of street lighting may influence the probability of wildlife mortality in collisions with vehicles. It seems logical that most types of lighting will make animals more visible to drivers and thus reduce risk of mortality by giving the driver more time to react. There is no research supporting this idea, however, and Reed (1995), Reed et al. (1979), and Reed and Woodward (1981) concluded that increased highway illumination was not effective at reducing deer–vehicle accidents in the United States.

Some artificial night lighting makes it difficult for nocturnal mammals to avoid collisions with vehicles if the animal experiences a rapid shift in illumination. Many nocturnal species are using only the rod system, and bright lighting saturates their retinas. Although many nocturnal mammals have a rudimentary cone system and can switch over to it within a couple seconds, during those seconds they are blinded. Once they switch to the cone system, areas illuminated to lower levels become black, and the animal may become disoriented, unable to see the dark area across the road and unwilling to flee into the unseeable shadows whence it came. This is not solely a problem for a rod-dominated visual system because even a cone-dominated system is ineffective when a small part of the visual field is many orders of magnitude brighter than the remaining field. This glare phenomenon is familiar to any backcountry camper who has been temporarily blinded by a companion's flashlight. Finally, if the animal is in the lighted area long enough to saturate its rod system, it will be at a distinct disadvantage for 10–40 minutes after returning to darkness.

The lowest possible lighting level consistent with human safety is the best for mammals crossing roads. There is no advantage to using lighting that is closer to the sunlight spectrum for these cone-poor animals. Indeed, low pressure sodium lights, with emission at 589 nm, provide reasonably effective vision for human drivers, who have mixed cone and rod vision, while interfering least, of the available lamp types, with the dominant rod-based vision of nocturnal mammals. Because the rod system has peak sensitivity near 496 nm, low pressure sodium lights should appear about one-tenth as bright to a rod-dominated retina as to a human retina.

Little ecological research, and a modest amount of human and engineering research, is needed on the issue of designing highway lighting to minimize roadkill mortality. Our knowledge of mammalian vision is sufficient to conclude that, from the animal's perspective, less is better. Research should focus on the straightforward issue of determining the lowest level of illumination that increases the ability of human drivers to see a large animal, thus allowing the driver to avoid collision, without disabling the rod-dominated retina of mammals, thus allowing them to escape into the darkness. Other technical questions, relevant not only to roadkill but also to biological clocks, predation risk, and foraging behavior, include developing cost-effective designs to confine lighting to the roadway and balancing them with a lamp height and beam pattern that reduces effects on the sensitive central part of the driver's retina.

Disruption of Dispersal Movements and Corridor Use

With increasing emphasis on providing biotic connectivity at the landscape scale, there is an increased need for information on how various factors influence the utility of a connective area. It follows from the preceding that street lighting negatively affects a mammal's ability and willingness to cross a road or to move through any area with artificial night lighting. Although planners and conservationists have focused on the issue of how wide a corridor should be, it is obvious that the answer depends on how bright it is.

Only two studies attempted to address how a mammal, moving at night through unfamiliar terrain, might react to natural or artificial light or otherwise use visual information to find suitable habitat. A study of dispersing puma (*Puma concolor*) in urban southern California noted several exploratory movements that did not follow favored topography or vegetation patterns (Beier 1995). Beier speculated that the pumas were moving away from the urban glow and navigating toward the darkest horizon. Beier also noted instances in which an animal, exploring new habitat for the first time, stopped during the night at a lighted highway crossing its direction of travel with unlit terrain beyond. In several instances, the animal would bed down until dawn, selecting a location where it could see the terrain beyond the highway after sunrise. The next evening, the puma would attempt to cross the road if wildland lay beyond or would turn back if industrial land lay beyond.

Another study revealed that white-footed mice (*Peromyscus leucopus*) are capable of a similar "look now and move later" strategy (Zollner and Lima 1999). Zollner and Lima experimentally released woodland mice in bare agricultural fields at night under dark or moonlit conditions and at various distances from a single woodland patch, which was suitable habitat for the mouse, in the area. Under dark conditions, the mice were incapable of perceiving and orienting to the woodland patch at distances of 30 m (98 ft) or more. Full moonlight extended the perceptual range to 60 m (197 ft), and mice given a twilight look at the landscape before sunset were able to orient from 90 m (295 ft) away. Thus, if mice were not deterred by psychology, activity pattern, and predation risk, interpatch dispersal by mice would be more successful under daylight illumination. The study demonstrates, however, that mice are able to assess the landscape under full light and use that information to move successfully in the dark.

Zollner and Lima (1999) also open a new realm of research, namely empirically determining the perceptual range of an animal, or the distance at which habitat patches can be perceived. Goodwin et al. (1999) provide helpful suggestions for sound statistical analyses and alternative approaches. Such research, using species for which corridors are designed, may provide a scientific basis for designing corridors and determining how animals use vision to explore new terrain.

Although perceptual range of mice increased in moonlight, there are two reasons that artificial night lighting may not similarly increase perceptual range and help animals find new patches. First, by saturating an animal's rod system, artificial night lighting plunges most of the landscape into darkness. Second, because a dispersing animal can anticipate this effect, it may orient away from the lights.

Movement in connective areas can be affected by adjacent lights of recreational fields, industrial parks, service stations, and housing. In southern California, where the South Coast Missing Linkages effort is attempting to maintain and restore landscape linkages between fifteen pairs of large wildlands, three riparian corridors are lined with homes sitting atop a low manufactured slope, and all fifteen linkages are crossed by lighted freeways (Beier et al. in press). Efforts to maintain and restore these landscape linkages should incorporate the general rule that less light is better for animal movement.

Research Issues

The literature on the effects of light on foraging behavior, predation risk, and biological clocks consists of two distinct approaches with little overlap. One approach is to study effects of moonlight on behavior of individual wild mammals; the other is to study the effects of artificial light on animals in laboratories. The discussion in this chapter underscores the need for studies using artificial lights on natural populations. Substantial expertise already exists, and productive collaborations between ecologists and laboratory physiologists could result in rapid progress.

Population-Level Research

A simple fusion of the two approaches will fall short of the mark unless at least some research efforts focus above the level of the response of individual animals. For instance, if research were to confirm that artificial night lighting increased numbers of tumors in wild mice by 25% or increased predation risk by 15%, this finding still would not address the issue of effects on the wild population. Conceivably, the induced tumors could shorten the lifespan of affected mice by only a few weeks or days, or predation mortality could act in a compensatory fashion with other types of mortality to reduce greatly the net effect on survival rates of animals living in the light-polluted zone. This effect could be further diluted if the light-polluted zone were part of a larger habitat, most of which was not directly affected by light, in which case the polluted zone may be a small population sink. Conversely, interactions between individuals from the polluted zone with neighbors in dark zones, such as dissolution of the synchrony of estrus and parturition, could amplify the effect. Only careful, whole-population studies can address these more important questions.

A critical element in study design is to include both treatment populations and control populations. Ideally, studies will include both replication with more than one treatment and control population and observations in both treatment and control populations before light pollution (Stewart-Oaten et al. 1986). This paired before–after–control–impact study design also is appropriate for situations in which replication is not possible. Although this design lacks random allocation of treatments to experimental units, it can provide meaningful answers to important applied questions (Beier and Noss 1998). It is far better to have an approximate answer to the right question than a precise answer to the wrong question.

Equivalence Testing

In the study of individuals or populations, the statistical analysis of the effects of artificial night lighting should use equivalence testing (Patel and Gupta 1984, McBride et al. 1993), in which the null hypothesis is "artificial night lighting has biologically meaningful negative effects on mammals," rather than the traditional null hypothesis of "no effect." Failure to reject the traditional null hypothesis typically leads to complacency, even if the failure to reject resulted from undersampling or other design flaws. The burden of proof falls, inappropriately, on the most plausible point of view. In contrast, in an equivalence test, failure to reject the null hypothesis lends continued support to the most plausible state of nature, namely that there is an effect, and shifts the burden of proof to proponents of the idea that there is no biologically significant effect. Equivalence testing therefore is appropriate in all situations in which related studies and known cause–effect relationships suggest an environmental impact. Because the procedure requires the analyst to specify the direction and magnitude of a biologically meaningful effect, rejection of the null hypothesis is by definition a biologically, as well as statistically, significant outcome. This is in marked contrast to tests of traditional null hypotheses, in which the "insignificance of significance testing" (Johnson 1999) is an intractable issue.

Conclusion

For small, nocturnal, herbivorous mammals, artificial night lighting increases risk of being killed by a predator and decreases food consumption. Such lighting probably also disrupts circadian rhythms and melatonin production of mammals. Most research, however, has documented the response of individual wild animals to moonlight or of laboratory animals to artificial light. Research on how artificial lights affect wild mammals at the population level is lacking. Significant progress relevant to management decisions will require collaboration between ecologists and laboratory physiologists and assessment of population-level responses (e.g., rates of survival and reproduction) as well as individual behavioral and physiological responses (e.g., food consumption, avoidance of lighted areas, and melatonin levels). I recommend an experimental design that includes observation on paired control (dark) and treatment (lighted) landscapes both before and after installation of artificial night lighting. Given the preponderance of evidence from previous studies and known cause–effect relationships, statistical procedures should test the null hypothesis that artificial night lighting has a biologically significant negative effect on survival and reproduction, appropriately placing the burden of proof on proponents of the idea that such lighting is benign.

Night lighting also may increase roadkill of animals and can disrupt mammalian dispersal movements and corridor use. Most research on these issues is a straightforward matter of determining an intensity, spectral output, and physical arrangement of lighting fixtures that enhances human safety while minimally affecting the rod-dominated visual system of nocturnal mammals. In addition, experiments to determine the perceptual range of mammals (i.e., the distance at which habitat patches can be discerned by an animal exploring new terrain) may, for example, enhance significantly a land manager's ability to locate artificial night lighting adjacent to wildlife linkages such that it minimizes interference with perception of habitat patches by species to be served by the linkage.

Acknowledgments

Christian B. Luginbuhl (U.S. Naval Observatory, Flagstaff, Arizona) and an anonymous reviewer provided useful comments on this chapter.

Literature Cited

Alkon, P. U., and D. Saltz. 1988. Influence of season and moonlight on temporal-activity patterns of Indian crested porcupines (*Hystrix indica*). *Journal of Mammalogy* 69:71–80.

Barry, R. E. Jr., and E. N. Francq. 1982. Illumination preference and visual orientation of wild-reared mice, *Peromyscus leucopus*. *Animal Behaviour* 30:339–344.

Bartness, T. J., and B. D. Goldman. 1989. Mammalian pineal melatonin: a clock for all seasons. *Experientia* 45:939–945.

Beier, P. 1995. Dispersal of juvenile cougars in fragmented habitat. *Journal of Wildlife Management* 59:228–237.

Beier, P., and D. R. McCullough. 1990. Factors influencing white-tailed deer activity patterns and habitat use. *Wildlife Monographs* 109:1–51.

Beier, P., and R. F. Noss. 1998. Do habitat corridors provide connectivity? *Conservation Biology* 12:1241–1252.

Beier, P., K. L. Penrod, C. Luke, W. D. Spencer, and C. Cabañero. In press. South Coast Missing Linkages: restoring connectivity to wildlands in the largest metropolitan area in the United States. In K. R. Crooks and M. A. Sanjayan (eds.), *Connectivity and conservation*. Cambridge University Press, Cambridge.

Bender, D. J., E. M. Bayne, and R. M. Brigham. 1996. Lunar condition influences coyote (*Canis latrans*) howling. *American Midland Naturalist* 136:413–417.

Bird, B. L., L. C. Branch, and D. L. Miller. 2004. Effects of coastal lighting on foraging behavior of beach mice. *Conservation Biology* 18:1435–1439.

Blake, D., A. M. Hutson, P. A. Racey, J. Rydell, and J. R. Speakman. 1994. Use of lamplit roads by foraging bats in southern England. *Journal of Zoology, London* 234:453–462.

Bowers, M. A. 1988. Seed removal experiments on desert rodents: the microhabitat by moonlight effect. *Journal of Mammalogy* 69:201–204.

Brainard, G. C., F. M. Barker, R. J. Hoffman, M. H. Stetson, J. P. Hanifin, P. L. Podolin, and M. D. Rollag. 1994. Ultraviolet regulation of neuroendocrine and circadian physiology in rodents. *Vision Research* 34:1521–1533.

Brainard, G. C., M. D. Rollag, and J. P. Hanifin. 1997. Photic regulation of melatonin in humans: ocular and neural signal transduction. *Journal of Biological Rhythms* 12:537–546.

Brillhart, D. B., and D. W. Kaufman. 1991. Influence of illumination and surface structure on space use by prairie deer mice (*Peromyscus maniculatus bairdii*). *Journal of Mammalogy* 72:764–768.

Bronson, F. H. 1989. *Mammalian reproductive biology*. University of Chicago Press, Chicago.

Butynski, T. M. 1984. Nocturnal ecology of the springhare, *Pedetes capensis*, in Botswana. *African Journal of Ecology* 22:7–22.

Clarke, J. A. 1983. Moonlight's influence on predator/prey interactions between short-eared owls (*Asio flammeus*) and deermice (*Peromyscus maniculatus*). *Behavioral Ecology and Sociobiology* 13:205–209.

Cresswell, W. J., and S. Harris. 1988. The effects of weather conditions on the movements and activity of badgers (*Meles meles*) in a suburban environment. *Journal of Zoology, London* 216:187–194.

Daly, M., P. R. Behrends, M. I. Wilson, and L. F. Jacobs. 1992. Behavioural modulation of predation risk: moonlight avoidance and crepuscular compensation in a nocturnal desert rodent, *Dipodomys merriami*. *Animal Behaviour* 44:1–9.

Dark, J., T. S. Kilduff, H. C. Heller, P. Licht, and I. Zucker. 1990. Suprachiasmatic nuclei influence hibernation rhythms of golden-mantled ground squirrels. *Brain Research* 509:111–118.

Dauchy, R. T., L. A. Sauer, D. E. Blask, and G. M. Vaughan. 1997. Light contamination during the dark phase in "photoperiodically controlled" animal rooms: effect on tumor growth and metabolism in rats. *Laboratory Animal Science* 47:511–518.

DeCoursey, P. J. 1986. Light-sampling behavior in photoentrainment of a rodent circadian rhythm. *Journal of Comparative Physiology A* 159:161–169.

Dice, L. R. 1945. Minimum intensities of illumination under which owls can find dead prey by sight. *American Naturalist* 79:385–416.

Elangovan, V., and G. Marimuthu. 2001. Effect of moonlight on the foraging behaviour of a megachiropteran bat *Cynopterus sphinx*. *Journal of Zoology, London* 253:347–350.

Emmons, L. H., P. Sherman, D. Bolster, A. Goldizen, and J. Terborgh. 1989. Ocelot behavior in moonlight. Pages 233–242 in K. H. Redford and J. F. Eisenberg (eds.), *Advances in neotropical mammalogy*. Sandhill Crane Press, Gainesville, Florida.

Erkert, H. G. 2000. Bats: flying nocturnal mammals. Pages 253–272 in S. Halle and N. C. Stenseth (eds.), *Activity patterns in small mammals: an ecological approach* (Ecological Studies, 141). Springer, Berlin.

Erkert, H. G. 2004. Extremely low threshold for photic entrainment of circadian activity rhythms in molossid bats (*Molossus molossus*; Chiroptera–Molossidae). *Mammalian Biology* 69:361–374.

Falkenberg, J. C., and J. A. Clarke. 1998. Microhabitat use of deer mice: effects of interspecific interaction risks. *Journal of Mammalogy* 79:558–565.

Foster, R. G., and I. Provencio. 1999. The regulation of vertebrate biological clocks by light. Pages 223–243 in S. N. Archer, M. B. A. Djamgoz, E. R. Loew, J. C. Partridge, and S. Vallerga (eds.), *Adaptive mechanisms in the ecology of vision*. Kluwer Academic Publishers, Dordrecht, The Netherlands.

Gerkema, M. P., G. A. Groos, and S. Daan. 1990. Differential elimination of circadian and ultradian rhythmicity by hypothalamic lesions in the common vole, *Microtus arvalis*. *Journal of Biological Rhythms* 5:81–95.

Gilbert, B. S., and S. Boutin. 1991. Effect of moonlight on winter activity of snowshoe hares. *Arctic and Alpine Research* 23:61–65.

Goodwin, B. J., D. J. Bender, T. A. Contreras, L. Fahrig, and J. F. Wegner. 1999. Testing for habitat detection distances using orientation data. *Oikos* 84:160–163.

Gwinner, E. 1986. *Circannual rhythms: endogenous annual clocks in the organization of seasonal processes*. Springer-Verlag, Berlin.

Halle, S., and N. C. Stenseth (eds.). 2000. *Activity patterns in small mammals: an ecological approach* (Ecological Studies, 141). Springer, Berlin.

Hecker, K. R., and R. M. Brigham. 1999. Does moonlight change vertical stratification of activity by forest-dwelling insectivorous bats? *Journal of Mammalogy* 80:1196–1201.

Hock, R. J. 1955. Photoperiod as stimulus for onset of hibernation. *Federation Proceedings* 14:73–74.

[IESNA] Illuminating Engineering Society of North America. 2000. *American national standard practice for roadway lighting*. ANSI/IESNA RP-8-00. IESNA, New York.

Johnson, D. H. 1999. The insignificance of statistical significance testing. *Journal of Wildlife Management* 63:763–772.

Joshi, D., and M. K. Chandrashekaran. 1985. Spectral sensitivity of the photoreceptors responsible for phase shifting the circadian rhythm of activity in the bat, *Hipposideros speoris*. *Journal of Comparative Physiology A* 156:189–198.

Julien-Laferrière, D. 1997. The influence of moonlight on activity of woolly opossums (*Caluromys philander*). *Journal of Mammalogy* 78:251–255.

Kaufman, D. W., and G. A. Kaufman. 1982. Effect of moonlight on activity and microhabitat use by Ord's kangaroo rat (*Dipodomys ordii*). *Journal of Mammalogy* 63:309–312.

Kie, J. G. 1996. The effects of cattle grazing on optimal foraging in mule deer (*Odocoileus hemionus*). *Forest Ecology and Management* 88:131–138.

Kotler, B. P. 1984. Effects of illumination on the rate of resource harvesting in a community of desert rodents. *American Midland Naturalist* 111:383–389.

Kramer, K. M., and E. C. Birney. 2001. Effect of light intensity on activity patterns of Patagonian leaf-eared mice, *Phyllotis xanthopygus*. *Journal of Mammalogy* 82:535–544.

Law, B. S. 1997. The lunar cycle influences time of roost departure in the common blossom bat, *Syconycteris australis*. *Australian Mammalogy* 20:21–24.

Lee, T. M., M. S. Carmichael, and I. Zucker. 1986. Circannual variations in circadian rhythms of ground squirrels. *American Journal of Physiology* 250:R831–R836.

Lockard, R. B., and D. H. Owings. 1974. Moon-related surface activity of bannertail (*Dipodomys spectabilis*) and Fresno (*D. nitratoides*) kangaroo rats. *Animal Behaviour* 22:262–273.

Lythgoe, J. N. 1979. *The ecology of vision*. Clarendon Press, Oxford.

McBride, G. B., J. C. Loftis, and N. C. Adkins. 1993. What do significance tests really tell us about the environment? *Environmental Management* 17:423–432.

McFarland, W., T. S. C. Wahl, and F. McAlary. 1999. The behaviour of animals around twilight with emphasis on coral reef communities. Pages 583–628 in S. N. Archer, M. B. A. Djamgoz, E. R. Loew, J. C. Partridge, and S. Vallerga (eds.), *Adaptive mechanisms in the ecology of vision*. Kluwer Academic Publishers, Dordrecht, The Netherlands.

Molenaar, J. G. de, R. J. H. G. Henkens, C. ter Braak, C. van Duyne, G. Hoefsloot, and D. A. Jonkers. 2003. *Road illumination and nature, IV. Effects of road lights on the spatial behaviour of mammals*. Alterra, Green World Research, Wageningen, The Netherlands.

Morrison, D. W. 1978. Lunar phobia in a neotropical fruit bat, *Artibeus jamaicensis* (Chiroptera: Phyllostomidae). *Animal Behaviour* 26:852–855.

Murray, M. G. 1982. The rut of impala: aspects of seasonal mating under tropical conditions. *Zeitschrift für Tierpsychologie* 59:319–337.

Nash, L. T. 1986. Influence of moonlight level on travelling and calling patterns in two sympatric species of *Galago* in Kenya. Pages 357–367 in D. M Taub and F. A. King (eds.), *Current perspectives in primate social dynamics*. Van Nostrand Reinhold Company, New York.

Negraeff, O. E., and R. M. Brigham. 1995. The influence of moonlight on the activity of little brown bats (*Myotis lucifugus*). *Zeitschrift für Säugetierkunde* 60:330–336.

Patel, H. I., and G. D. Gupta. 1984. A problem of equivalence in clinical trials. *Biometrical Journal* 26:471–474.

Pengelley, E. T., and K. C. Fisher. 1963. The effect of temperature and photoperiod on the yearly hibernating behavior of captive golden-mantled ground squirrels (*Citellus lateralis tescorum*). *Canadian Journal of Zoology* 41:1103–1120.

Perlman, I., and R. A. Normann. 1998. Light adaptation and sensitivity controlling mechanisms in vertebrate photoreceptors. *Progress in Retinal and Eye Research* 17:523–563.

Price, M. V., N. M. Waser, and T. A. Bass. 1984. Effects of moonlight on microhabitat use by desert rodents. *Journal of Mammalogy* 65:353–356.

Reed, D. F. 1995. *Efficacy of methods advocated to reduce cervid–vehicle accidents: research and rationale in North America*. Presented at "Wildlife–Traffic Collisions," January 23–28, Sapporo, Japan [Proceedings in Japanese; updated from presentation at Colloque International "Route et Faune Sauvage," Strasbourg, Conseil de l'Europe, June 5–7, 1985].

Reed, D. F., and T. N. Woodard. 1981. Effectiveness of highway lighting in reducing deer–vehicle accidents. *Journal of Wildlife Management* 45:721–726.

Reed, D. F., T. N. Woodard, and T. D. I. Beck. 1979. *Regional deer–vehicle accident research*. U.S. Department of Transportation Federal Highway Administration Report No. FHWA-RD-79-11. National Technical Information Service, Springfield, Virginia.

Rogowitz, G. L. 1997. Locomotor and foraging activity of the white-tailed jackrabbit (*Lepus townsendii*). *Journal of Mammalogy* 78:1172–1181.

Rydell, J., and H. J. Baagøe. 1996. Bats & streetlamps. *Bats* 14:10–13.

Shanahan, T. L., J. M. Zeitzer, C. A. Czeisler. 1997. Resetting the melatonin rhythm with light in humans. *Journal of Biological Rhythms* 12:556–567.

Sharma, V. K., M. K. Chandrashekaran, and P. Nongkynrih. 1997. Daylight and artificial light phase response curves for the circadian rhythm in locomotor activity of the field mouse *Mus booduga*. *Biological Rhythm Research* 28(Supplement):39–49.

Skinner, J. D., and A. S. van Jaarsveld. 1987. Adaptive significance of restricted breeding in southern African ruminants. *South African Journal of Science* 83: 657–663.

Stewart-Oaten, A., W. W. Murdoch, and K. R. Parker. 1986. Environmental impact assessment: "pseudoreplication" in time? *Ecology* 67:929–940.
Subbaraj, R., and J. Balasingh. 1996. Night roosting and "lunar phobia" in Indian false vampire bat *Megaderma lyra*. *Journal of the Bombay Natural History Society* 93:1–7.
Topping, M. G., J. S. Millar, and J. A. Goddard. 1999. The effects of moonlight on nocturnal activity in bushy-tailed wood rats (*Neotoma cinerea*). *Canadian Journal of Zoology* 77:480–485.
Travers, S. E., D. W. Kaufman, and G. A. Kaufman. 1988. Differential use of experimental habitat patches by foraging *Peromyscus maniculatus* on dark and bright nights. *Journal of Mammalogy* 69:869–872.
Vásquez, R. A. 1994. Assessment of predation risk via illumination level: facultative central place foraging in the cricetid rodent *Phyllotis darwini*. *Behavioral Ecology and Sociobiology* 34:375–381.
Vaughan, T. A. 1978. *Mammalogy*. Second edition. Saunders College Publishing, Philadelphia.
Vickery, W. L., and J. R. Bider. 1981. The influence of weather on rodent activity. *Journal of Mammalogy* 62:140–145.
Walls, G. L. 1942. *The vertebrate eye and its adaptive radiation*. Cranbrook Institute of Science Bulletin No. 19, Bloomfield Hills, Michigan.
Wolfe, J. L., and C. T. Summerlin. 1989. The influence of lunar light on nocturnal activity of the old-field mouse. *Animal Behaviour* 37:410–414.
Wright, P. C. 1981. The night monkeys, genus *Aotus*. Pages 211–240 in A. F. Coimbra-Filho and R. A. Mittermeier (eds.), *Ecology and behavior of neotropical primates*, Volume 1. Academia Brasileira de Ciências, Rio de Janeiro.
Zollner, P. A., and S. L. Lima. 1999. Illumination and the perception of remote habitat patches by white-footed mice. *Animal Behaviour* 58:489–500.
Zucker, I., M. Boshes, and J. Dark. 1983. Suprachiasmatic nuclei influence circannual and circadian rhythms of ground squirrels. *American Journal of Physiology* 244:R472–R480.

Chapter 3

Bats and Their Insect Prey at Streetlights

Jens Rydell

Bats have long been observed feeding on insects attracted to artificial light sources. Indeed, before the advent of high-tech bat observation equipment such as ultrasound detectors and infrared-sensitive video cameras in the 1980s, artificially lit places provided otherwise rare opportunities to observe bats hunting insects (Griffin 1958, Roeder 1967, Shields and Bildstein 1979), and this research tradition has continued ever since (Belwood and Fullard 1984, Schnitzler et al. 1987, Barak and Yom-Tov 1989, Hickey and Fenton 1990, Dunning et al. 1992, Acharya and Fenton 1999, Fullard 2001).

Experimental ultraviolet lights set up in otherwise dark areas rapidly attract bats, which feed on the insects that accumulate around the light (Fenton and Morris 1976, Bell 1980). The habit of feeding at artificial lights is now so common and widespread among bats that it must be considered part of the normal life habit of many species. Some bat species obtain a large part, perhaps even most, of their food at lights. Nevertheless, what bats and insects do in lit places is a special case. Lighted areas are not representative of the conditions to which bats and insects have become evolutionarily adapted. The phenomenon is new in an evolutionary sense, having increased rapidly only over the last century, following the spread of artificial lighting.

Many populations and species of bats are considered threatened or endangered. This is often thought to be because of habitat degradation and landscape fragmentation, which have negative effects on the availability of roosts for bats or on their insect prey (Hutson et al. 2001). Conservation of bats and their habitats is now an important issue in Europe and North America, and interest in bat conservation is spreading rapidly to other parts of the world as knowledge about tropical and subtropical bat populations improves (Hutson et al. 2001). Because bats often feed on insects attracted to lights, it seems possible or even likely that lights indirectly influence the survival and reproductive performance, and hence the conservation status, of both bats and insects. Therefore, the relationship between streetlights, bats, and insects is an important conservation issue.

Lights affect bats not only by providing food but probably also in other more and less subtle ways. For example, lights may interfere with navigation of bats during nocturnal migrations or commuting flights (Buchler and Childs 1982) or disrupt their circadian clocks (Erkert 1982). Bats feeding at lights may be exposed to increased predation risk from visually oriented raptorial birds such as bat hawks (*Macaerhamphus* spp.; Hartley and Hustler 1993). Such issues, which are largely unexplored, are not considered in this chapter. Instead, I deal with feeding by bats and behavior of their insect prey at artificial lights, particularly streetlights. I begin with an introduction to bat biology and a discussion of the flight patterns and morphology of insectivorous bats. This is followed by an explanation of the methods of measuring habitat use by bats, an evaluation of the use of streetlights for bat foraging, and an assessment of the consequences for bats, their insect prey, and their competitive interactions.

A Brief Introduction to Bats

The order Chiroptera (bats) represents about one-quarter of all mammal species. Corbet and Hill (1991) recognize 977 species of bats belonging to two suborders: the Macrochiroptera, or Old World fruit bats (162 species), most of which are large and principally nonecholocating and which feed on fruits, flowers, and nectar, and the Microchiroptera (815 species), most of which are small (usually 5–20 g body mass), use ultrasonic echolocation, and feed predominantly on insects.

The suborder Microchiroptera consists of 17 families that differ substantially in morphology, behavior, and ecology. In general, they occur worldwide except in the polar regions and on some isolated oceanic islands. They are found in nearly all kinds of terrestrial habitats, from forests to deserts and lakes. The diversity of bats is highest in the tropics and subtropics. All Macrochiroptera, as well as the plant-feeding and the carnivorous Microchiroptera, are tropical or subtropical, and the insectivorous Microchiroptera are most abundant and diverse near the equator as well. The insectivorous Microchiroptera also occur throughout the temperate zones, although their diversity varies widely between different regions and habitats and generally decreases rapidly toward the poles. Findley (1993) and Altringham (1996) provide good reviews of the diversity and distribution of the Chiroptera. This chapter addresses the insectivorous Microchiroptera exclusively.

Flight and Morphology of Insectivorous Bats

The size and weight of bats that hunt insects in the air by echolocation are strongly constrained by the need to maneuver rapidly in the air; bats need to be small to intercept and catch their prey. They need to approach and catch prey within a fraction of a second after it is detected because the detection range is very short (usually no more than 1–5 m [3–16 ft]; Barclay and Brigham 1991, Waters et al. 1995). The detection range is so short because ultrasound suffers high (frequency-dependent) attenuation and spreading loss in air, and the reflective power of an insect-sized target is small (Lawrence and Simmons 1982, Kick 1982).

Despite the general need to be small, insectivorous bats vary widely in size and wing morphology. These properties in turn correlate with the type of echolocation call used by a species. Together, size, wing morphology, and echolocation call characteristics are strong determinants of the flight speed, maneuverability, and prey detection capabilities of a bat species (Norberg and Rayner 1987) and therefore to a large extent determine habitat use and diet (Jones and Rydell 2003). As a gross generalization, large bats fly faster and are less maneuverable than small bats, and because the flight style is also influenced by the shape of the wings, long-winged bats generally fly faster and are less maneuverable than broad-winged bats. Furthermore, fast-flying bats use echolocation calls that permit longer prey detection ranges but are inefficient for detection of small prey. As a consequence, they tend to feed mostly on larger insects, including moths (Lepidoptera; Barclay 1985, 1986).

Big, fast-flying bats are confined to open airspace, such as that around and above streetlights, whereas smaller and slower bats also can exploit spatially (and acoustically) more complex habitats (Neuweiler 1984). For our purpose, it is relevant to consider which bats come to lights and how the different types of bats use the airspace around the light source.

Figure 3.1 represents a rough categorization of a typical street scene in a tropical area, in the city of Mérida in Yucatán, Mexico (Rydell et al. 2002, unpublished observations). The situation is similar elsewhere, although the bat species involved obviously are different. The aim here is to set the scene by illustrating how the airspace near a light source can be divided between different categories of bats. This partitioning of the space is to some extent independent of the geographic location and the particular species involved. At higher latitudes, however, fewer species and categories of bats are usually involved.

First, species of large (about 30–100 g), fast-flying bats, usually belonging to the free-tailed bats (e.g., *Eumops perotis* and other larger species of the family Molossidae), often roost in city buildings and hunt at high elevation over and outside the city (Bowles et al. 1990), typically exploiting insects over extensive areas with many lights. They often are

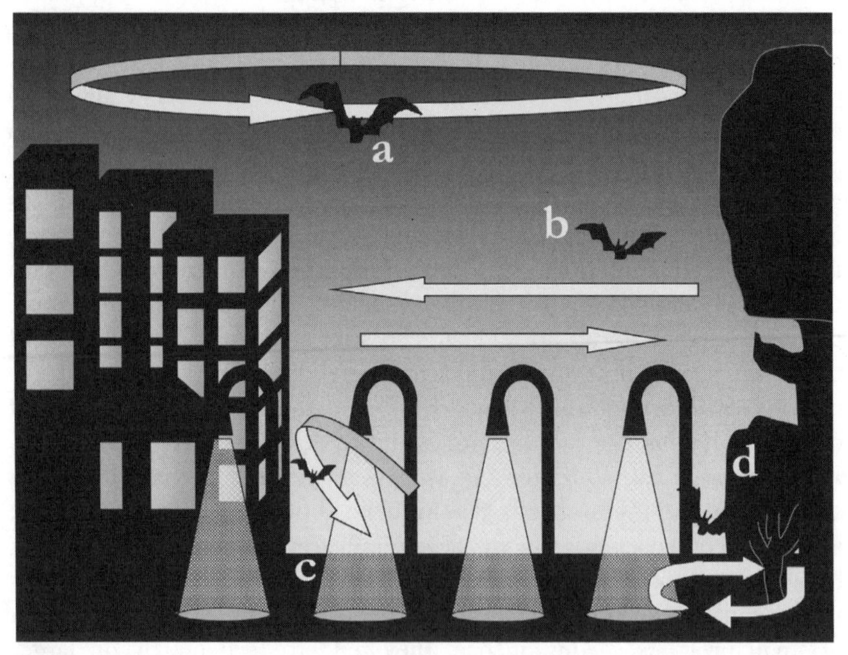

Figure 3.1. A street scene in Mexico: characteristic bat behaviors at and around streetlamps. *(a)* Large, fast-flying bats; *(b)* medium-sized, fast-flying bats; *(c)* small, fast-flying bats; *(d)* broad-winged, slow-flying but highly maneuverable bats.

found over lit sports fields and airports and similar places, sometimes in large numbers (Gould 1978, Bowles et al. 1990), and they even search for insects in the sky above the center of big cities, including Mexico City (Avila-Flores 2002). Large free-tailed bats almost never fly low enough to be seen from the ground, so their presence usually can be appreciated only by using an ultrasound detector. In the Old World, apart from the large free-tailed bats (e.g., *Otomops martiensseni*; Fenton et al. 2002), some large species of sheath-tailed bats (e.g., *Taphozous* spp.; Emballonuridae) behave in a similar way (Gould 1978).

Second, species of medium-sized (about 10–30 g) but still fast-flying bats, usually belonging to the Vespertilionidae (e.g., *Lasiurus* spp.), typically fly back and forth in straight flight along rows of streetlights and similar places. They typically patrol the street from well above the lights and can be seen only as they dive toward insects in the light cone. Again, their abundance usually can be appreciated only by using an ultrasound detector. This behavior is perhaps the most characteristic and universal for bats that hunt at lights and is representative of many typical "streetlight bats" all over the world, such as the red and hoary bats (*Lasiurus* spp.), big brown bats and serotines (*Eptesicus* spp.), noctules (*Nyctalus* spp.), and larger pipistrelles (*Pipistrellus* spp.), all belonging to the Vespertilionidae (Belwood and Fullard 1984, Geggie and Fenton 1985, Haffner and Stutz 1985/1986, Schnitzler et al. 1987, Kronwitter 1988, Barak and Yom-Tov 1989, Rydell 1992, Gaisler et al. 1998).

Third, several species of smaller (less than 10 g) but fast-flying bats are maneuverable enough to forage around single light posts or below the lights. Yucatán examples of this category include *Rhogeessa* spp., *Eptesicus furinalis* (Vespertilionidae), and *Pteronotus davyi* (Mormoopidae). Small *Pipistrellus* spp. (Vespertilionidae) represent this category in the Old World (Haffner and Stutz 1985/1986, Rydell and Racey 1995).

Fourth, at least one Yucatán bat species, the moustached bat *Pteronotus parnellii* (Mormoopidae), typically searches for food within the cover of vegetation rather than over lit roads but nevertheless sometimes passes in the light cone along the ground, presumably searching for insects. This species represents many broad-winged and slow-flying but highly maneuverable bats, including the so-called gleaners and flutter-detectors, most of which are seldom or never observed at lights (Rydell and Racey 1995). Nevertheless, there are exceptions. For example, the large horseshoe bat *Rhinolophus philippinensis* (Rhinolophidae), a flutter-detector, has been reported to feed regularly on insects at lights in Queensland, Australia (Pavey 1999).

In sum, insects at streetlights often are exploited by fast-flying bats of various sizes. These species use high-intensity echolocation calls and otherwise feed on flying insects that they find in more or less open habitats such as, for example, that above the forest canopy; along forest edges, rivers, or lakeshores; and in smaller gaps. Because these bats are equipped to feed in habitats that are structurally and acoustically relatively simple, they can be considered preadapted to make use of urban or suburban habitats with little vegetation, particularly if insects are attracted to such habitats by lights (Rydell 1992).

Evaluating the Importance of Streetlights for Bats

To evaluate the importance of streetlights for bats we need to know whether bats actually prefer to forage at streetlights and how the quality and quantity of food obtained by bats differ between foraging at streetlights and in other habitats.

Methods of Estimating Habitat Use of Bats

Two methods are available to study the habitat use of bats. Presence of bats in the air can be estimated by using an ultrasound detector or "bat detector" that transforms the ultrasonic echolocation calls to audible signals. Because different bat species use different types of signals, a bat detector also can be used to identify bat species (Ahlén 1981). The type of echolocation sound to some extent indicates what the bat is doing. For example, bursts of short pulses merging into a buzz indicate that the bat is closing in on its prey and therefore show unambiguously that the bat is feeding (Griffin et al. 1960). The absolute number of bats in an area is hard or impossible to estimate acoustically. Instead, the number of bat passes per unit time normally is used as an estimate of relative bat activity. Ultrasonic detectors can be used either automatically or manually and either from fixed positions or by a person moving along transects on foot (Gaisler et al. 1998) or from a bicycle or a car (Ahlén 1981). Counting bat passes from a car (moving at 30–50 km/h [20–30 mph]) is particularly efficient for estimating the relative abundance of bats along streets and roads (Ahlén 1981, Jüdes 1990, Rydell 1991, 1992).

The time that individual bats spend in particular places or habitats can be quantified by recording their positions through radiotelemetry. There are low-weight (less than 1 g) radio transmitters that bats can carry for extended periods (e.g., Kronwitter 1988, Catto 1993).

Use of Streetlights by Bats

Streetlights have a strong effect on the local distribution of several bat species, as shown by the methods described earlier (Rydell 1991, 1992, Rydell and Racey 1995). Evidence also indicates that bats that come to light do so primarily because they are attracted to the lights directly, not, for example, because they prefer to roost in nearby buildings (Blake et al. 1994).

Currently, three main kinds of lamps are used as streetlights: mercury vapor lamps, which emit a bluish white light that includes ultraviolet; low pressure sodium vapor lamps, which emit a nearly monochromatic yellow light (no ultraviolet); and high pressure sodium vapor lights, which emit mostly orange light but also some at shorter and longer wavelengths, including some ultraviolet. Because of these characteristics, mercury vapor lamps and, to a lesser extent, high pressure sodium vapor lamps attract insects and hence bats. Low pressure sodium vapor lamps appear to have no significant attractive effect on insects and bats (Rydell 1992).

Radiotracking and bat detector monitoring conducted in Europe and North America have shown that several bat species, all belonging to the fast-flying, aerial-hawking categories, feed around streetlights frequently and in preference to other habitats (Table 3.1). Concentrations of bats at lights may be large compared with those in surrounding habitats. For example, studies on the northern bat (*Eptesicus nilssonii*) and the parti-

Table 3.1. Bat species of Europe and North America that have been suggested to prefer (+) or avoid (–) foraging at lights.

Bat Species	*Location*	*Preference*	*References*
Eptesicus fuscus	Quebec, Canada	+	Geggie and Fenton 1985
	Ontario, Canada	+	Furlonger et al. 1987
Eptesicus serotinus	England	+	Catto 1993
Eptesicus nilssonii	Sweden	+	Rydell 1991, 1992
Vespertilio murinus	Sweden	+	Rydell 1992
Nyctalus noctula	Germany	+	Kronwitter 1988
Lasiurus borealis	Ontario, Canada	+	Furlonger et al. 1987
Lasiurus cinereus	Ontario, Canada	+	Furlonger et al. 1987
Pipistrellus pipistrellus	Switzerland	+	Haffner and Stutz 1985/1986
	Scotland	+	Rydell and Racey 1995
	England	+	Blake et al. 1994
Pipistrellus kuhlii	Switzerland	+	Haffner and Stutz 1985/1986
Myotis spp.	Ontario, Canada	–	Furlonger et al. 1987
	Sweden	–	Rydell 1992
Plecotus auritus	Sweden	–	Rydell 1992

colored bat (*Vespertilio murinus*; Vespertilionidae) along roads in Scandinavia have shown that densities of bats are 3–20 times higher in road sections with streetlights than in similar sections without lights (or in sections with low pressure sodium vapor lights). There were 2–5 bats, occasionally up to 20 bats, per kilometer of lit road, in the latter case equaling about one bat per lamppost (Rydell 1991, 1992). There is no reason to think that these figures are extreme in any way.

Other species, mostly slow-flying gleaners and flutter-detectors, avoid lights (Table 3.1), and so it seems likely that whereas lights affect some bats positively, they affect others negatively. For example, extensive radio-tracking has shown that the greater horseshoe bat (*Rhinolophus ferrumequinum*; Rhinolophidae) in England does not use lights, although roads with mercury vapor lights occur in abundance within their normal feeding range (Jones and Morton 1992, Jones et al. 1995). The same situation applies to the spotted bat (*Euderma maculatum*; Vespertilionidae), a long-eared and broad-winged North American species (Woodsworth et al. 1981, Leonard and Fenton 1983, Wai-Ping and Fenton 1989). There are also instances in which avoidance of lit areas by some European and North American species of mouse-eared bats (*Myotis* spp.) and long-eared bats (*Plecotus* spp.; Vespertilionidae) has been shown statistically (Furlonger et al. 1987, Rydell 1992).

Why some bat species usually avoid streetlights, however, is not clear. For example, *Myotis* spp. (Vespertilionidae) seldom turn up at streetlights, although these bats come to single lights set up for experimental purposes in desert scrub (Fenton and Morris 1976, Bell 1980). This suggests that they do not necessarily avoid streetlights directly, but perhaps they avoid the openness (treelessness) that prevails over most illuminated streets and roads.

Generally, the density of bats at streetlights usually is highest in suburban or rural areas and declines toward the town or city centers (Rydell 1992, Gaisler et al. 1998). This is probably because insect density declines drastically in urban areas, where concrete has replaced most or all of the natural vegetation (see Chapter 13, this volume). There also may be a dilution effect toward town and city centers, that is, fewer insects remaining per lamp when there are many lamps (Robinson and Robinson 1950). Nevertheless, as we already know, the centers of towns and cities are used to some extent by bats, even if the bats do not necessarily breed there.

An unusual example of an urban bat is the parti-colored bat (*Vespertilio murinus*; Vespertilionidae), a species known for its audible display songs that echo between buildings in Danish and Scandinavian towns and cities in late autumn. The females of this species migrate from their maternity colonies in the countryside to towns and cities in the autumn, where they mate and spend the winter in multistory buildings (Rydell and Baagøe 1994). It is not known whether these bats come to urban areas primarily because of the insects attracted by the lights or because of the potential roosts provided in the buildings.

Do Streetlights Enhance the Feeding of Bats?

There are usually more insects around streetlights than in surrounding, unlit areas (Hickey and Fenton 1990, Rydell 1992, Blake et al. 1994), and we may therefore assume that lights benefit bats by providing unusually rich food patches. Even away from lights, however, bats typically feed on insects that have accumulated for various reasons, including mating swarms and insects that have become concentrated in sheltered places by the wind. Therefore, to compare streetlights with other habitats in terms of quality as feeding habitat for bats, we need to estimate the average food intake rate of bats at lights in comparison with bats that feed in other habitats. By observing a population of the northern bat (*Eptesicus nilssonii*; Vespertilionidae) feeding in various rural habitats in Sweden and weighing the insects that occurred in each habitat, Rydell (1992) concluded that the food intake at lights typically was higher than the food intake in other habitats. Although the bats' capture rate of insects at lights was low, the insects that they caught there were mostly moths (Lepidoptera), and these were much larger than insects (flies and beetles) that they typically caught in other habitats. Therefore, the size of the average insects caught at lights more than compensated for the low capture rate in terms of food intake.

Although this study was limited to one species and one locality, circumstantial evidence exists from other parts of the world that the situation is similar elsewhere. For example, in Canada big brown bats (*Eptesicus fuscus*; Vespertilionidae) feeding at lights in urban areas have a low capture rate compared with bats in rural habitats without lights (Geggie and Fenton 1985). A low capture rate may indicate either that their food intake was low or, as suggested earlier, that the prey items caught were large and therefore required a long handling time. The literature abounds with observations suggesting that bats feeding at lights typically prey on large insects, usually moths (Belwood and Fullard 1984, Hickey and Fenton 1990, Dunning et al. 1992, Acharya and Fenton 1992, 1999, Fullard 2001).

An additional advantage of streetlights, from the bats' point of view, could be that they supply food on a temporally and spatially more predictable basis than other habitats. This aspect has not yet been investigated, however. In summary, although the evidence suggesting that streetlights enhance the feeding efficiency of bats is far from conclusive, the information we have indicates that this is generally true. In fact, it is possible that the benefit could be substantial.

No conclusive evidence exists that bat populations have increased as a result of streetlights, although this seems likely for several species such as *Pipistrellus pipistrellus* (Arlettaz et al. 2000) and *Eptesicus nilssonii* (Vespertilionidae; Rydell 1993) in Europe. Unfortunately, reliable population estimates for common bats such as these are not easy to obtain.

Bat–Bat Interactions at Streetlights

As already mentioned, aerial-hawking bats often feed on insects that occur in swarms or that have accumulated in a limited area for other reasons. Streetlights presumably provide easily located food, assuming that bats can see the lights over long distances. This seems likely because at least some bats (i.e., the big brown bat [*Eptesicus fuscus*]) are able to use distant light sources for orientation and navigation (Childs and Buchler 1981, Buchler and Childs 1982). The lights also may provide food at highly predictable sites and for extended periods of time.

Given that streetlights probably are profitable feeding sites and also are easy to locate, it is not surprising that bats often accumulate in areas with lights. Bats are also able to evaluate the quality of a particular feeding site and eavesdrop on other bats by listening to their echolocation calls from a distance. Bats that feed successfully thus attract other bats (Barclay 1982, Balcombe and Fenton 1988). Competition for feeding space may in turn lead to territorial behavior in the form of chases and audible vocalizations, resulting in either exclusion of individuals from the feeding site or subdivision of the feeding site between individuals (Rydell 1986). What appears to be aggressive behavior has been observed among bats at lights, for example, in red bats (*Lasiurus borealis*; Vespertilionidae), although in this instance there was no evidence of territoriality (Hickey and Fenton 1990), suggesting that when there are many bats feeding in the same place, they may compete directly for the same food items. On the other hand, it has also been suggested that Kuhl's pipistrelles (*Pipistrellus kuhlii*; Vespertilionidae) actually forage cooperatively on insects at lights in Israel (Barak and Yom-Tov 1989).

Baagøe (1986) has suggested, although speculatively, that competition for feeding space also may be interspecific and result in exclusion of one species by another from preferred feeding sites. For example, the serotine bat (*Eptesicus serotinus*; Vespertilionidae) and the parti-colored bat (*Vespertilio murinus*; Vespertilionidae) have allopatric distributions on Zealand, Denmark (Baagøe 1986). The two species are similar in size and feeding habits, and both often feed at lights. One possible mechanism behind the allopatry may be competitive exclusion of the smaller *Vespertilio murinus* by *Eptesicus serotinus* at lights. A similar situation has been suggested to explain why the common pipistrelle (*Pipistrellus pipistrellus*) and Kuhl's pipistrelle (*Pipistrellus kuhlii*; Vespertilionidae), both of which typically feed in large numbers along lit roads in Switzerland, almost never occur together (Haffner and Stutz 1985/1986).

A dramatic example of a bat population that may be negatively affected by streetlights is the European lesser horseshoe bat (*Rhinolophus hipposideros*; Rhinolophidae), a species that is declining rapidly in many areas. This species, which does not use streetlights, became locally extinct in several mountain valleys in Switzerland after streetlights were installed in these valleys. At the same time, the valleys were invaded by large numbers of pipistrelle bats (*Pipistrellus pipistrellus*), an expanding species that presumably was attracted by the lights. Because the two species are of similar size and feed on the same kind of insects, it seems possible that the local extinctions of the horseshoe bat were caused by competitive exclusion (Arlettaz et al. 2000).

Unfortunately, these examples are not yet supported by conclusive evidence. Competition is notoriously difficult to demonstrate in wild animals, and interspecific competition has not even been shown to exist among bats. Nevertheless, competition in bats in general and at lights in particular is an interesting research issue that deserves further study.

Bat–Insect Interactions at Streetlights

As a result of the coevolutionary "arms race" between bats and nocturnal insects, many insects have evolved ultrasonic detectors in the form of simple ears (Roeder 1967, Fullard 1998, Miller and Surlykke 2001). These ears form part of the primary defense against bats. Detection of ultrasound, indicating presence of a bat, normally triggers various kinds of defensive behavior, including evasive maneuvers or cessation of flight (Miller and Olesen 1979) and sometimes warning sounds (Dunning and Roeder 1965). The ultrasound-based defensive behavior has been most intensively studied in moths. Interactions between bats and moths have been studied frequently at streetlights, where direct observation of the behavior is fairly easy (Belwood and Fullard 1984, Acharya and Fenton 1992).

At least some bats that feed at lights increase their consumption of moths dramatically compared with what they catch in other habitats (Rydell 1992), suggesting that moths become easier to catch at lights. There could be several reasons for this. It has been shown that the evasive response of moths is switched off under bright light conditions such as near streetlights. Bright light presumably indicates diurnal conditions to a moth, under which an evasive response to ultrasound is likely to be maladaptive. During the day ultrasound is unlikely to come from bats (Svensson and Rydell 1998, Svensson et al. 2003, Acharya and Fenton 1999). It also seems possible that the effect of moths' evasive flights are limited in the vicinity of streetlights because of the lack of protective vegetation underneath.

Surprisingly, it has not yet been investigated whether and how bat predation at lights affects the size of moth populations. This is a pressing research need. Frank (1988; Chapter 13, this volume) has speculated that a substantial depletion of moths could be caused by lights, particularly in heavily lit urban and suburban areas, although this may not necessarily result from bat predation alone. Predation by bats at lights affects male moths much more than female moths (Acharya 1995), the consequences of which are unknown. This bias probably exists because males, which usually fly to search for mating opportunities, spend much more time in the air than females (Svensson 1996). This generalization does not necessarily apply to all species, however (Kolligs 2000).

Conclusion

Replacement of mercury vapor lamps with high pressure sodium vapor lamps, which do not attract insects to the same extent, is positive both for insects and bats. Such replacements, however, probably result in lower food intake and presumably lower reproductive success for some populations of aerial-hawking bats. It probably will have little or no direct effect on species that seldom or never feed at insect concentrations at lights, such as the horseshoe bats (Rhinolophidae), the long-eared bats (*Plecotus* spp.; Vespertilionidae), and most of the mouse-eared bats (*Myotis* spp.; Vespertilionidae). It seems likely, however, that the disuse of mercury vapor streetlamps may have indirect positive effects on some of these species. As discussed in this chapter, there is circumstantial evidence that in the presence of insect-attracting lights species that do not forage at streetlights, such as the European lesser horseshoe bat (*Rhinolophus hipposideros*), can be displaced by interspecific competition from species that exploit the lights (Arlettaz et al. 2000). On the other hand, some rare species are known to use streetlights in some areas but not in others, such as the European barbastelle (*Barbastella barbastellus*; Vespertilionidae; Sierro 1999), and for this and perhaps other species the overall effect of the disuse of white lights is harder to predict. In summary, the disuse of streetlights should be encouraged whenever possible for many reasons related to conservation of animals. Given that at least some streetlights are here to stay, the replacement of mercury vapor lights by sodium vapor lights, preferably low pressure sodium vapor, should be encouraged at least for bat and insect conservation.

The likely outcomes of predator–prey interactions vary in space and time and depend on the prevailing circumstances. For bats and nocturnal moths, for example, it seems as if artificial lighting has changed the likely outcome in favor of the bats by immobilizing the moths' sophisticated and highly efficient bat defense system. Lights thereby initiate an interesting "natural" experiment by changing the direction of a long-coevolved pathway between prey and predator (Rydell et al. 1995). At the same time, large-scale lighting probably results in conservation problems for both.

Acknowledgments

I thank Travis Longcore and Catherine Rich for their invitation to contribute to this volume and Johan Eklöf for making the figure and for comments on the manuscript. I was funded by the Swedish Research Council. This chapter is an update and extension of an earlier review of the subject (Rydell and Racey 1995).

Literature Cited

Acharya, L. 1995. Sex-biased predation on moths by insectivorous bats. *Animal Behaviour* 49:1461–1468.

Acharya, L., and M. B. Fenton. 1992. Echolocation behaviour of vespertilionid bats (*Lasiurus cinereus* and *Lasiurus borealis*) attacking airborne targets including arctiid moths. *Canadian Journal of Zoology* 70:1292–1298.

Acharya, L., and M. B. Fenton. 1999. Bat attacks and moth defensive behaviour around street lights. *Canadian Journal of Zoology* 77:27–33.

Ahlén, I. 1981. *Identification of Scandinavian bats by their sounds*. Department of Wildlife Ecology, Report 6. Swedish University of Agricultural Science, Uppsala.

Altringham, J. D. 1996. *Bats: biology and behaviour*. Oxford University Press, Oxford.

Arlettaz, R., S. Godat, and H. Meyer. 2000. Competition for food by expanding pipistrelle bat populations (*Pipistrellus pipistrellus*) might contribute to the decline of lesser horseshoe bats (*Rhinolophus hipposideros*). *Biological Conservation* 93:55–60.

Avila-Flores, R. 2002. *Habitat use by insectivorous bats in a mega-urban environment*. Paper presented at 32nd Annual North American Symposium on Bat Research, November 6–9, Burlington, Vermont.

Baagøe, H. J. 1986. Summer occurrence of *Vespertilio murinus* Linné, 1758 and *Eptesicus serotinus* (Schreber, 1780) (Chiroptera, Mammalia) on Zealand, Denmark, based on records of roosts and registrations with bat detectors. *Annalen des Naturhistorischen Museums in Wien B* 88/89:281–291.

Balcombe, J. P., and M. B. Fenton. 1988. Eavesdropping by bats: the influence of echolocation call design and foraging strategy. *Ethology* 79:158–166.

Barak, Y., and Y. Yom-Tov. 1989. The advantage of group hunting in Kuhl's bat *Pipistrellus kuhli* (Microchiroptera). *Journal of Zoology, London* 219:670–675.

Barclay, R. M. R. 1982. Interindividual use of echolocation calls: eavesdropping by bats. *Behavioral Ecology and Sociobiology* 10:271–275.

Barclay, R. M. R. 1985. Long- versus short-range foraging strategies of hoary (*Lasiurus cinereus*) and silver-haired (*Lasionycteris noctivagans*) bats and the consequences for prey selection. *Canadian Journal of Zoology* 63:2507–2515.

Barclay, R. M. R. 1986. The echolocation calls of hoary (*Lasiurus cinereus*) and silver-haired (*Lasionycteris noctivagans*) bats as adaptations for long- versus short-range foraging strategies and the consequences for prey selection. *Canadian Journal of Zoology* 64:2700–2705.

Barclay, R. M. R., and R. M. Brigham. 1991. Prey detection, dietary niche breadth, and body size in bats: why are aerial insectivorous bats so small? *American Naturalist* 137:693–703.

Bell, G. P. 1980. Habitat use and response to patches of prey by desert insectivorous bats. *Canadian Journal of Zoology* 58:1876–1883.

Belwood, J. J., and J. H. Fullard. 1984. Echolocation and foraging behaviour in the Hawaiian hoary bat, *Lasiurus cinereus semotus*. *Canadian Journal of Zoology* 62:2113–2120.

Blake, D., A. M. Hutson, P. A. Racey, J. Rydell, and J. R. Speakman. 1994. Use of lamplit roads by foraging bats in southern England. *Journal of Zoology, London* 234:453–462.

Bowles, J. B., P. D. Heideman, and K. R. Erickson. 1990. Observations on six species of free-tailed bats (Molossidae) from Yucatan, Mexico. *Southwestern Naturalist* 35:151–157.

Buchler, E. R., and S. B. Childs. 1982. Use of the post-sunset glow as an orientation cue by big brown bats (*Eptesicus fuscus*). *Journal of Mammalogy* 63:243–247.

Catto, C. M. C. 1993. *Aspects of the ecology and behaviour of the serotine bat (*Eptesicus serotinus*)*. Ph.D. thesis, University of Aberdeen, Scotland.

Childs, S. B., and E. R. Buchler. 1981. Perception of simulated stars by *Eptesicus fuscus* (Vespertilionidae): a potential navigational mechanism. *Animal Behaviour* 29:1028–1035.

Corbet, G. B., and J. E. Hill. 1991. *A world list of mammalian species*. Third edition. Oxford University Press, Oxford.

Dunning, D. C., L. Acharya, C. B. Merriman, and L. Dal Ferro. 1992. Interactions between bats and arctiid moths. *Canadian Journal of Zoology* 70:2218–2223.

Dunning, D. C., and K. D. Roeder. 1965. Moth sounds and the insect-catching behavior of bats. *Science* 147:173–174.

Erkert, H. G. 1982. Ecological aspects of bat activity rhythms. Pages 201–242 in T. H. Kunz (ed.), *Ecology of bats*. Plenum Press, New York.

Fenton, M. B., and G. K. Morris. 1976. Opportunistic feeding by desert bats (*Myotis* spp.). *Canadian Journal of Zoology* 54:526–530.

Fenton, M. B., P. J. Taylor, D. S. Jacobs, E. J. Richardson, E. Bernard, S. Bouchard, K. R. Debaeremaeker, H. ter Hofstede, L. Hollis, C. L. Lausen, J. S. Lister, D. Rambaldini, J. M. Ratcliffe, and E. Reddy. 2002. Researching little-known species: the African bat *Otomops martiensseni* (Chiroptera: Molossidae). *Biodiversity and Conservation* 11:1583–1606.

Findley, J. S. 1993. *Bats: a community perspective*. Cambridge University Press, Cambridge.

Frank, K. D. 1988. Impact of outdoor lighting on moths: an assessment. *Journal of the Lepidopterists' Society* 42:63–93.

Fullard, J. H. 1998. The sensory coevolution of moths and bats. Pages 279–326 in R. R. Hoy, A. N. Popper, and R. R. Fay (eds.), *Comparative hearing: insects*. Springer-Verlag, New York.

Fullard, J. H. 2001. Auditory sensitivity of Hawaiian moths (Lepidoptera: Noctuidae) and selective predation by the Hawaiian hoary bat (Chiroptera: *Lasiurus cinereus semotus*). *Proceedings of the Royal Society of London. Series B, Biological Sciences* 268:1375–1380.

Furlonger, C. L., H. J. Dewar, and M. B. Fenton. 1987. Habitat use by foraging insectivorous bats. *Canadian Journal of Zoology* 65:284–288.

Gaisler, J., J. Zukal, Z. Rehak, and M. Homolka. 1998. Habitat preference and flight activity of bats in a city. *Journal of Zoology, London* 244:439–445.

Geggie, J. F., and M. B. Fenton. 1985. A comparison of foraging by *Eptesicus fuscus* (Chiroptera: Vespertilionidae) in urban and rural environments. *Canadian Journal of Zoology* 63:263–266.

Gould, E. 1978. Opportunistic feeding by tropical bats. *Biotropica* 10:75–76.

Griffin, D. R. 1958. *Listening in the dark: the acoustic orientation of bats and men*. Yale University Press, New Haven, Connecticut.

Griffin, D. R., F. A. Webster, and C. R. Michael. 1960. The echolocation of flying insects by bats. *Animal Behaviour* 8:141–154.

Haffner, M., and H. P. Stutz. 1985/1986. Abundance of *Pipistrellus pipistrellus* and *Pipistrellus kuhlii* foraging at street-lamps. *Myotis* 23/24:167–172.

Hartley, R., and K. Hustler. 1993. A less-than-annual breeding cycle in a pair of African bat hawks *Machaeramphus alcinus*. *Ibis* 135:456–458.

Hickey, M. B. C., and M. B. Fenton. 1990. Foraging by red bats (*Lasiurus borealis*): do intraspecific chases mean territoriality? *Canadian Journal of Zoology* 68:2477–2482.

Hutson, A. M., S. P. Mickleburgh, and P. A. Racey. 2001. *Microchiropteran bats: global status survey and conservation action plan*. IUCN, Gland, Switzerland.

Jones, G., P. L. Duvergé, and R. D. Ransome. 1995. Conservation biology of an endangered species: field studies of greater horseshoe bats. Pages 309–324 in P. A. Racey and S. M. Swift (eds.), *Ecology, evolution and behaviour of bats*. Symposia of the Zoological Society of London No. 67. Clarendon Press, Oxford.

Jones, G., and M. Morton. 1992. Radio-tracking studies on habitat use by greater horseshoe bats (*Rhinolophus ferrumequinum*). Pages 521–537 in I. G. Priede and S. M. Swift (eds.), *Wildlife telemetry: remote monitoring and tracking of animals*. Ellis Horwood Ltd., New York.

Jones, G., and J. Rydell. 2003. Attack and defense: interactions between echolocating bats and their insect prey. Pages 301–345 in T. H. Kunz and M. B. Fenton (eds.), *Bat ecology*. University of Chicago Press, Chicago.

Jüdes, U. 1990. Analysis of the distribution of flying bats along line transects. Pages 311–318 in V. Hanák, I. Horácek, and J. Gaisler (eds.), *European bat research 1987*. Charles University Press, Prague.

Kick, S. A. 1982. Target-detection by the echolocating bat, *Eptesicus fuscus*. *Journal of Comparative Physiology A* 145:431–435.

Kolligs, D. 2000. Ökologische Auswirkungen künstlicher Lichtquellen auf nachtaktive Insekten, insbesondere Schmetterlinge (Lepidoptera) [Ecological effects of artificial light sources on nocturnally active insects, in particular on moths (Lepidoptera)]. *Faunistisch–Ökologische Mitteilungen* Supplement 28:1–136.

Kronwitter, F. 1988. Population structure, habitat use and activity patterns of the noctule bat *Nyctalus noctula* Schreb., 1774 (Chiroptera: Vespertilionidae) revealed by radio-tracking. *Myotis* 26:23–85.

Lawrence, B. D., and J. A. Simmons. 1982. Measurements of atmospheric attenuation at ultrasonic frequencies and the significance for echolocation by bats. *Journal of the Acoustical Society of America* 71:585–590.

Leonard, M. L., and M. B. Fenton. 1983. Habitat use by spotted bats (*Euderma maculatum*, Chiroptera: Vespertilionidae): roosting and foraging behaviour. *Canadian Journal of Zoology* 61:1487–1491.

Miller, L. A., and J. Olesen. 1979. Avoidance behavior in green lacewings. I. Behavior of free flying green lacewings to hunting bats and ultrasound. *Journal of Comparative Physiology A* 131:113–120.

Miller, L. A., and A. Surlykke. 2001. How some insects detect and avoid being eaten by bats: tactics and countertactics of prey and predator. *BioScience* 51:570–581.

Neuweiler, G. 1984. Foraging, echolocation and audition in bats. *Naturwissenschaften* 71:446–455.

Norberg, U. M., and J. M. V. Rayner. 1987. Ecological morphology and flight in bats (Mammalia; Chiroptera): wing adaptations, flight performance, foraging strategy and echolocation. *Philosophical Transactions of the Royal Society of London. Series B, Biological Sciences* 316:335–427.

Pavey, C. R. 1999. Foraging ecology of the two taxa of large-eared horseshoe bat, *Rhinolophus philippinensis*, on Cape York Peninsula. *Australian Mammalogy* 21:135–138.

Robinson, H. S., and P. J. M. Robinson. 1950. Some notes on the observed behaviour of Lepidoptera in flight in the vicinity of light-sources together with a description of a light-trap designed to take entomological samples. *Entomologist's Gazette* 1:3–20.

Roeder, K. D. 1967. *Nerve cells and insect behavior*. Revised edition. Harvard University Press, Cambridge, Massachusetts.

Rydell, J. 1986. Feeding territoriality in female northern bats, *Eptesicus nilssoni*. *Ethology* 72:329–337.

Rydell, J. 1991. Seasonal use of illuminated areas by foraging northern bats *Eptesicus nilssoni*. *Holarctic Ecology* 14:203–207.

Rydell, J. 1992. Exploitation of insects around streetlamps by bats in Sweden. *Functional Ecology* 6:744–750.

Rydell, J. 1993. *Eptesicus nilssonii*. *Mammalian Species* 430:1–7.

Rydell, J., H. T. Arita, M. Santos, and J. Granados. 2002. Acoustic identification of insectivorous bats (order Chiroptera) of Yucatan, Mexico. *Journal of Zoology, London* 257:27–36.

Rydell, J., and H. J. Baagøe. 1994. *Vespertilio murinus*. *Mammalian Species* 467:1–6.

Rydell, J., G. Jones, and D. Waters. 1995. Echolocating bats and hearing moths: who are the winners? *Oikos* 73:419–424.

Rydell, J., and P. A. Racey. 1995. Street lamps and the feeding ecology of insectivorous bats. Pages 291–307 in P. A. Racey and S. M. Swift (eds.), *Ecology, evolution and behaviour of bats*. Symposia of the Zoological Society of London No. 67. Clarendon Press, Oxford.

Schnitzler, H.-U., E. Kalko, L. Miller, and A. Surlykke. 1987. The echolocation and hunting behavior of the bat, *Pipistrellus kuhli*. *Journal of Comparative Physiology A* 161:267–274.

Shields, W. M., and K. L. Bildstein. 1979. Bird versus bats: behavioral interactions at a localized food source. *Ecology* 60:468–474.

Sierro, A. 1999. Habitat selection by barbastelle bats (*Barbastella barbastellus*) in the Swiss Alps (Valais). *Journal of Zoology, London* 248:429–432.

Svensson, A. M., J. Eklöf, N. Skals, and J. Rydell. 2003. Light dependent shift in the anti-predator response of a pyralid moth. *Oikos* 101:239–246.

Svensson, A. M., and J. Rydell. 1998. Mercury vapour lamps interfere with the bat defence of tympanate moths (*Operophtera* spp.; Geometridae). *Animal Behaviour* 55:223–226.

Svensson, M. 1996. Sexual selection in moths: the role of chemical communication. *Biological Reviews* 71:113–135.

Wai-Ping, V., and M. B. Fenton. 1989. Ecology of spotted bat (*Euderma maculatum*) roosting and foraging behavior. *Journal of Mammalogy* 70:617–622.

Waters, D. A., J. Rydell, and G. Jones. 1995. Echolocation call design and limits on prey size: a case study using the aerial-hawking bat *Nyctalus leisleri*. *Behavioral Ecology and Sociobiology* 37:321–328.

Woodsworth, G. C., G. P. Bell, and M. B. Fenton. 1981. Observations of the echolocation, feeding behaviour, and habitat use of *Euderma maculatum* (Chiroptera: Vespertilionidae) in southcentral British Columbia. *Canadian Journal of Zoology* 59:1099–1102.

Part II

Birds

Night, Maine Woods

Going out at night conjures up many allures, and thoughts of nocturnal excursions resonate with a glow of the mystery that makes life the adventure it ought to be instead of the drudgery it can be. I had early inkling of the night life. An hour or so after dark my father took me along into the woods to check his small mammal traps. I recall hearing the bugle of the elk and an occasional crashing in the woods. And then early in the morning we checked on his carefully dug pit traps to find shrews, mice, and carabid beetles that had an hour or two earlier been wandering about under the cover of darkness.

For visually oriented creatures, the unseen night world can be tinged with apprehension. I was not yet ten years old then, and I still half believed in gnomes of the forest living in the hollowed-out moundlike cushions of a pale green moss that grew in our dark woods. My sister and I placed white-flecked red amanita mushrooms in front of our gnome homes to provide shelter from the rain. I don't recall if we really believed in these fantasies, but they were alluring precisely because we had not yet established clear boundaries between the real and the imagined. Anything still seemed possible, and night blurred distinctions.

When we came to western Maine we were, as complete strangers, welcomed into the company of a lively and large family on what would in these days disparagingly be called a run-down farm. Given the traffic, it should have been. It was populated by a dozen cows, some pigs, dogs, cats, honeybees, many free-ranging bantam chickens, and a pet skunk. As for me, I had a cage full of caterpillars of various sorts; I was raising moths and butterflies instead of livestock. The barn had, along with its other inhabitants, a sizable population of mice, swallows, and starlings. The boundaries between the real and the imagined were by now to me more clearly defined, but those between wild nature and domesticity of human and beast were not. To me it seemed we'd been delivered to paradise. As folks at all the other nearby farms, we did the bare minimum to survive. The rest of the time was spent lining bees to find wild bee trees, fishing, coon hunting with the hounds, and deer hunting. In that order and, except for the first, mostly at night.

The white perch at the nearby Pease Pond reputedly bit best at night. I can't vouch for the truth of that assertion, except that they took to our worm-baited hooks reliably enough for Leona, the wonder woman who was the main civilizing influence of the place, to serve fish chowder with home-grown corn and pork rind on a regular basis. Every meal at the red-patterned linoleum-covered table next to the big black kitchen stove was a reminder of an excursion with Floyd and the boys, making our way through the woods in the evening, untying the boat by the alder bushes, pushing out past the lily pads, and then rowing and trolling lazily over the dark waters over which bats fluttered to catch insects and where fish rose to leave concentric circles on glassy surface. The fireflies were flashing from the tall black hemlocks leaning over from shore, and from up in the woods by the apple orchard that we had passed through, we heard the whip-poor-will giving its energetic refrain. Dog-tired, we returned with a pailful of perch.

A couple of years later we had our own, even more economically challenged farm. It was located on the other side of the same pond. After I had fallen in love with everything local, I sought to expand my contact in as many ways as possible; I deliberately went out to experience a blizzard in our woods, or a pounding rainstorm, or a dark overcast summer night.

It was wind-still and clear on my night out. And after I crossed the pasture and entered the maple–ash–beech woods, the dense leaf canopy shut out most of the starlight. Almost immediately I no longer saw the familiar tree trunks that usually served as landmarks, and I stumbled along like a blind person holding a hand in front of me. My only bearing was the direction of the barking dog at a distant farmstead. I sat down, propped myself up against a tree, and listened.

It was early in the summer (probably June) because the silence was not yet shattered by the orthopteran concert, the crickets and katydids that are diagnostic of late summer nights both here in New England and in the tropics. At first I heard nothing but the dog, but after awhile other night life revealed itself. I had, in the forests of Africa and New Guinea, been impressed as the incredibly diverse ringing voices of the birds during the day changed almost abruptly to the soft flutter of bats in the evening, accompanied by the pulsing of katydids and other orthopterans. The shrill and steady orthopteran sound pulses unceasingly, unvaryingly. Yet there are many species, each with its own pitch and cadence. The volume is at first almost overwhelming, but as with the steady roar of my Maine hometown woolen mill with hundreds of looms working simultaneously, we eventually filter out steady sounds to be alert to the new. Thus, we also filter out the sometimes most amazing, which is not necessarily the rare.

Listening during my night out in the Maine summer woods I at first also noted a steady sound. But it was a faint, barely perceptible patter that was noticeable only because it was amplified in my brain, focused by the still darkness. It could have been steady but sparse raindrops on the leaves, except that the night sky was clear. By the process of elimination I eventually decided I was hearing caterpillar feces falling on leaves. And not just those of any caterpillars. These were those primarily of moths, like those of most Catacola, some sphingids, and some of the large-bodied geometrids. All would have been camouflaged in the daytime because of their disguises involving shape, color, and behavior. Most of these animals lie motionless and without feeding all day long, when they are disguised as sticks and bark. Only at night do they venture forth out on the branches to feed, where they then gain an edge of safety from the relentlessly searching sharp-eyed birds, who are alert to the slightest movement of their prey. With the sound of caterpillar rain as a steady backdrop, I then heard the slow lumbering of a porcupine. I heard squeaks. I heard the faint peep of a bird in the canopy. Possibly this bird had been dreaming. But not all day-active songbirds sleep soundly at night. I have often been surprised in the spring that our common ground-dwelling ovenbird, which gives its familiar and monotonous "teacher, teacher, teacher" from the forest floor in the breeding season, will at night occasionally flutter up through and over the forest canopy and sail over it to burst forth in a musical lark-like refrain that bears not the slightest resemblance to its song in the daytime. Some animals do things differently at night; they then become extraordinary.

Bernd Heinrich

Chapter 4

Effects of Artificial Night Lighting on Migrating Birds

Sidney A. Gauthreaux Jr. and Carroll G. Belser

Many hundreds of species of birds typically migrate at night, and it is well known that fires and artificial lights attract birds during migration, particularly when the sky is cloudy and the ceiling is low. Romanes (1883) was first to discuss the similarities of the attraction of insects to a flame at night, birds to lighthouses, and fish to lanterns. In some instances, humans have exploited the attraction of migrating and local birds to lighted buildings, floodlights, and spotlights. In one early example, hunters used a simple reflecting lamp to attract shorebirds at night. "[T]he birds came all around and about them—like chickens when called to feed," reported the St. Augustine *Press* (quoted in Hallock 1874:150). In Jatinga, a small village on a ridge in the North Cachar Hills district of Assam in northeastern India, from August to October on moonless, foggy nights with south winds and drizzle, villagers use searchlights and lanterns to attract, capture, and kill hundreds of local birds for food (Dubey 1990). Up to fifty species have been collected, with herons and egrets being some of the largest victims and pittas and kingfishers representing some of the smaller species. In Africa the attraction of nocturnal migrants to artificial lights at lodges and to automobile headlights has been used to enhance ecotourism (Backhurst and Pearson 1977, Nikolaus 1980, Nikolaus and Pearson 1983).

As human populations expand geographically, artificial lighting also expands, and it is now almost impossible to find areas that are free from its influence. Verheijen (1981b) was first to apply the term *photopollution* to situations in which artificial light has adverse effects on wildlife. His 1985 review elaborates on the concept of photopollution and highlights incidents involving birds and sea turtles, natural and artificial light fields, orientation issues, and remedies (Verheijen 1985). All evidence indicates that the increasing use of artificial light at night is having an adverse effect on populations of birds, particularly those that typically migrate at night.

In this chapter we provide a review of the literature on the attraction of birds to light at night. We first examine how and why birds are attracted to light and the mechanisms of avian vision. We then review examples of this attraction, organized by the type of lighting: lighthouses and lightships, floodlights and ceilometers, city lights and horizon glows, fires and flares, and broadcast and communication towers. We then report our observations of the response of migratory birds to lights on communication towers. We conclude with some specific recommendations to minimize light attraction of migrating birds and reduce the associated mortality.

Mechanisms of Bird Attraction to Artificial Light

Little is known about how birds are attracted to light at night (Verheijen 1985). It has been suggested that when a bird flies into lights at night it loses its visual cues to the horizon, and the bird uses the lights as a visual reference, resulting in spatial disorientation (Herbert 1970). According to Herbert (1970), birds using a light on a tower as a horizon cue would circle the tower, which may be a factor in bird attraction to lights on tall communication towers. Exposure to a light field at night causes alteration of a straight flight path (e.g., hovering, slowing down, shifting direction, or circling), and the change in flight path would keep the bird near the light source longer than if the flight path remained straight (Gauthreaux and Belser 1999, unpublished data). Under such circumstances *attraction* may not be as appropriate a term for the behavioral response as *capture* (Verheijen 1958). It is also likely in some cases that the intensity of the light bleaches visual pigments so that the birds are in effect blinded and can no longer see visual details that they could detect when dark adapted (Verheijen 1985).

There is also evidence that horizon glows from cities may influence the orientation behavior of caged migratory birds. It is well established that caged migratory birds often orient toward horizon glows produced by the lights of cities (Kramer 1949, 1951). Immature migratory birds may be more susceptible to the disruptive influences of artificial night lighting than adults (Gauthreaux 1982). In two different experiments that examined the age-dependent orientation of caged migratory birds, it was found that birds of the year responded to the sky glow of a city, whereas adults did not (Gauthreaux 1982). In the first experiment performed by Williams (1978) during the spring of 1978, 8 immature and 14 adult white-crowned sparrows (*Zonotrichia leucophrys leucophrys*) were tested in six circular, automatic orientation cages once the birds exhibited migratory restlessness (*Zugunruhe*). All field tests were conducted under relatively clear, starry skies with no more than a thin crescent moon and essentially calm winds. Each bird was tested for a total of four nights, except for one immature that was tested for an additional night. Both age classes were tested each night, starting just before sunset and ending about 7 A.M. local time.

Figure 4.1 shows the distribution of the summed activity in minutes for each treadle in the orientation cage for adults and immatures and the distribution of horizon glow around the test cages. The orientation of the adult and immature groups is significantly different ($p = 0.05$) from a uniform distribution (Rayleigh test), and the mean direction of the adult group is significantly different ($p = 0.05$) from that of the immature group (F test of Watson and Williams). The circular distribution of horizon glow as measured with a sensitive photometer also is shown in Figure 4.1. The two longest radii at azimuths 90° and 105° have the maximum horizon glow intensity of 0.0096 lux. The vector resultant of the horizon glow is 80° ($r = 0.455$).

The mean direction of the activity of the adult sparrows was largely seasonally appropriate, but most of the activity of the immatures was oriented toward the direction of maximum horizon glow. The results show that age is a factor in the influence of horizon glows on the orientation of caged migratory birds, and although this is a study of caged birds it suggests that free-flying birds could respond similarly.

In a related study by Beacham (1982), 12 immature and 12 adult indigo buntings (*Passerina cyanea*) were tested in six circular, automatic orientation cages once the birds exhibited migratory restlessness in the spring of 1981 (Gauthreaux 1982). Each bird was tested for a minimum of four nights and a maximum of seven nights. Six birds from each age group (12 total) were tested in Emlen funnels (newsprint funnels with an

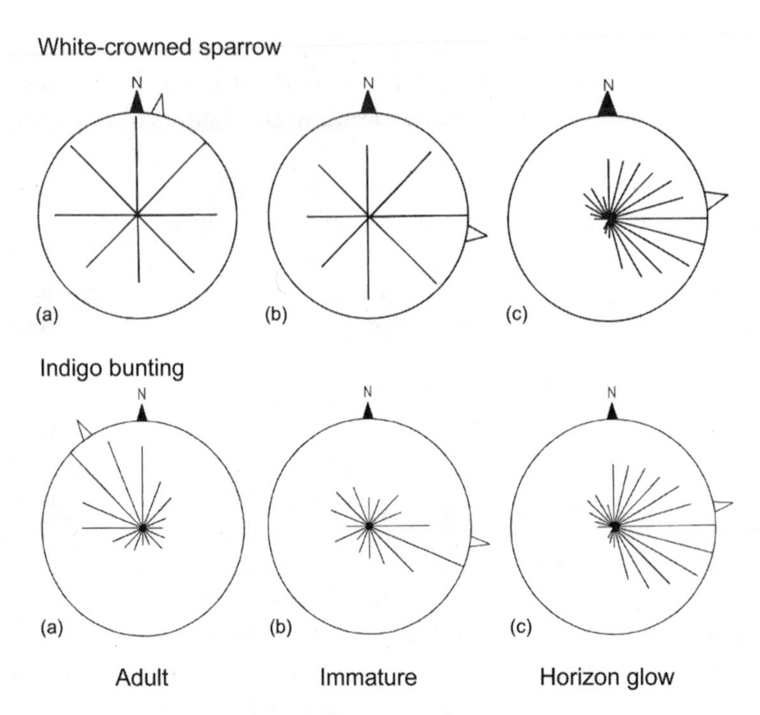

Figure 4.1. Distribution of migratory restlessness of *(a)* adult and *(b)* immature white-crowned sparrows and indigo buntings and *(c)* pattern of horizon glow around orientation cages. For white-crowned sparrows the longest radius is 17,015 minutes of activity toward 45° for adults and 8,265 minutes toward 90° for immatures. The resultant vector (mean direction) of all activity is 12° (r = 0.097) in *(a)* and 100° (r = 0.138) in *(b)*. For indigo buntings the longest radius is 5,221 hops toward 315° for adults and 4,402 hops toward 112.5° for immatures. The resultant vector (mean direction) of all hops is 327.6° (r = 0.392) in *(a)* and 98.5° (r = 0.181) in *(b)*.

ink-soaked sponge at the base). The funnel was divided into 16 sectors for analysis, and the number of hops in a sector was computed by comparing the density of hops with a scale of density patterns containing known numbers of hops. Figure 4.1 shows the distribution of the activity for adults and immatures as well as the pattern of horizon glow around the test funnels. The resultant vectors are significantly different (p = 0.05, Watson and Williams test) for the two age groups. As with the white-crowned sparrow, the greatest number of hops in the immature group is directed toward the greatest intensity of horizon glow, whereas the activity of the adults is oriented in a seasonally appropriate direction. Experimental studies of the migratory orientation of caged European robins

(*Erithacus rubecula*) also found that the birds oriented toward the vector resultant of the distribution of illumination in the night sky (Katz and Vilks 1981).

Birds have five different types of visual pigment and seven different types of photoreceptor: rods, double (uneven twin) cones, and four types of single cones (Hart 2001). Birds have a four-cone system and therefore broader spectral sensitivity than humans with a three-cone system (Wessels 1974, Graf and Norren 1974, Norren 1975). The extra cone type of birds is responsive to wavelengths in the ultraviolet range of the spectrum. In addition, bird eyes have oil droplets of different colors that narrow receptor sensitivities (Partridge 1989, Vorobyev et al. 1998). Because of these differences birds likely see their environment differently than do humans, which makes it difficult to speculate about the mechanism of how light pollution affects migrating birds at night. Another possible influence of artificial lighting on the behavior of night-migrating birds relates to the magnetic compass that several species use for direction finding during migration, as discussed in greater detail later in this chapter.

Sources of Light That Attract Birds

The tendency of birds to move toward lights at night when migrating and their reluctance to leave the sphere of light influence once encountered has been well documented. We review the various contexts in which birds have been attracted to lights and illustrate those that cause bird mortality. Although death or injury from collisions with structures is the most obvious adverse effect for migrating birds, attraction to lights may have other adverse consequences such as reducing energy stores necessary for migration because of delays and altered migration routes, mortality from collisions with glass during the daytime, and delayed arrival at breeding or wintering grounds.

Lighthouses and Lightships

Lighthouses and lightships (Figure 4.2) have attracted migrating birds since they were first operated (Dutcher 1884, Miller 1897, Hansen 1954), and this attraction was the basis of early, detailed studies of bird migration (e.g., Barrington 1900, Clarke 1912). In the 1800s lighthouse keepers noted that birds struck the lanterns most often on dark, cloudy nights with haze, fog, or rain, and that bird strikes on clear nights were extremely rare (Brewster 1886, Dixon 1897:268–274). Early studies supported the

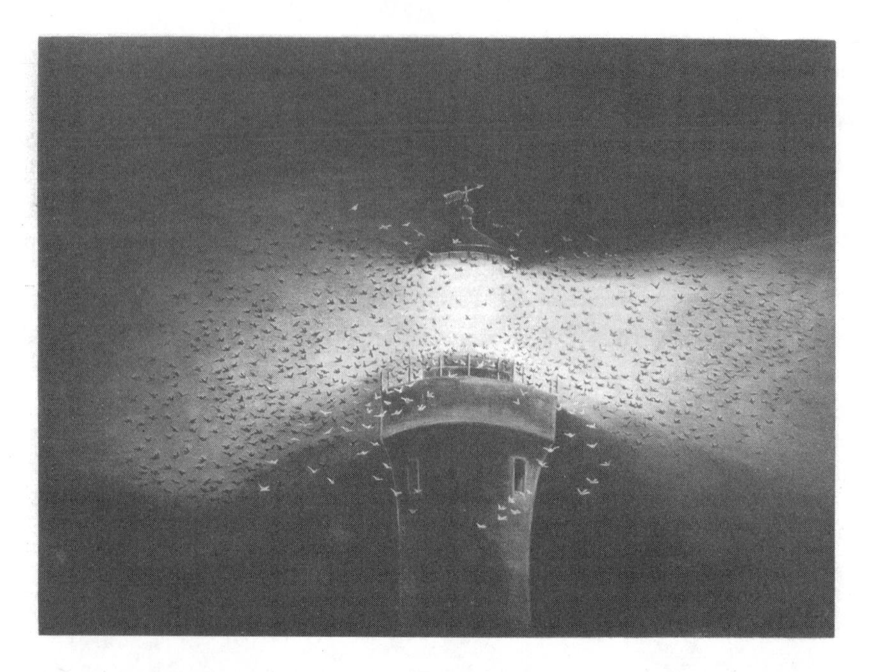

Figure 4.2. Lighthouses and lightships have attracted migrating birds at night since they were first operated. Above, an illustration of the Eddystone Lighthouse southwest of Plymouth, England, with nocturnal migrants milling around the lantern. Below, the Kentish Knock Lightship in the North Sea northeast of the coast of Kent, England. Illustrations are by M. E. Clarke and appear as frontispieces in volumes 1 and 2, respectively, of Clarke (1912).

notion that lightships attracted more birds than lighthouses on headlands and islands (Dixon 1897), and in more recent times lights on ships continue to attract nocturnal migrants crossing expanses of water (e.g., lighted ship in Gulf of Mexico, Bullis 1954; Cross Rip lightship in Nantucket Sound, Massachusetts, Bagg and Emery 1960; ore transport steamer on Lake Superior, Green and Perkins 1964).

An early survey of 24 lighthouses in North America showed that destruction of migrant birds was greatest at lighthouses in the Carolinas, Florida, and Louisiana, lower in the Northeast, and not present at two lighthouses in San Francisco Bay (Allen 1880). In a study of the destruction of migrating birds at lighthouses along the California coast, Squires and Hanson (1918) found relatively little collision mortality and attributed this to fewer migration waves in the western United States and less migration along the coastline. In contrast, a survey of 45 lighthouses on the coast of British Columbia by Munro (1924) found that nine reported high mortality rates, and overall annual mortality was more than 6,000 birds. Tufts (1928) surveyed lighthouses in the Canadian maritime provinces and Newfoundland and found that 45 of 197 experienced bird mortalities. Another survey of lighthouse keepers along the North American coasts and 225 lighthouse keepers in Central and South America and in the West Indies indicated that bird collisions were annual occurrences on the coasts of North America and along both coasts of South America and in the West Indies, especially along the northern coast of Cuba (Merriam 1885).

The literature on how migrant birds react to different types of lights at lighthouses and lightships is confusing. An analysis of birds that were killed (collided or shot) or caught at 58 light stations from 1881 through 1897 indicated that fixed lights were more attractive to migrants than rotating or blinking lights and that white lights were more attractive than red lights (Barrington 1900). Dixon (1897:269) also noted that "fixed white lights were more deadly than revolving or coloured lights." Thomson (1926:333–334) commented that it was known that "coloured lights do not attract the birds as white ones so fatally do." He added, "It therefore seems not unreasonable to ask that serious consideration should be given to the question of the gradual substitution of coloured for white lights at stations where great destruction commonly occurs and when the change may not be found to be impracticable for navigation reasons." Similarly, Cooke (1915) noted that a fixed white light beam attracted birds, a flashing light frightened them away, and red light was avoided. When the light beam of the lighthouse at Long Point, Lake Erie, Ontario, Canada was made narrower and dimmer in 1989, a dramatic reduction in avian mortality occurred (Jones and Francis 2003). From 1960 through 1989 mean annual kills were 200 birds in spring and 393 in autumn, and from 1990–2002 mean annual kills dropped to 18.5 birds in spring and 9.6 in autumn.

The above findings, however, are not consistent with those reported by other authors. Munro (1924) reported that flashing and rotating lights caused more mortality than fixed lights, and the results of a survey of 135 lighthouse keepers showed that the responses of birds to different types of lighting at lighthouses varied widely (Lewis 1927), with flashing white lights causing the greatest mortality and fixed beacons and red lights attracting fewer birds. In a survey of bird mortality at lighthouses in the maritime provinces of Canada and Newfoundland, 152 lighthouses recorded no losses, of which 131 had fixed lights and 21 had flashing or rotating lights (Tufts 1928). Approximately half of the 45 lighthouses reporting some mortality had fixed lights, and half had flashing lights (Tufts 1928). The confusing results about the responses of birds to different types of lighting at lighthouses and lightships probably are related to the characteristics of the individual lamps, such as the wavelength or intensity, because the responses of the birds changed when the type of lamp was changed at a station.

In the early 1900s, when gas and kerosene lanterns at lighthouses were replaced with electric lamps, collision mortality decreased (Hansen 1954). When the original revolving white beacon at the Dungeness Light in Great Britain was replaced with a xenon-filled lamp that produced a bluish beam that flashed for one second in ten, attraction and mortality of migrants were eliminated (Baldwin 1965). When foghorns were placed near lighted structures, the number of birds striking the lights decreased dramatically (Dixon 1897).

Floodlights and Ceilometers

In an attempt to further reduce collision mortality at lighthouses, some were floodlit, and this practice produced mixed results. Illuminating a lighthouse in Denmark with a floodlight increased mortality (Hansen 1954). A similar increase in mortality followed illumination of the rotating beam Long Point lighthouse in Ontario (Baldwin 1965). In contrast, when five lighthouses in England were floodlit the number of collisions declined (Baldwin 1965). There is now considerable evidence to indicate that illuminating chimneys, buildings, bridges, and monuments with floodlights attracts and kills migrating birds, particularly on nights during the fall and spring that are misty with a low cloud layer (e.g., the 169-m [555-ft] Washington Monument in the District of Columbia [Overing 1938]; Bluff's lodge on the Blue Ridge Parkway, Wilkes County, North

Carolina [Lord 1951]; the Long Point lighthouse [Baldwin 1965]; buildings on Holston Mountain, Tennessee [Herndon 1973]; and the 108-m [354-ft] Perry International Peace Monument and nuclear power plant cooling towers [Jackson et al. 1974]). Searchlights can also influence the flight behavior of migrating birds at night. Bruderer et al. (1999) switched on and off a strong searchlight mounted parallel to a radar antenna while tracking single migrants at night. The light beam caused a wide variation in shifts of flight direction (an average of 8° in the first 10 seconds and an average of 15° in the third 10-second interval). The mean velocity of the birds was reduced by 2–3 m (7–10 ft) per second (15–30% of normal airspeed), and climbing rate showed a slight increase, a possible response to escape the light beam. These effects declined with distance from the light source, and Bruderer et al. (1999) calculated that no reactions to the light should occur beyond 1 km (0.6 mi).

In the late 1940s meteorologists began using very bright (more than 1,000,000 candela), fixed-beam, vertically pointing spotlights, called ceilometers, to measure the height of the cloud ceiling at airports and weather stations. A rotating sensor measured the angle of the light spot on the cloud layer from a fixed distance from the base of the vertically pointing light beam, and the instrument computed the ceiling height using a simple trigonometric function. From 1948 through 1964, on overcast, misty nights these instruments were responsible for great losses of migrating birds. One of the earliest reports of mortality at a ceilometer was recorded at Nashville, Tennessee on the night of September 9–10, 1948 (Spofford 1949). On this night low clouds and misty conditions were present, and migrating birds congregated where the intense beam illuminated the cloud layer. Some circled, some collided, and some even dived into the ground. The largest kill of migrating birds ever recorded at a ceilometer, approximately 50,000, occurred October 6–8, 1954 at Warner Robins Air Force Base near Macon, Georgia, when a cold front moved over the Southeast (Chamberlain 1955, Johnston 1955). Howell et al. (1954) summarized the weather conditions that caused the buildup of migrants at ceilometers, the behavior of birds in and around the light beam, and the causes of injuries and mortality. In many instances the mortality of migrants at light beams would have been greater had it not been for meteorologists turning off ceilometers when birds began to accumulate in and around the beam (Ferren 1959, Fobes 1956, Green 1963, 1965).

Two changes to ceilometers greatly reduced and eventually eliminated the attraction and mortality of migrants. The first change involved filtering the wavelength of the light. When the longer wavelengths of ceilometer lamps were filtered so that mainly ultraviolet light remained, the attraction was greatly reduced and mortalities were essentially eliminated (Laskey 1960, Terres 1956). The second change made in the early 1960s by the National Weather Service was the replacement of fixed-beam ceilometers with rotating beam units (Velie 1963). In the new units the ultraviolet beam rotated and the detector was stationary, and the rotating beam did not attract birds (Avery et al. 1980:5).

City Lights and Horizon Glows

In 1886, Gastman reported that nearly 1,000 migratory birds were killed around electric light towers in Decatur, Illinois on the evening of September 28 (Gastman 1886). Kumlien (1888) provided detailed accounts of migrating birds striking a Milwaukee building a year later, from September 22 to 29, 1887. The building had a tower 61 m (200 ft) above street level and was illuminated by four floodlights. Since these early reports we have seen steady increases in the number of streetlights, the number and sizes of cities, the heights of office buildings, and the number of offices with lights on after dark. In 1951 after a single stormy spring night, 2,421 dead migrants of 39 species (mostly warblers) were gathered beneath light poles on Padre Island, Texas (James 1956). The mortality of birds attracted to the lights of tall buildings has also increased. The hazards of lighted structures and windows to migrating birds are well documented in a report published by World Wildlife Fund Canada and the Fatal Light Awareness Program (Evans Ogden 1996). The executive summary concludes, "The collision of migrating birds with human-built structures and windows is a world-wide problem that results in the mortality of millions of birds each year in North America alone" (Evans Ogden 1996:2). This publication contains valuable references on the subject of light attractions. Two additional publications contain annotated bibliographies of avian mortality at human-made structures. A bibliography of bird kills as a result of attraction to lighted structures and possible solutions to the problem can be found in Weir (1976), and a bibliography of 1,042 references on this subject has been compiled by Avery et al. (1980). The latter effort has a very useful subject index, taxonomic index, geographic index, and author index.

Fires and Flares

Fires on the ground can attract birds during nocturnal migration. In March 1906 migrating birds were attracted to a large lumberyard fire in Philadelphia (Stone 1906). Stone noted that the birds did not change their flight direction as they flew over, but they appeared to lower their flight altitude. Some 30 birds were burned to death when they came too close to the flames. Gas flares on offshore oil and gas platforms and at oil refineries also pose a threat to migrating birds at night (see Chapter 5, this volume). Numerous reports of mass mortality of migrating songbirds at gas flares on oil platforms in the North Sea have been reported (Sage 1979), and Tornielli (1951) has reported an incident in Italy. Birds congregate around the flares on misty and foggy nights, and as they fly near and through the flames they are burned to death. Newman (1960) reports on an event after midnight on April 30, 1960 at an oil refinery in Baton Rouge, Louisiana. More than 1,000 migrants of 17 species were killed when they were attracted to a 76.2-m (250-ft) gas flame illuminating a low overcast sky. A similar incident occurred in late May 1980 in northwest Alberta (Bjorge 1987). On this evening approximately 3,000 birds of 26 species were found dead within 75 m (246 ft) of a 104-m (341-ft) flare stack.

Broadcast and Communication Towers

Television and FM radio station towers have steadily increased in height above ground level since they were first constructed in the late 1940s (Aldrich et al. 1966, Malakoff 2001, Manville 2001, in press). In the mid-1960s it was estimated that television towers in the United States killed more than a million birds per year (Aldrich et al. 1966). By the mid-1970s, 26 towers were between 580 m and 630 m (1,902–2,067 ft) above ground level in the United States, heights that penetrate the altitudinal layer where songbirds typically migrate. Taller towers need more stabilizing guylines and warning lights for aircraft, and it is documented that aircraft warning lights (Cochran and Graber 1958, Avery et al. 1976) and guylines (Brewer and Ellis 1958) on such towers are responsible for the deaths of hundreds of thousands of birds during nocturnal migration. Over the years many individuals have collected dead and injured birds at the base of broadcast towers and documented their findings in local and regional ornithological journals and newsletters. Some of the more extensive studies of tower kills have emerged as classics (e.g., Brewer and Ellis 1958, Stoddard 1962, Stoddard and Norris 1967, Crawford 1974). Only a few studies have continued into the 1990s (e.g., Kemper 1996, Nehring 1998, Morris et al. 2003), and these studies indicate a significant decline in the number of tower fatalities over the last 20 years (Figure 4.3). More work is needed to distinguish between the roles of evolutionary adaptation, behavioral habituation, declining populations of migratory birds, changing weather conditions, and changes in tower lighting systems as possible explanations for such declines (Morris et al. 2003, Clark et al. 2005).

Bird kills at tall lighted structures in the United States and at Dutch lighthouses show similar lunar periodicity (Verheijen 1980, 1981a,

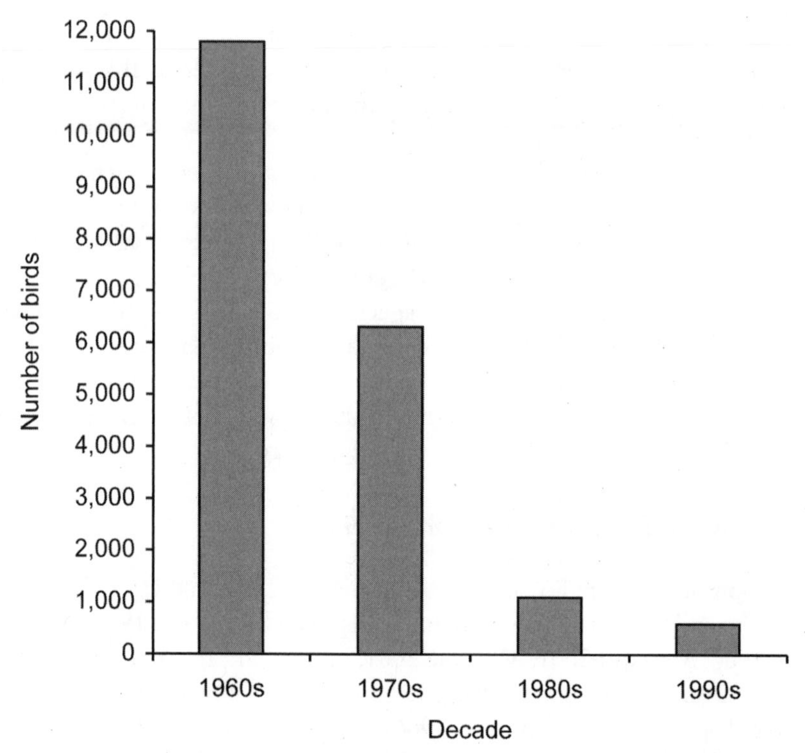

Figure 4.3. Long-term (38-year) trend in casualties of migratory birds at the WSMV-TV tower in Nashville, Tennessee. Redrawn from Nehring (1998).

1981b). None of the 229 events compiled by Verheijen occurred near a full moon, and most were clustered around the new moon period. When Crawford (1981) examined data from 1956–1980 from the WCTV tower in Leon County, Florida, however, he identified 683 nights on which ten or more birds were killed. Of these nights 40% occurred when the moon was 0–30% illuminated, while 28% occurred when the moon was 71–100% illuminated. If one examines the numbers of birds killed, the largest kill (2,127 birds) occurred on a night when the moon was 97% illuminated. The inverse relationship between lunar illumination and tower mortality is not as universal as Verheijen (1980, 1981a, 1981b) would suggest, but Crawford's (1981) analysis also suggests that more mortality events occur on nights when the moon is new or only slightly illuminated (notwithstanding the largest kill occurring during the full moon).

The flight behavior of migrant birds near tall towers with aircraft warning lights has been described on several occasions (e.g., Cochran and Graber 1958). We (SAG) observed the flight paths with an infrared scope near the

WBRZ television tower in Baton Rouge, Louisiana on two occasions when winds were blowing from the same direction but at different wind speeds (Figure 4.4). When wind velocity was low to moderate, 3–8 m (10–26 ft) per second, some birds circled the tower, hovered, and accumulated on the leeward side of the tower (Figure 4.4a). When wind velocity increased, few birds hovered on the leeward side of the tower, and most passed the tower with only minor deflections in their flight path (Figure 4.4b).

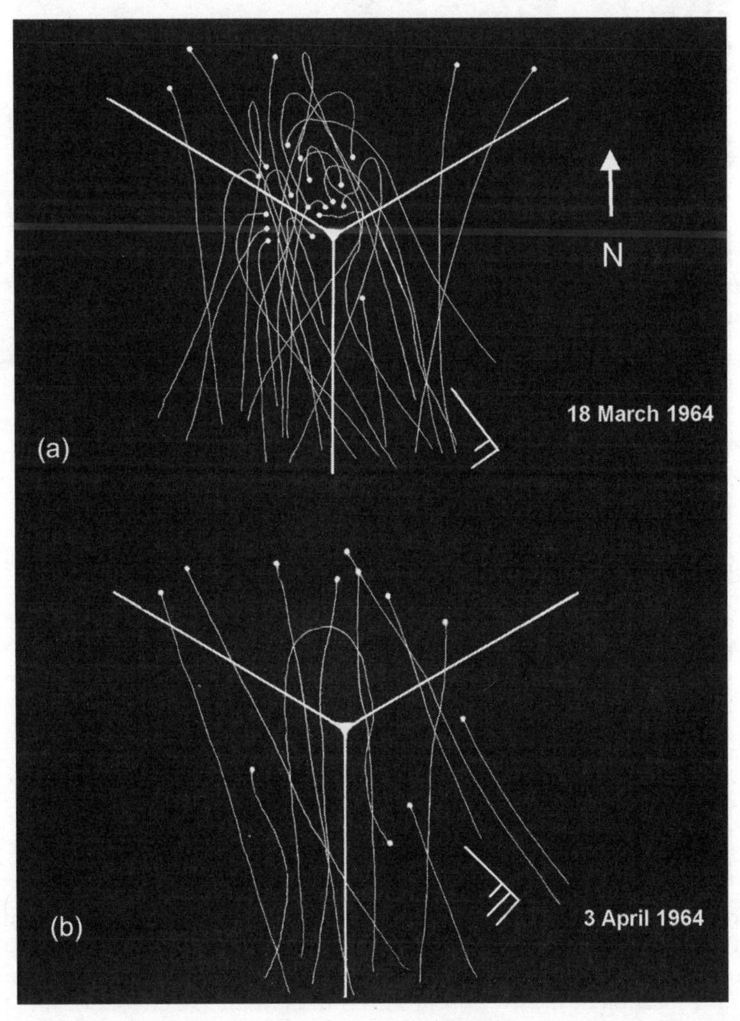

Figure 4.4. Flight paths of migrants near the WBRZ television tower in Baton Rouge, Louisiana in the spring of 1964. *(a)* March 18: winds are from the southeast at 7.5 m per second (15 knots). *(b)* April 3: winds are 12.5 m per second (25 knots). The tower is supported by three sets of guylines.

Portable tracking radar has been used to record the circular paths of birds flying near a broadcast tower in a sparsely populated area of the Upper Peninsula of Michigan on the night of September 9–10, 1983, when low clouds surrounded the tower (Larkin and Frase 1988). Figure 4.5 shows

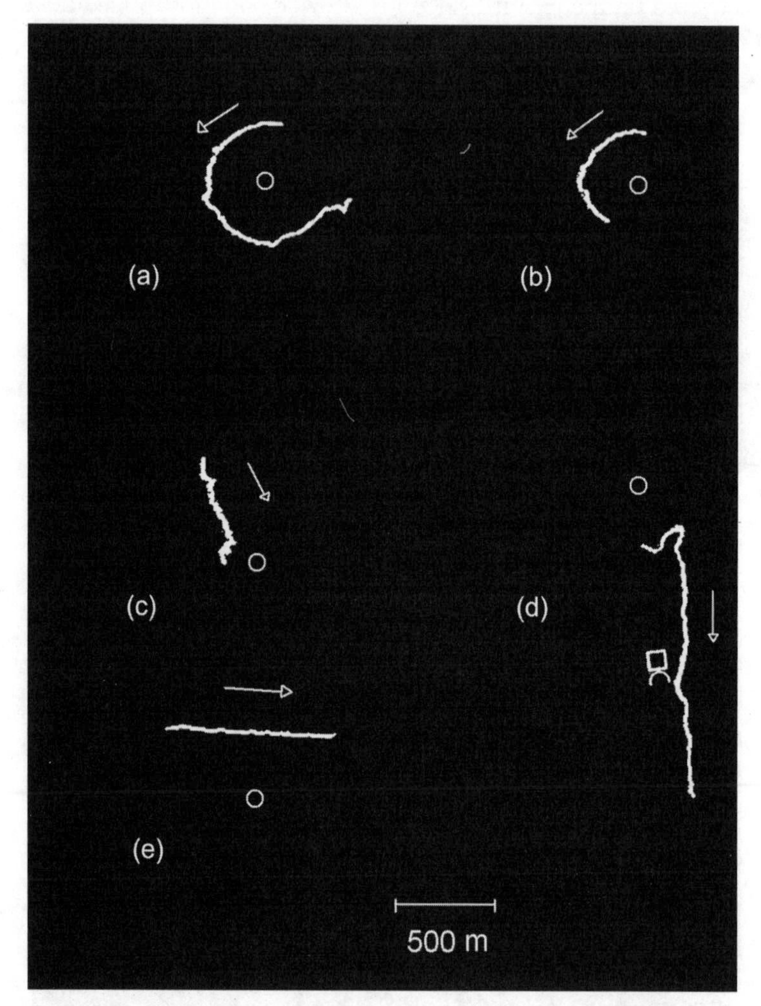

Figure 4.5. Flight paths of birds recorded by a tracking radar near a 308-m (1,010-ft) broadcast tower in the Upper Peninsula of Michigan on the night of September 9–10, 1983, when clouds surrounded the tower. The open circle shows the location of the tower, and the position of the radar is shown in *(d)*. Arrows indicate the direction of flight. *(a)* Time start 06:37 Universal Time Coordinate (UTC), duration 164 seconds. *(b)* 08:57 UTC, 94 seconds. *(c)* 09:04 UTC, 124 seconds. *(d)* 08:42 UTC, 186 seconds. *(e)* 05:31 UTC, 41 seconds. Redrawn from Larkin and Frase (1988).

the flight paths of nocturnal migrants as they were migrating near the tower. Arcs and circular paths were centered on the tower and had radii greater than 100 m (328 ft). Birds closer to the tower could not be tracked because the tracking radar would lock on the tower, the stronger target. When skies were clear or high overcast was present, the circling and curving behavior was not observed.

Influence of Lighting Type on Bird Behavior Near Tall Communication Towers

Tall communication towers historically have had arrays of incandescent red lighting as aircraft warning lights. This type of array includes continuously illuminated red lights alternating with slowly blinking red lights at different intervals along the length of the tower. Since the 1970s arrays of white strobe lights have been used increasingly as aircraft warning lights on tall communication towers. On towers with white strobe lights, the lights at different heights on the tower pulse (i.e., strobe) in unison or in sequence. Currently both types of lighting arrays are widely used. In an effort to understand why birds are attracted to lights and to assess the influences of different types of arrays of aircraft warning lights on towers, we examined the behavior of nocturnal migrants flying near tall towers with different types of lighting (Gauthreaux and Belser 1999, unpublished data).

During spring migration of 1986 from April 27 through May 15 we monitored flight behavior of migrating birds near the WNGC FM radio tower with strobe lights in northeast Georgia near the settlement of Neese and over a control area to the northeast (no tower present) for ten evenings (only nine with migration). We also collected data at this tower on the evenings of October 6 and 7, 1986. The FM radio tower was 366 m (1,200 ft) tall, with three sets of ten guylines and strobe lights located every 91.4 m (300 ft). The strobe lights pulsed at a rate of 40–46 pulses per minute. We used a vertically pointing image intensifier (AN/TVS 5, Varo Inc., Midland, Texas, 7× magnification) to monitor birds flying overhead near the tower (Figure 4.6) and coded the flight paths of migrants into the following categories: linear flight (straight) and nonlinear flight (pause or hover, curved, or circling). For statistical analysis the response variables were RATE (total paths per 20 minutes), PLIN (linear flight paths per 20 minutes), and PNON (nonlinear flight paths per 20 minutes). The explanatory variables were DATE and LOCATION (strobe tower or control site). We used a SAS general linear models procedure (SAS Institute 1999) for statistical analysis.

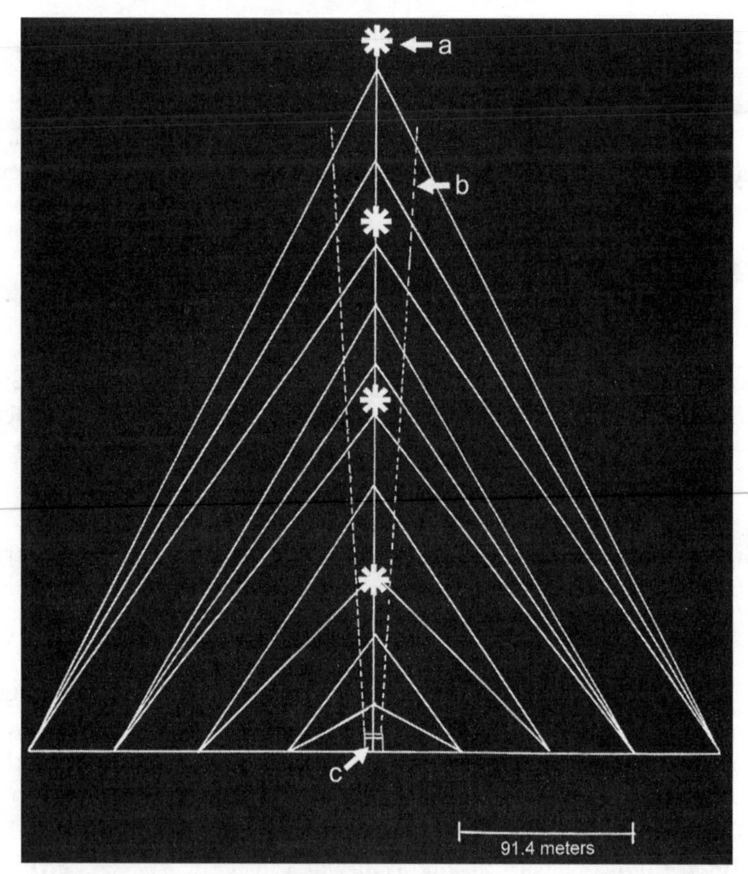

Figure 4.6. Communication tower and cone of observation. *(a)* Strobe aircraft warning lights and *(b)* cone of observation through the *(c)* image intensifier (AN/TVS 5). Only two of the three sets of ten guylines are shown in this illustration.

During the spring study in Georgia, the number of birds showing nonlinear flight near the tower with white strobe lights was significantly greater than at the control site, but the number of birds recorded at each site was not significantly different. For the response variables PLIN and PNON the explanatory variable DATE was not significant ($F_{8,8} = 1.43$, $p < 0.3116$), but the explanatory variable LOCATION was significant ($F_{1,8} = 5.78$, $p < 0.0429$). The rate of straight line flight paths was significantly greater at the control site, and the rate of flight paths that curved or showed hovering was significantly greater at the strobe light tower (Table 4.1). For the response variable RATE we found that the explanatory variables DATE and LOCATION were not significant ($F_{8,8} = 1.23$, $p < 0.3882$

Table 4.1. Results of *t*-tests (least significant difference) for response variables RATE, LRATE, PLIN, and PNON for Neese, Georgia and Garris (Moore's) Landing, South Carolina studies.

T Grouping	*Mean*	*N*	*Light*
	Neese, Georgia		
	RATE (flight paths per 20 min)		
A*	8.593	9	Control
A	5.521	9	Strobe
	PLIN (linear flight paths per 20 min)		
A	0.974	9	Control
B	0.856	9	Strobe
	PNON (nonlinear flight paths per 20 min)		
A	0.026	9	Control
B	0.144	9	Strobe
	Garris (Moore's) Landing, South Carolina		
	LRATE (log transformed flight paths per 20 min)		
A	2.674	14	Red
B	1.976	14	Control
B	1.390	13	Strobe
	PLIN (linear flight paths per 20 min)		
A	0.701	14	Red
B	0.979	14	Control
C	0.861	13	Strobe
	PNON (nonlinear flight paths per 20 min)		
A	0.299	14	Red
B	0.021	14	Control
C	0.139	13	Strobe

*Means with the same letter are not significantly different.

and $F_{1,8} = 2.33$, $p < 0.1657$, respectively), indicating that the date when the samples were taken and the location (near or away from the strobe tower) had no effect on the number of flight paths per 20 minutes (Table 4.1).

During the fall migration of 1986 we monitored the flight behavior of migrating birds near a television tower with red lights, near a television tower with white strobe lights, and over a control area that had no tower, all in the vicinity of Garris (Moore's) Landing, South Carolina (Figure

4.7). The two towers were 1.3 km (0.8 mile) apart, and both were 9 km (5.6 miles) southwest of the control site. One tower, operated by television station WTAT (Channel 24), was 508.1 m (1,667 ft) tall, with six continuously burning sets of red lights at 49, 146, 245, 340, 437, and 487 m (161, 479, 804, 1,115, 1,433, and 1,598 ft) that alternated with five slowly blinking red lights at 98, 195, 294, 392, and 508 m (322, 640, 965, 1,286, and 1,667 ft). The second tower, operated jointly by television stations WCSC and WCIV (Channels 4 and 5), was 614 m (2,016 ft) tall with nine sets of white strobe lights positioned approximately every 60 m (197 ft) starting at 76 m (249 ft). All strobes flashed synchronously. We used the same observation and data recording procedures as in the spring study. We recorded data on 14 evenings from October 2 to November 8, when conditions were favorable for bird migration. The rate data were log transformed (LRATE) to meet the requirements of normality, and we used the same SAS statistical procedures to analyze the fall data as we had used to analyze the spring data. The response variables were the same as in the spring study (LRATE, PLIN, and PNON), and the explanatory variables included RED (red light tower), STROBE (strobe light tower), and CONTROL.

During the fall study the log transformed flight paths per 20 minutes (LRATE) differed significantly by DATE ($F_{13,25} = 6.42, p < 0.0001$)

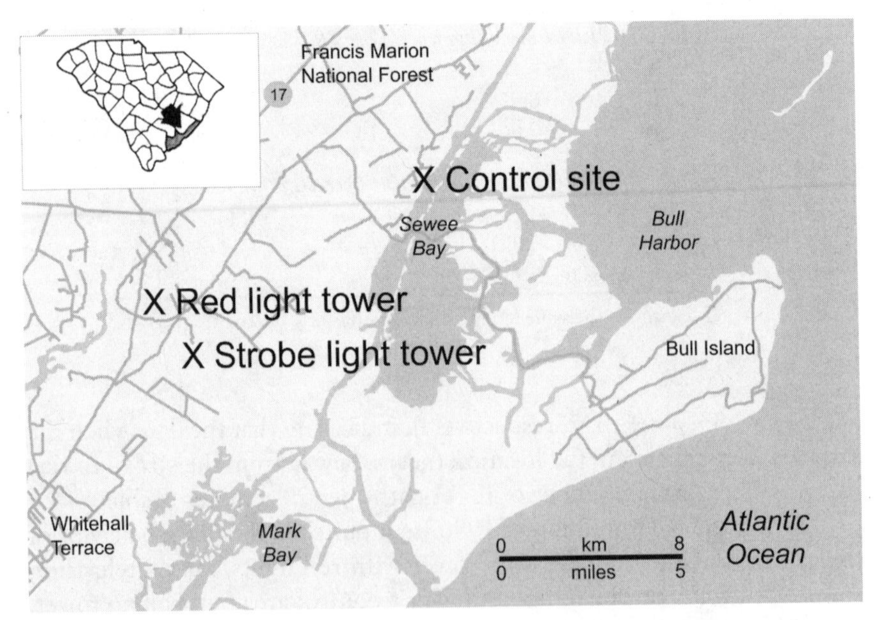

Figure 4.7. Map showing sites for the South Carolina tower lighting study.

and by light type ($F_{2,25} = 8.02, p < 0.0020$). The rate of linear flight paths did not change as a function of date ($F_{13,25} = 1.70, p < 0.1237$) but was significantly related to the type of lighting ($F_{2,25} = 14.29, p < 0.0001$). Nonlinear flight behavior was significantly greater near the tower with red lighting than near the tower with white strobes, and nonlinear flight behavior was significantly greater near the tower with white strobes than over the control site (Table 4.1). When the number of flight paths per 20 minutes recorded at each site is considered, significantly more birds were recorded flying near the tower with red lights than were recorded flying near the tower with white strobes and over the control site (Table 4.1). The number of birds detected flying near the tower with white strobes did not differ significantly from the number recorded over the control site, a result similar to that found in the Georgia spring migration study.

The type of lighting system on broadcast and communication towers influences the flight behavior of migrating birds at night. At both study sites the flight behavior of migrants differed significantly between the strobe light towers and the control sites, but the rate of flight paths per 20 minutes did not differ between the sites. In contrast, in the South Carolina fall migration study the number of flight paths per 20 minutes was greater at the tower with the red light array than at the tower with strobe lights and at the control site. Migrating birds at the red light tower showed significantly more nonlinear and hovering flight than birds passing the strobe tower and flying over the control site, and it is likely that nonlinear flight behavior over time contributed to the concentration of migrants at the tower with red lights. The greater number of birds near the tower with red lights probably is the result of attraction to the constantly illuminated lights on towers with red light arrays and the proportion of the time the birds showed nonlinear flight behavior. Whereas birds in linear flight spend only a brief instant near the tower and leave the area, birds showing curved, circling, or hovering behavior spend more time near the tower and thus build concentrations of migrants in the vicinity of the tower. Once concentrations build, the birds themselves may become collision hazards to other birds.

Our results showing greater disorientation from red lights are consistent with recent studies of the spectral sensitivity of birds. It has been demonstrated in the last decade that certain wavelengths of light appear to influence the magnetoreception of compass information by migratory birds (Wiltschko et al. 1993, Deutschlander et al. 1999, Wiltschko and Wiltschko 1999). Three passerine bird species have shown normal orientation of migratory restlessness under dim monochromatic light from the blue-green range of the spectrum, whereas they were disoriented under yellow and red light (Wiltschko and Wiltschko 2002). If red light disrupts the magnetic compass used by birds during migration, then this could be an additional factor contributing to the aberrant flight behavior of migrating birds near towers with red warning light arrays. The absence of compass information could be the reason birds alter a straight flight path by hovering, slowing down, changing direction, or circling. We do not know how quickly red light affects the compass or whether birds are actually using the magnetic compass once a direction has been selected at the beginning of a migratory flight. Much more research is needed to answer these questions and those related to the mechanisms of how migratory birds are influenced by artificial lighting. If bird conservation is a goal, we must in the meantime develop and follow policies that minimize the types of lighting with which bird mortality and behavioral disruption have been observed.

Conclusion

Efforts are needed to avoid the adverse effects of lighting on birds, and if circumstances prevent such action, then appropriate mitigation measures should be developed. It would be impossible to detail all design and mitigation measures here; each project will require its own unique considerations. Our current understanding of bird behavior suggests the following general guidelines.

The intense glow of city lights can be reduced by making certain that all light is directed toward the ground whenever possible. Streetlights should be shielded so that the pattern of illumination is below the horizontal plane of the light fixture. Floodlights on the ground that point upward to illuminate buildings, bridges, and monuments are harmful and should be avoided. Such architectural lighting often is hazardous to migrating birds, particularly on nights that are misty with a low overcast ceiling. If such lighting designs must be used, then they should be turned off during migration seasons when weather conditions could contribute to attraction and mortality.

The Fatal Light Awareness Program has developed a Bird-Friendly Building Program that is aimed at building managers, building owners, office tenants, and employees of skyscrapers in Toronto. Their program has been effective in reducing bird mortality at participating buildings (Table 4.2). A similar program should be instituted in cities throughout the world.

Billboards often are lighted by floodlights that point upward, illumi-

Table 4.2. Fatal Light Awareness Program (FLAP) Bird-Friendly Building (BFB) Program (from www.flap.org).

EDUCATION

Implement the following educational strategies that carry the message about reducing bird collisions with your building:

- Elevator news.
- Lobby signage.
- E-mail migration alerts to tenants and staff in spring and fall.
- Educational displays.

LIGHTING CONTROL STRATEGIES

Program building's lighting system to achieve a measurable reduction in night lighting from 11 P.M. to 7 A.M. or, ideally, ensure that all lights are extinguished during that period.

Extinguish all exterior vanity lighting (e.g., rooftop floods, perimeter spots) during the migration periods.

When lights must be left on at night, examine and adopt alternatives to bright, all-night, floorwide lighting.

Options include the following:

- Installing motion-sensitive lighting.
- Using desk lamps and task lighting.
- Reprogramming timers.
- Adopting lower-intensity lighting.

TENANT RELATIONS

Work with reluctant tenants to ensure that they comply with BFB guidelines. Coordinate meetings between FLAP and tenants, establish guidelines for tenants, and offer incentives to reward positive action.

Participate in the Bird Action Group Stations (BAGS) program by setting up stations where concerned tenants and staff can pick up bags, nets, gloves, and literature that enable them to rescue birds.

BUILDING A SAFER ENVIRONMENT

Implement measures (window film, netting, or other) to prevent birds from hitting windows at ground level in high-collision areas.

Eliminate use of chemical fertilizers and pesticides on grounds in favor of natural methods of pest control.

STAYING ON COURSE

Report to FLAP on your progress every spring and fall using the e-mailed questionnaire sent out early in the season. (This includes providing copies of all educational tools used and supplying FLAP with light energy consumption data.)

Strive for a built environment safe for birds and people by implementing these measures and developing partnerships with neighboring towers to ensure safe passage for night-migrating birds.

nating not only the billboard but also the sky above. A better design would be to mount floodlights on the top of the billboard pointing down. Alternatively, billboards could be illuminated from within, using less energy. Flashing lights do not attract migrating birds as much as constant lights (Avery et al. 1976). When floodlighting was replaced by strobe lighting at utility plants in Ontario, Canada, injury and mortality of birds at stacks and towers was nearly eliminated (Evans Ogden 1996:29).

Our studies and other evidence indicate that changing red warning lights to white strobes on broadcast towers may reduce the mortality of migrating birds (Gauthreaux and Belser 1999, unpublished data, M. Avery, personal communication, 2003). This solution, however, poses an additional problem. People living in the vicinity of strobe light towers complain about the flashing lights, particularly on overcast, misty nights. They report that it is like living in a thunderstorm with constant lightning and no thunder. Perhaps red strobe lights would be better, but no work has been completed on red strobes to see whether they are less likely than solid lights to attract migrating birds and less likely than white strobe lights to draw the attention of residents living near towers.

Populations of migratory birds are declining throughout the world, and the decline can be attributed to several different factors, including migration mortality, habitat change, and habitat destruction. By eliminating or controlling light pollution we can reduce one of the factors responsible for mortality during migration.

Acknowledgments

The work on the influence of different lighting arrays on the flight behavior of migrating birds near tall communication towers was supported by a grant from the Air Force Engineering and Services Center of the Air Force Systems Command to examine the behavioral responses of migrating birds to aircraft strobe lights.

We thank the personnel of Channel 24 in Charleston, South Carolina for access to the tower with the red light array and Channels 4 and 5 in Charleston, South Carolina for permission to work at the tower with the strobe light array. We thank Marianne B. Willey, Stuart R. Reitz, and David S. Mizrahi for their assistance during and after the tower lighting study. We also appreciate the detailed review by A. M. Manville II of an earlier draft of this manuscript.

Literature Cited

Aldrich, J. W., R. R. Graber, D. A. Munro, G. J. Wallace, G. C. West, and V. H. Cahalane. 1966. Report of the Committee on Bird Protection, 1965. *Auk* 83:457–467.

Allen, J. A. 1880. Destruction of birds by light-houses. *Bulletin of the Nuttall Ornithological Club* 5:131–138.

Avery, M., P. F. Springer, and J. F. Cassel. 1976. The effects of a tall tower on nocturnal bird migration: a portable ceilometer study. *Auk* 93:281–291.

Avery, M. L., P. F. Springer, and N. S. Dailey. 1980. *Avian mortality at man-made structures: an annotated bibliography* (revised). U.S. Fish and Wildlife Service, Biological Services Program, FWS/OBS-80/54.

Backhurst, G. C., and D. J. Pearson. 1977. Ethiopian region birds attracted to the lights of Ngulia Safari Lodge, Kenya. *Scopus* 1:98–103.

Bagg, A. M., and R. P. Emery. 1960. Fall migration: northeastern maritime region. *Audubon Field Notes* 14:10–17.

Baldwin, D. H. 1965. Enquiry into the mass mortality of nocturnal migrants in Ontario: final report. *Ontario Naturalist* 3:3–11.

Barrington, R. M. 1900. *The migration of birds as observed at Irish lighthouses and lightships*. R. H. Porter, London and Edward Ponsonby, Dublin.

Beacham, J. L. 1982. *Phototaxis and age dependent migratory orientation in the indigo bunting,* Passerina cyanea. M.S. thesis, Clemson University, Clemson, South Carolina.

Bjorge, R. R. 1987. Bird kill at an oil industry flare stack in northwest Alberta. *Canadian Field-Naturalist* 101:346–350.

Brewer, R., and J. A. Ellis. 1958. An analysis of migrating birds killed at a television tower in east-central Illinois, September 1955–May 1957. *Auk* 75:400–414.

Brewster, W. 1886. Bird migration. Part 1. Observations on nocturnal bird flights at the light-house at Point Lepreaux, Bay of Fundy, New Brunswick. *Memoirs of the Nuttall Ornithological Club* 1:5–10.

Bruderer, B., D. Peter, and T. Steuri. 1999. Behaviour of migrating birds exposed to X-band radar and a bright light beam. *Journal of Experimental Biology* 202:1015–1022.

Bullis, H. R. Jr. 1954. Trans-Gulf migration, spring 1952. *Auk* 71:298–305.

Chamberlain, B. R. 1955. Fall migration: southern Atlantic coast region. *Audubon Field Notes* 9:17–18.

Clark, A. R., C. E. Bell, and S. R. Morris. 2005. Comparison of daily avian mortality characteristics at two television towers in western New York, 1970–1999. *Wilson Bulletin* 117:35–43.

Clarke, W. E. 1912. *Studies in bird migration*. Volumes I and II. Gurney & Jackson, London.

Cochran, W. W., and R. R. Graber. 1958. Attraction of nocturnal migrants by lights on a television tower. *Wilson Bulletin* 70:378–380.

Cooke, W. W. 1915. Bird migration. *United States Department of Agriculture Bulletin* 185:1–48.

Crawford, R. L. 1974. Bird casualties at a Leon County, Florida, TV tower: October 1966–September 1973. *Bulletin of the Tall Timbers Research Station* 18:1–27.

Crawford, R. L. 1981. Bird kills at a lighted man-made structure: often on nights close to a full moon. *American Birds* 35:913–914.

Deutschlander, M. E., J. B. Phillips, and S. C. Borland. 1999. The case for light-dependent magnetic orientation in animals. *Journal of Experimental Biology* 202:891–908.

Dixon, C. 1897. *The migration of birds: an attempt to reduce avine season-flight to law.* Windsor House, London.

Dubey, R. M. 1990. Some studies on Jatinga bird mystery. *Tiger Paper* (Bangkok) 17(4):20–25.

Dutcher, W. 1884. Bird notes from Long Island, N.Y. *Auk* 1:174–179.

Evans Ogden, L. J. 1996. *Collision course: the hazards of lighted structures and windows to migrating birds.* World Wildlife Fund Canada and the Fatal Light Awareness Program, Toronto, Canada.

Ferren, R. L. 1959. Mortality at the Dow Air Base ceilometer. *Maine Field Naturalist* 15:113–114.

Fobes, C. B. 1956. Bird destruction at ceilometer light beam. *Maine Field Naturalist* 12:93–95.

Gastman, E. A. 1886. Birds killed by electric light towers at Decatur, Ill. *American Naturalist* 20:981.

Gauthreaux, S. A. Jr. 1982. Age-dependent orientation in migratory birds. Pages 68–74 in F. Papi and H. G. Wallraff (eds.), *Avian navigation.* Springer-Verlag, Berlin.

Gauthreaux, S. A. Jr., and C. G. Belser. 1999. The behavioral responses of migrating birds to different lighting systems on tall towers [abstract]. In W. R. Evans and A. M. Manville II (eds.), *Avian mortality at communication towers.* Transcripts of proceedings of the Workshop on Avian Mortality at Communication Towers, August 11, Cornell University, Ithaca, New York. Online: http://migratorybirds.fws.gov/issues/towers/agenda.html.

Graf, V. A., and D. V. Norren. 1974. A blue sensitive mechanism in the pigeon retina: λ_{max} 400 nm. *Vision Research* 14:1203–1209.

Green, J. C. 1963. Destruction of birdlife in Minnesota: September 1963. III. Notes on kills at Duluth on September 18/19. *Flicker* 35:112–113.

Green, J. C. 1965. Fall migration: western Great Lakes region. *Audubon Field Notes* 19:37–41, 44.

Green, J. C., and J. P. Perkins. 1964. Some notes on the fall 1964 migration of vireos and warblers. *Loon* 36:127–129.

Hallock, C. (ed.). 1874. Shot gun and rifle: game in season for October. *Forest and Stream* 3:149–150.

Hansen, L. 1954. Birds killed at lights in Denmark 1886–1939. *Videnskabelige Meddelelser fra Dansk Naturhistorisk Forening* 116:269–368.

Hart, N. S. 2001. The visual ecology of avian photoreceptors. *Progress in Retinal and Eye Research* 20:675–703.

Herbert, A. D. 1970. Spatial disorientation in birds. *Wilson Bulletin* 82:400–419.

Herndon, L. R. 1973. Bird kill on Holston Mountain. *Migrant* 44:1–4.

Howell, J. C., A. R. Laskey, and J. T. Tanner. 1954. Bird mortality at airport ceilometers. *Wilson Bulletin* 66:207–215.

Jackson, W. B., E. J. Rybak, and S. H. Vessey. 1974. Vertical barriers to bird migration. Pages 279–287 in S. A. Gauthreaux Jr. (ed.), *A conference on the biological aspects of the bird/aircraft collision problem*. Clemson University, Clemson, South Carolina.

James, P. 1956. Destruction of warblers on Padre Island, Texas, in May 1951. *Wilson Bulletin* 68:224–227.

Johnston, D. 1955. Mass bird mortality in Georgia, October, 1954. *Oriole* 20:17–26.

Jones, J., and C. M. Francis. 2003. The effects of light characteristics on avian mortality at lighthouses. *Journal of Avian Biology* 34:328–333.

Katz, Y. B., and I. K. Vilks. 1981. The role of distribution of illumination in the stellar picture in a planetarium for the orientation of redbreasts robins (*Erithacus rubecula*) in round cages [Russian with English abstract]. *Zoologicheskii Zhurnal* 60:1222–1230.

Kemper, C. 1996. A study of bird mortality at a west central Wisconsin TV tower from 1957–1995. *Passenger Pigeon* 58:219–235.

Kramer, G. 1949. Über Richtungstendenzen bei der nächtlichen Zugunruhe gekäfigter Vögel [On the directional tendencies of nocturnal *Zugunruhe* of caged birds]. Pages 269–283 in E. Mayr and E. Schüz (eds.), *Ornithologie als biologische Wissenschaft*. Carl Winter, Heidelberg.

Kramer, G. 1951. Eine neue Methode zur Erforschung der Zugorientierung und die bisher damit erzielten Ergebnisse [A new method to study migratory orientation and the results thereby obtained thus far]. Pages 269–280 in S. Hörstadius (ed.), *Proceedings of the Xth International Ornithological Congress, Uppsala June 1950*. Almqvist & Wiksell, Uppsala.

Kumlien, L. 1888. Observations on bird migration at Milwaukee. *Auk* 5:325–328.

Larkin, R. P., and B. A. Frase. 1988. Circular paths of birds flying near a broadcasting tower in cloud. *Journal of Comparative Psychology* 102:90–93.

Laskey, A. R. 1960. Bird migration casualties and weather conditions, autumns 1958–1959–1960. *Migrant* 31:61–65.

Lewis, H. F. 1927. Destruction of birds by lighthouses in the provinces of Ontario and Quebec. *Canadian Field-Naturalist* 41:55–58, 75–77.

Lord, W. G. 1951. Bird fatalities at Bluff's Lodge on the Blue Ridge Parkway, Wilkes County, N.C. *Chat* 15:15–16.

Malakoff, D. 2001. Faulty towers. *Audubon* 103(5):78–83.

Manville, A. M. II. 2001. The ABCs of avoiding bird collisions at communication towers: next steps. Pages 85–103 in R. L. Carlton (ed.), *Avian interactions with utility and communication structures*. Proceedings of a workshop held in Charleston, SC, USA. December 2–3, 1999. Electric Power Research Institute Technical Report 1005180.

Manville, A. M. II. In press. Bird strikes and electrocutions at power lines, communication towers, and wind turbines: state of the art and state of the science —next steps toward mitigation. In C. J. Ralph and T. D. Rich (eds.), *Bird conservation implementation and integration in the Americas: proceedings of the third international Partners in Flight conference 2002*. GTR-PSW-191. USDA Forest Service, Albany, California.

Merriam, C. H. 1885. Preliminary report of the committee on bird migration. *Auk* 2:53–65.

Miller, G. S. Jr. 1897. Winge on birds at the Danish lighthouses. *Auk* 14:415–417.

Morris, S. R., A. R. Clark, L. H. Bhatti, and J. L. Glasgow. 2003. Television tower mortality of migrant birds in western New York and Youngstown, Ohio. *Northeastern Naturalist* 10:67–76.

Munro, J. A. 1924. A preliminary report on the destruction of birds at lighthouses on the coast of British Columbia. *Canadian Field-Naturalist* 38:141–145, 171–175.

Nehring, J. D. 1998. *Assessment of avian population change using migration casualty data from a television tower in Nashville, Tennessee*. M.S. thesis, Middle Tennessee State University, Murfreesboro, Tennessee.

Newman, R. J. 1960. Spring migration: central southern region. *Audubon Field Notes* 14:392–397.

Nikolaus, G. 1980. An experiment to attract migrating birds with car headlights in the Chyulu Hills, Kenya. *Scopus* 4:45–46.

Nikolaus, G., and D. J. Pearson. 1983. Attraction of nocturnal migrants to car headlights in the Sudan Red Sea Hills. *Scopus* 7:19–20.

Norren, D. V. 1975. Two short wavelength sensitive cone systems in pigeon, chicken and daw. *Vision Research* 15:1164–1166.

Overing, R. 1938. High mortality at the Washington Monument. *Auk* 55:679.

Partridge, J. C. 1989. The visual ecology of avian cone oil droplets. *Journal of Comparative Physiology A* 165:415–426.

Romanes, G. J. 1883. *Mental evolution in animals*. Kegan, Paul, Trench & Co., London.

Sage, B. 1979. Flare up over North Sea birds. *New Scientist* 81:464–466.

SAS Institute, Inc. 1999. *The SAS system for Windows*. SAS Institute, Inc., Cary, North Carolina.

Spofford, W. R. 1949. Mortality of birds at the ceilometer of the Nashville airport. *Wilson Bulletin* 61:86–90.

Squires, W. A., and H. E. Hanson. 1918. The destruction of birds at the lighthouses on the coast of California. *Condor* 20:6–10.

Stoddard, H. L. Sr. 1962. Bird casualties at a Leon County, Florida TV tower, 1955–1961. *Bulletin of Tall Timbers Research Station* 1:1–94.

Stoddard, H. L. Sr., and R. A. Norris. 1967. Bird casualties at a Leon County, Florida TV tower: an eleven-year study. *Bulletin of Tall Timbers Research Station* 8:1–104.

Stone, W. 1906. Some light on night migration. *Auk* 23:249–252.

Terres, J. K. 1956. Death in the night. *Audubon Magazine* 58:18–20.

Thomson, A. L. 1926. *Problems of bird-migration*. H. F. & G. Witherby, London.

Tornielli, A. 1951. Comportamento di migratori nei riguardi di un pozzo metanifero in fiamme [Behavior of migratory birds under the influence of a burning natural gas well]. *Rivista Italiana di Ornitologia* II 21:151–162.

Tufts, R. W. 1928. A report concerning destruction of bird life at lighthouses on the Atlantic coast. *Canadian Field-Naturalist* 42:167–172.

Velie, E. D. 1963. Report of a survey of bird casualties at television towers, ceilometers, and other obstructions. *Flicker* 35:79–84.

Verheijen, F. J. 1958. The mechanisms of the trapping effect of artificial light sources upon animals. *Archives Néerlandaises de Zoologie* 13:1–107.

Verheijen, F. J. 1980. The moon: a neglected factor in studies on collisions of nocturnal migrant birds with tall lighted structures and with aircraft. *Die Vogelwarte* 30:305–320.

Verheijen, F. J. 1981a. Bird kills at lighted man-made structures: not on nights close to a full moon. *American Birds* 35:251–254.

Verheijen, F. J. 1981b. Bird kills at tall lighted structures in the USA in the period 1935–1973 and kills at a Dutch lighthouse in the period 1924–1928 show similar lunar periodicity. *Ardea* 69:199–203.

Verheijen, F. J. 1985. Photopollution: artificial light optic spatial control systems fail to cope with. Incidents, causations, remedies. *Experimental Biology* 44:1–18.

Vorobyev, M., D. Osorio, A. T. D. Bennett, N. J. Marshall, and I. C. Cuthill. 1998. Tetrachromacy, oil droplets and bird plumage colours. *Journal of Comparative Physiology A* 183:621–633.

Weir, R. D. 1976. *Annotated bibliography of bird kills at man-made obstacles: a review of the state of the art and solutions.* Department of Fisheries and the Environment, Environmental Management Service, Canadian Wildlife Service, Ontario Region, Ottawa.

Wessels, R. H. A. 1974. *Tetrachromatic vision in the daw (*Corvus monedula *L.).* Ph.D. dissertation, State University of Utrecht, The Netherlands.

Williams, J. N. 1978. *Age dependent factors in the migratory behavior of the white-crowned sparrow,* Zonotrichia leucophrys leucophrys. M.S. thesis, Clemson University, Clemson, South Carolina.

Wiltschko, W., U. Munro, H. Ford, and R. Wiltschko. 1993. Red light disrupts magnetic orientation of migratory birds. *Nature* 364:525–527.

Wiltschko, W., and R. Wiltschko. 1999. The effect of yellow and blue light on magnetic compass orientation in European robins, *Erithacus rubecula. Journal of Comparative Physiology A* 184:295–299.

Wiltschko, W., and R. Wiltschko. 2002. Magnetic compass orientation in birds and its physiological basis. *Naturwissenschaften* 89:445–452.

Chapter 5

Influences of Artificial Light on Marine Birds

William A. Montevecchi

The nocturnal activities of many animals have been changed by artificial lighting. Ambient light influences the reproductive physiology, migration, foraging, and hence parental behavior of many species. Perhaps more than any other vertebrates, birds are intimately and inextricably linked with the light features of their environments (e.g., Farner 1964).

Nocturnal oceans are essentially flat, dark environments in which marine birds negotiate their lives. Some seabirds exploit coastal and nearshore habitats, and others are pelagic, ranging over vast ocean expanses. Many seabirds are nocturnally active, in part to avoid diurnal avian predators, primarily gulls. Many of these nocturnal birds also prey on vertically migrating and bioluminescent prey.

Somewhat paradoxically perhaps, many nocturnal seabird species are highly attracted to artificial light. The attraction to light by nocturnal-feeding petrels has been hypothesized to result from their adaptations and predisposition to exploit bioluminescent prey (Imber 1975) and from a predilection to orient to specific star patterns (Reed et al. 1985). In these instances, artificial light sources might be perceived as attractive "supernormal" stimuli. Well before the age of electric lighting, humans used light from fires to attract nocturnal birds for exploitation (Maillard 1898, Murphy 1936, Murie 1959).

Migratory birds move seasonally over tens of degrees of latitude and longitude, often exhibiting movements of hemispheric proportions. These creatures are especially vulnerable to increasing sources and extents of artificial lighting. Light-associated mortality of nocturnal avian migrants involving collisions of hundreds or thousands or more birds with lights and lighted structures has been well documented for well more than a century (Allen 1880, Brewster 1886, Kumlien 1888, Johnston and Haines 1957, Evans 1968; see Chapter 4, this volume). Considering that mortality during migration is more than an order of magnitude higher than during energy-demanding breeding and winter seasons (Sillett and Holmes 2002), the population effects of additive mortality associated with artificial lighting could be profound.

Increasing risks associated with artificial lighting cumulate with other sources of environmental modification, degradation, and change, including deforestation, pollution, overfishing, and global climate change (e.g., Vitousek et al. 1997, Hughes 2000). For example, because global fish stocks are being overexploited, more fishery effort is directed at invertebrates on lower levels of marine food webs (Pauly et al. 1998). As a consequence, light-induced fisheries for squid are increasing in capacity and ocean coverage, with unknown influences on marine ecosystems (Rodhouse et al. 2001).

Given the dramatic influence of artificial lighting on marine organisms in the instances that have been documented, a general effect on marine birds, mammals, fishes, and invertebrates can be expected. Birds that spend most of their lives at sea are often highly influenced by artificial lighting in coastal areas and in dark, two-dimensional ocean environments. Except for coastal areas, oceanscapes tend to have less artificial lighting than terrestrial environments. Much artificial lighting on the ocean occurs at intense source points that can attract marine birds from very large catchment areas (Rodhouse et al. 2001, Wiese et al. 2001).

This chapter reviews the major sources of artificial illumination in the marine environment and their direct and indirect influences on seabirds. The cumulative effects of artificial lighting with other sources of environmental risk are considered. Different species and age classes of marine birds exhibit different degrees of attraction, and hence vulnerability, to artificial lighting. Mortality associated with artificial lighting threatens populations of endangered and rare species. Current levels of mitigative action are nonexistent or inadequate to address problems posed by artificial lighting for marine organisms. Environmentally sound and ecologically precautionary broad-scale and long-term adaptive planning programs are needed to address current and future problems.

Sources of Artificial Light in the Marine Environment

The major sources of artificial light in marine environments include vessels, lighthouses, light-induced fisheries, and oil and gas platforms. Vessels have plied the seas for as long as humans have inhabited coastal environments, though most widely and prolifically during the last few centuries. Vessel numbers, sizes, and lights have increased exponentially throughout this period. Yet the more recent changes associated with lighthouses, marine gas and oil platforms, and light-induced fisheries are likely having the most significant influences on marine birds.

Lighthouses and Coastal Lighting

Lighthouse beacons have been an important aspect of coastal navigation for centuries, with their proliferation probably peaking in the late nineteenth century. Rotational beams identified landfall and specific sites for mariners. At times, lightships have been moored at sea and at coastal sites with treacherous navigation. Because of improved navigational aids such as sonar and global positioning systems aboard vessels, the number of active lighthouses decreased dramatically in the late twentieth century, a trend that will continue over the next decades.

As large segments of human populations moved to coastal areas for housing, recreation, and leisure, the extent of artificial lighting along coasts spread throughout the twentieth century. Moreover, artificial illumination increased in power and intensity, as well as proliferating during this period.

Oil and Gas Platforms at Sea

The intense flares at offshore hydrocarbon platforms, undoubtedly the most lethal light there is (Terres 1956, Bourne 1979, Sage 1979, Hope-Jones 1980, Wallis 1981), can be detected easily on satellite images (Muirhead and Cracknell 1984). These flares relieve pressures associated with natural gas from drilled wells and can reach up to 40 m (131 ft). Flares tend to burn most intensely during the initial operational phases of drilling and when hydrocarbon is not offloaded to vessels during extreme sea conditions (Burke et al. 2005). Hydrocarbon platforms are being constructed and deployed at remote ocean sites, where they impose novel artificial light sources, such as the shelf edge of the Grand Banks of eastern Canada. Both the intensity and oceanographic novelty of the light source could have a cumulative effect on the attraction and mortality of seabirds.

Light-Induced Fisheries

Many fisheries use intense artificial lighting to attract, concentrate, and facilitate prey capture (e.g., Vojkovich 1998, Arcos and Oro 2002; see Chapter 11, this volume). Rodhouse et al. (2001) estimated that 63–89% of the world catch of squid is caught using lights that can be mapped using satellite imagery. Small artisanal vessels fishing squid often use a single light, whereas large vessels may use 150 lamps, with about 300 kW of illumination power (Rodhouse et al. 2001), and several vessels often work in the same area. Squid species that have large, well-developed eyes are attracted to the intense lights. The highest concentrations of light-induced fisheries for squid (also octopus and cuttlefish) are pursued in the Kuroshio Current on the China Sea Shelf southwest of Japan and along the Sunda–Arafura Shelf primarily in the Gulf of Thailand. Other major light-induced fisheries for squid are carried out around New Zealand, in the southwest Atlantic, and in the California and Humboldt currents.

Influences of Ambient Light, Lunar Phase, and Season on Avian Attraction to Artificial Lighting

Attraction to and mortality at lighted structures is influenced by visibility, ambient light conditions, and lunar phase. Birds are more attracted to light during low cloud cover and overcast skies, especially foggy, drizzly conditions that are pervasive in many ocean regions (Brewster 1886, Kemper 1964, Aldrich et al. 1966, Weir 1976, Hope-Jones 1980, Wallis 1981, Telfer et al. 1987). Moisture droplets in the air refract light and greatly increase illuminated volumes (i.e., catchment basins), whereas concentrated beams of light act as bright corridors in the darkness into which birds fly (Weir 1976). Birds entrained in intense artificial light often circle the source for hours to days, especially during overcast conditions, when they are reluctant to fly outside of the sphere of illumination into darkness (Avery et al. 1976, Wallis 1981). Also, seabirds and marine waterfowl fly closer to land during foggy conditions (Chaffey

2003; see also Weir 1976, Blomqvist and Peterz 1984), increasing their chances of encountering and being affected by coastal lighting.

Seabird vulnerability to artificial light is influenced by lunar cycles. There is significantly less attraction to artificial lighting on bright, clear nights with a full moon (Verheijen 1980, 1981, Telfer et al. 1987). In these conditions, breeding nocturnal seabirds exhibit less activity at colonies (Warham 1960, Harris 1966, Boersma et al. 1980, Watanuki 1986, 2002, Bryant 1994). Conversely, more birds are attracted to, stranded at, and killed at artificial lights during new moon phases, when activity at breeding colonies is also greater.

Autumn and spring migratory periods are critical times for mortality associated with artificial lighting at coastal and offshore sources. In autumn high proportions of relatively easily disoriented young-of-the-year are on the wing, and during both spring and autumn seabirds move in large numbers across oceans and hemispheres. In the northwest Atlantic, for example, tens of millions of birds move into the region from breeding areas in the Arctic in the fall and the Southern Hemisphere in the spring.

Direct Influences of Artificial Light on Seabirds

Marine birds are attracted to and often collide with lighthouses (Evans 1968, Crawford 1981, Verheijen 1981, Roberts 1982), coastal resorts (Reed et al. 1985), offshore hydrocarbon platforms (Ortego 1978, Hope-Jones 1980, Tasker et al. 1986, Baird 1990, Wiese et al. 2001, Burke et al. 2005), and vessels that use intense artificial lighting to attract and catch squid and other fish (Dick and Davidson 1978, Arcos and Oro 2002).

Mass collisions of birds with lighted structures can result in high levels of mortality. In one documented incident, the lights of a fishing vessel were estimated to attract about 6,000 crested auklets (*Aethia cristatella*) weighing 1.5 metric tons, which nearly capsized the boat (Dick and Davidson 1978). Mass collisions and incidences of hundreds, thousands, and tens of thousands of circling birds have been reported at coastal and offshore artificial light sources (Bourne 1979, Wiese et al. 2001). Seabirds are attracted to the flares of offshore oil and gas platforms and can be killed by intense heat, by collisions with structures, and by oil on and around brightly lit platforms (Figure 5.1; Wood 1999, Wiese et al. 2001, Burke et al. 2005; see also Newman 1960).

Mortality associated with flaring and artificial lighting is episodic, which probably explains why some observers report hundreds and even tens of thousands of birds killed by flares (Sage 1979), and others report

Figure 5.1. Hibernia oil platform and its flare by night on the edge of the Grand Banks of eastern Canada on February 18, 2003. Photo courtesy of C. Burke.

many birds attracted to platforms with little or no associated mortality (Hope-Jones 1980, Wallis 1981). These apparently discrepant findings also provide a rationale for the necessity of having dedicated independent observers rather than casual industry observers on offshore hydrocarbon facilities (Wiese et al. 2001) and on light-induced nocturnal fishing boats. Observer independence is needed to ensure validity and transparency of the process, as is true for observers on fishing vessels to monitor catches and bycatches (Weimerskirch et al. 2000, Melvin and Parrish 2001).

Indirect Influences of Artificial Light on Seabirds

Much of the mortality associated with artificial lighting is indirect and difficult to document. For instance, migrating passerines have been observed to circle platforms continuously for hours to days and to fall on the ocean or, less often, to land on platforms exhausted and emaciated (Hope-Jones 1980, Wallis 1981). This holding or trapping effect (Verheijen 1981) of intense light can deplete the energy reserves of migrating birds, rendering them incapable of making it to nearest landfalls. Although migratory seabirds do not use landfalls, the energetic costs associated with such diversions could have severe consequences for winter survival or subsequent reproduction.

Offshore hydrocarbon platforms develop rapidly into artificial reefs that create marine communities. These reefs attract, concentrate, and proliferate flora, crustaceans, fishes, and squids (Carlisle et al. 1964, Duffy 1975, Sonnier et al. 1976, Ortego 1978, Wolfson et al. 1979, Hope-Jones 1980, de Groot 1996). Lighting attracts invertebrates, fishes, and birds, and organisms at higher trophic levels are in turn attracted to lower ones as well as to the lighting.

Many species of marine birds have been recorded feeding in artificial nocturnal lighting. Most of these events have been recorded in coastal situations, but feeding at lights has also been observed for terrestrial waterbirds (e.g., Brown et al. 1982) and at offshore fishing vessels and hydrocarbon platforms (Hope-Jones 1980, Burke et al. 2005). Conversely, some nocturnally migrating crustaceans and associated predators might be less likely to migrate toward surface waters that are artificially illuminated.

Purse seine fisheries for small clupeids in the Mediterranean Sea use lights to attract and concentrate fishes (Arcos and Oro 2002). These nocturnal fisheries also attract threatened Audouin's gulls (*Larus audouinii*) that capture fish during hauling. The fisheries thus might be considered as providing a short-term benefit for the gulls but could also be changing their distributions at sea and potentially depleting their prey.

Many of the squids taken by albatrosses are dead ones that are scavenged (Weimerskirch et al. 1986). Squid species that have positive buoyancy after death ("floaters"; Lipinski and Jackson 1989, McNeil et al. 1993) are the ones most often scavenged by procellariiform seabird species (e.g., Rodhouse et al. 1987). Some of the dead squid contain hooks that can injure or kill avian scavengers. Albatrosses and other procellariid avian species may be those most attracted to offshore squid fisheries.

Light-induced nocturnal fisheries at times are conducted near islands where nocturnal seabirds nest. These light levels could facilitate predation by night-hunting gulls and could also reduce visitation rates by burrow-nesting seabirds to mates, eggs, and chicks (Keitt 1998).

Cumulative Effects

The most complex indirect influences on populations often are those associated with cumulative effects that represent the interaction of a multiplicity of diverse causes (Clark and Leppert-Slack 1994, Duinker 1994). In such circumstances, a negative environmental or population effect might not be attributable to any single factor but rather to a multiplicity of cumulative interactions that are obscured from a causal analysis (e.g., Burke et al. 2005).

Many cumulative effects probably are associated with artificial lighting. For example, light and heat that facilitate marine plant growth attract invertebrates and fishes, and these in turn attract and concentrate feeding gulls and other seabirds at offshore hydrocarbon installations (Wiese et al. 2001, Burke et al. 2005). Wastewater discharged on site at these platforms fertilizes the artificial reefs and provides feeding opportunities that attract scavenging gulls, just as coastal sewage outflows do. Spilled oil and discharged oily drilling fluids at platforms also contaminate birds on site (Burke et al. 2005). Together, the cumulative attractive effects are likely synergistic and greater than the sum of the influences of light, food availability, heat, and structural effects.

Globally, cumulative natural (e.g., oceanographic and climate change) and anthropogenic changes (e.g., greenhouse gas emissions, overfishing) are having profound, long-term effects on the Earth's ecosystems (Vitousek et al. 1997). The proliferation of artificial light throughout the biosphere could act in synergistic and unknown ways with these other large-scale environmental changes. For example, overfishing of the world's fish stocks in recent decades has led to much fishing effort being directed at invertebrate prey, that is, fishing down marine food webs (Pauly et al. 1998). Consequently, light-induced squid fisheries are increasing in effort and extent (Rodhouse et al. 2001). Furthermore, as the fishing and oceanographic influences in particular areas of squid concentration produce stock collapse, such as off eastern Canada in the 1980s (Black et al. 1987, Montevecchi 1993), fishery efforts are concentrated in other hotspots, such as the southwestern Atlantic (Rodhouse et al. 2001).

Species Vulnerability

Many nocturnal seabirds have a preponderance of rods in their retinas, more rhodopsin, and often larger eyes than related diurnal species (McNeil et al. 1993). These species probably are more susceptible to the influences of artificial light. Many of the smaller planktivorous nocturnal species are highly sensitive to, and attracted to, night light (Imber 1975, Dick and Davidson 1978, Bretagnolle 1990). At least 21 species of procellariiform seabirds are known to be attracted to artificial lighting (Murphy

1936, Reed et al. 1985). For example, Leach's storm-petrels (*Oceanodroma leucorhoa*) are highly attracted to lighthouse beacons and to the illumination of offshore hydrocarbon platforms (Wiese et al. 2001). These storm-petrels have also been observed flying about lights at baseball fields in San Francisco (B. Sydeman, personal communication, 2004) and St. John's, Newfoundland (N. Montevecchi, personal communication, 2004).

Vulnerability to artificial light appears to be greatest among species that feed on bioluminescent prey and could have predispositions for light attraction. Many endangered and threatened species of marine birds therefore are at risk. Even some of the largest of marine birds, such as king penguins (*Aptenodytes patagonicus*), prey on bioluminescent myctophids often at low illumination levels (Cherel and Ridoux 1992). They likely also have keen sensitivity and possibly attraction to ambient and artificial light.

Age Vulnerability

Fledgling storm-petrels, petrels, shearwaters, and possibly some auks are more attracted to artificial light than are adults. This could result from disorientation associated with environmental inexperience or possibly from predispositions to find bioluminescent prey at sea (Imber 1975). Fledgling band-rumped storm-petrels (*Oceanodroma castro cryptoleucura*), dark-rumped petrels (*Pterodroma phaeopygia sandwichensis*), grey-faced petrels (*Pterodroma macroptera gouldi*), Barau's petrels (*Pterodroma baraui*), Newell's shearwaters (*Puffinus auricularis newelli*), wedge-tailed shearwaters (*Puffinus pacificus*), and Cory's shearwaters (*Calonectris diomedea*) incur considerable mortality as a result of their attraction to artificial lighting (Telfer et al. 1987, Bretagnolle 1990, Whittow 1997, Mougeot and Bretagnolle 2000, Day et al. 2003, J. Valerias, unpublished data). Many of these species are endangered or threatened, including band-rumped storm-petrels, dark-rumped petrels, and Newell's shearwaters. The varying age-class attraction of nocturnal species to light also suggests that some older birds may learn not to approach artificial light sources.

Among nocturnal seabirds, immature and nonbreeding birds appear to be more sensitive and vulnerable to the influences of lunar light than are breeding birds. This could be related to the greater vulnerability of immature and nonbreeding birds to visually hunting nocturnal predators when compared with breeders (Morse and Buchheister 1977, Huntington et al. 1996, Mougeot and Bretagnolle 2000, Stenhouse et al. 2000). In contrast to seabirds, adult passerines are more likely to be attracted to lighted coastal structures than are juveniles (Dunn and Nol 1980; but see Chapter 4, this volume).

Potential Population Effects

Wiese et al. (2001) suggested that artificial lighting at oil platforms on the Grand Banks could affect long-distance migrants from high latitudes in the Southern Hemisphere (shearwaters) and from the high Arctic (dovekies, murres) as well as from the world's largest populations of Leach's storm-petrels that breed locally. The species that are potentially most vulnerable to attraction to artificial lighting in marine environments, however, are nocturnal species that are at risk and endangered and whose populations are small and fragmented.

Endangered Species and Species of Concern

The small population sizes of some endangered and threatened species that are attracted to nocturnal light make them particularly vulnerable to artificial lighting. Barau's petrel, for example, an endangered endemic species that breeds on Réunion Island in the Indian Ocean, exhibits a very strong attraction to artificial lighting that leads to mortality (Le Corre et al. 2002). Very rare endangered Mascarene petrels (*Pseudobulweria aterrima*) are also killed by attraction to artificial lighting. Fledglings of two endemic Hawaiian seabirds, Newell's shearwater and dark-rumped petrel, suffer high mortality associated with artificial coastal lighting as they depart from inland nesting sites on their way to sea (Telfer et al. 1987, Ainley et al. 1997, 2001, Slotterback 2002, Day et al. 2003). Table 5.1 lists endangered, threatened, and rare species that experience mortality associated with artificial lighting.

Threatened Audouin's gulls feed on small clupeid fishes at nocturnal purse seine operations that use artificial lights to attract and concentrate fishes (Arcos and Oro 2002). Intense artificial lighting associated with commercial fisheries for squid exerted a negative influence on nesting Xantus's murrelets (*Synthliboramphus hypoleucus*; Carter et al. 2000, Pacific Seabird Group 2002), leading in part to their listing as a threatened species in California. The market-driven squid fishery has more than doubled the number of participating vessels from the 1970s to the 1990s, during which period catches increased about 4.5-fold (Vojkovich 1998). The fisheries are carried out just offshore from important nesting islands

Table 5.1. Marine bird species that are endangered, threatened, or of special concern and that are attracted to human light sources.

Species	*References*
Newell's shearwater (*Puffinus auricularis newelli*)	Telfer et al. 1987, Ainley et al. 1997, 2001, Day et al. 2003
Dark-rumped petrel (*Pterodroma phaeopygia sandwichensis*)	Telfer et al. 1987
Cahow (Bermuda petrel) (*Pterodroma cahow*)	Beebe 1935
Grey-faced petrel (*Pterodroma macroptera gouldi*)	Le Corre et al. 2002
Barau's petrel (*Pterodroma baraui*)	Le Corre et al. 2002
Mascarene petrel (*Pseudobulweria aterrima*)	Le Corre et al. 2002
Band-rumped storm-petrel (*Oceanodroma castro cryptoleucura*)	Telfer et al. 1987, Slotterback 2002
Audouin's gull (*Larus audouinii*)	Arcos and Oro 2002
Xantus's murrelet (*Synthliboramphus hypoleucus*)	Pacific Seabird Group 2002

for murrelets and black-vented shearwaters (*Puffinus opisthomelas*). Their lights have also facilitated nocturnal predation by barn owls (*Tyto alba*) and western gulls (*Larus californicus*) at colonies and possibly disrupted reproductive behavior, movement, and aggregations on the water, which leads to nest abandonment (Keitt 1998).

Methods to Reduce Effects of Artificial Light on Seabirds

About twenty-five years ago, Hope-Jones (1980) indicated the need for detailed study of the effects of hydrocarbon platforms on avian behavior and mortality. Despite the phenomenal proliferation of these platforms in the world's oceans and as surprising as it seems, these studies are still necessary (Montevecchi et al. 1999, Burke et al. 2005).

Working with Seasonal and Spatial Patterns of Avian Vulnerability

Peak fledging periods are highly concentrated during a few weeks in late summer in the Northern and Southern hemispheres. Minimizing coastal and offshore lighting at these times could significantly reduce unnecessary avian mortality. Moreover, some sites attract more birds than others. On the Hawaiian island of Kauai, for instance, the mortality of endangered shearwaters and petrels was highest at coastal sections near river mouths, apparently because fledglings of these species follow river valleys from inland mountain nesting sites to sea (Telfer et al. 1987). The Kauai Surf Hotel near the mouth of the Huleia River accounted for almost half of all the avian fallout documented during 1981 (Telfer et al. 1987). By shielding and eliminating skyward lighting at the hotel during fledging times, Reed et al. (1985) produced significant reductions in the mortality of these endangered endemic species. Such temporal mitigative strategies could also be applied profitably during periods of peak migratory movements. The County of Kauai initiated a program of insulating and shielding streetlights in 1980, and a Save Our Shearwaters program, aimed at recovering and releasing stranded young birds, has been in place since 1978 (Day et al. 2003).

The flares on offshore and land-based hydrocarbon facilities are periodically shut down for maintenance and refit. These downtimes should be scheduled to coincide with periods of greatest risk of avian mortality, that is, peak fledging and migration times.

Shielding, Extinguishing, and Modifying Light

Shielding lights to eliminate skyward illumination could greatly reduce the catch basin of light attraction for birds in or passing through a region. By shielding the upward projection of light, Reed et al. (1985) demonstrated experimental reductions of 30–50% of the landings of endangered endemic shearwater and petrel fledglings at a coastal Hawaiian resort. This approach indicates worthwhile opportunities for reducing coastal and offshore light pollution.

Some cites such as Tucson, Arizona and Prague, Czech Republic shield lights in their municipalities to reduce light pollution that interferes with astronomical observation. Light shielding also helps to direct more light downward, where it is intended. This action also benefits birds that are active and migrate at night. Shielding of lights at marine platforms must both eliminate the skyward projection of light and guard against increasing the incidence of light directed at the sea surface to avoid its attractiveness to fishes and invertebrates.

A practical but underused approach to reducing light pollution is simple conservation. Turning off unneeded exterior and interior lighting and covering windows at night could be extremely useful. In 2000, the

California Fish and Game Commission required that squid fishing vessels shield their lights and use no more than 30,000 W per boat. Observers are not required on these vessels, but they should be.

Different wavelengths of light have different attractiveness to animals; for example, red and blue appear to be less attractive than white light (Wiese et al. 2001; see also Weir 1976, Telfer et al. 1987). More compellingly, intermittent lights at lighthouses result in fewer bird losses compared with steady rotating beams (Weir 1976). Lighthouses in Canada and elsewhere still use rotating beams; these should be replaced with strobe or intermittent flashing signals.

Flaring at Offshore Hydrocarbon Platforms

Flaring cannot be shielded to prevent upward illumination, but it can be reduced and eventually eliminated by reinjecting gases into hydrocarbon basins. The technology is available to do this and should be implemented rapidly and universally.

During the initial operation of the Hibernia platform on the continental shelf of eastern Canada in 1998, there were reports of hundreds, thousands, and tens of thousands of seabirds circling the platform for hours. These reports have ceased, but because there are no dedicated independent observers or comprehensive protocols for collecting this type of information on this and other platforms, information is lacking on what has occurred and what is occurring. In the absence of information, it is impossible to assess the consequences of flaring and offshore artificial lighting. About a year after startup, potentially significant levels of seabird mortality were still ongoing at a sufficient level to be documented during a casual visit by a journalist (Wood 1999). Current protocols on offshore platforms are inadequate to detect significant episodic mortality (Burke et al. 2005).

Mandating Dedicated Independent Observers on Offshore Hydrocarbon Platforms and Light-Induced Fishery Vessels

Self-reporting does not always provide accurate or reliable assessments of activity, especially of negative, inappropriate, or illegal activity (Weimerskirch et al. 2000). Independent arm's-length monitoring is widely accepted as a more valid and reliable means of resource and environmental assessment because industries or individuals with vested interests in profits do not always self-regulate unless compelled to do so.

Long-term systematic observations by dedicated independent observers therefore are necessary to reliably document and understand the episodic nature of avian mortality at lighted structures at night (Montevecchi et al. 1999). Without such information, effective mitigation is essentially precluded. A program of independent, systematic observations throughout the year is necessary to detail the species present and times of greatest risk. Risk periods vary widely between species and between oceanographic regions, and an adaptive approach to mitigation is needed to implement different strategies in different circumstances.

Dedicated independent observers should be mandated as a legislative condition of operation of offshore hydrocarbon platforms in all jurisdictions. Observers are already required on fishing vessels because of the potential detrimental effects that biologically unsound fishing practices can have on populations of marine fishes (Stehn et al. 2001). The threats from lights and flares at offshore hydrocarbon platforms appear as severe and necessitate similar regulation.

Reducing Cumulative Effects

Light acts in concert with other environmental factors such as heat, structures, pollutants, and food to augment the risks to birds. For instance, seabirds attracted to offshore lights associated with squid fishing vessels or hydrocarbon platforms might also be killed by ingesting hooked prey or by oil on the water.

An example of an indirect cumulative influence relates to the unnecessary discharge of wastewater at offshore platforms. These wastes fertilize the developing reef below platforms and promote plant and crustacean growth that in turn attracts fish (Duffy 1975, Ortego 1978, Sonnier et al. 1976). The fishes in turn may be attracted to the surface waters by intense lighting, where they may be preyed on by birds at night (Burke et al. 2005). Retaining wastewater at platforms and recycling it at land-based facilities would prevent unnecessary fertilization and reduce the attraction of scavenging gulls.

Limiting the Expansion of Light-Induced Fisheries

Concerns have been expressed about the movement of light-induced squid fisheries into the Antarctic region and the consequences for squid-eating marine birds and mammals (Rodhouse et al. 2001). Quotas for squid in the Antarctic have been set conservatively on the basis of these concerns.

Limiting the Construction of New Lighted Structures

Artificial lighting is increasing globally, including in the marine environment (e.g., *Pipeline and Gas Journal* 2005). The most direct and effective mitigative measures to preserve darkness involve eliminating unnecessary illumination, reducing light intensity, and minimizing the skyward and seaward projection of artificial light.

Conclusion

Lighthouses, offshore and nearshore squid and other fisheries that use intense lighting to attract prey at night, and offshore oil and gas platforms and their brilliant gas flares are imposing new artificial light sources in heretofore dark nocturnal ocean environments. These developments attract, concentrate, and kill seabirds and other marine animals. The mortality of seabirds associated with these artificial sources is not monitored or studied effectively. To minimize these forms of mortality, it is essential to study their seasonal variation and species vulnerabilities. Some causes of this mortality are indirect (e.g., energy depletion from prolonged circling of light sources, increasing predation on nocturnal species by diurnal gulls hunting at night), and some are embedded in cumulative effects (e.g., offshore platforms create artificial reefs that attract crustaceans and fishes that in turn attract avian predators). Endangered, threatened, and rare species are at especially high risk for negative population effects. Fledglings making their initial flights to sea from nesting areas and migrating flocks are the most critically affected groups. Occurrences of light-associated mortality are episodic, so to document this mortality there is a compelling need to legislatively mandate dedicated independent observers on hydrocarbon platforms and light-enhanced nocturnal fishery vessels.

Acknowledgments

I am extremely grateful to Travis Longcore and Catherine Rich for involving me in this interesting and compelling environmental issue. Chantelle Burke provided the photograph of the Hibernia oil platform. Dr. Greg Robertson provided a bibliography report from the Canadian Wildlife Service. Brad Keitt provided information and material. Dr. Bill Sydeman provided a helpful review of the manuscript. My long-term research program with marine birds has been supported by the Natural

Sciences and Engineering Research Council of Canada. To each of these people and organizations, I am very thankful.

Literature Cited

Ainley, D. G., R. Podolsky, L. Deforest, G. A. Spencer, and N. Nur. 2001. The status and population trends of the Newell's shearwater on Kaua'i: insights from modeling. *Studies in Avian Biology* 22:108–123.

Ainley, D. G., T. C. Telfer, and M. H. Reynolds. 1997. Townsend's and Newell's shearwater (*Puffinus auricularis*). Pages 1–20 in A. Poole and F. Gill (eds.), *The birds of North America*, No. 297. The Academy of Natural Sciences, Philadelphia, Pennsylvania, and the American Ornithologists' Union, Washington, D.C.

Aldrich, J. W., R. R. Graber, D. A. Munro, G. J. Wallace, G. C. West, and V. H. Cahalane. 1966. Report of the Committee on Bird Protection, 1965. *Auk* 83:457–467.

Allen, J. A. 1880. Destruction of birds by light-houses. *Bulletin of the Nuttall Ornithological Club* 5:131–138.

Arcos, J., and D. Oro. 2002. Significance of nocturnal purse seine fisheries for seabirds: a case study off the Ebro Delta (NW Mediterranean). *Marine Biology* 141:277–286.

Avery, M., P. F. Springer, and J. F. Cassel. 1976. The effects of a tall tower on nocturnal bird migration: a portable ceilometer study. *Auk* 93:281–291.

Baird, P. H. 1990. Concentrations of seabirds at oil-drilling rigs. *Condor* 92:768–771.

Beebe, W. 1935. Rediscovery of the Bermuda cahow. *Bulletin of the New York Zoological Society* 38:187–190.

Black, G. A. P., T. W. Rowell, and E. G. Dawe. 1987. Atlas of the biology and distribution of the squids *Illex illecebrosus* and *Loligo pealei* in the northwest Atlantic. *Canadian Special Publication of Fisheries and Aquatic Sciences* 100:1–62.

Blomqvist, S., and M. Peterz. 1984. Cyclones and pelagic seabird movements. *Marine Ecology Progress Series* 20:85–92.

Boersma, P. D., N. T. Wheelwright, M. K. Nerini, and E. S. Wheelwright. 1980. The breeding biology of the fork-tailed storm-petrel (*Oceanodroma furcata*). *Auk* 97:268–282.

Bourne, W. R. P. 1979. Birds and gas flares. *Marine Pollution Bulletin* 10:124–125.

Bretagnolle, V. 1990. Effet de la lune sur l'activité des pétrels (classe Aves) aux îles Salvages (Portugal) [Effect of the moon on the activity of petrels (class Aves) on the Salvage Islands (Portugal)]. *Canadian Journal of Zoology* 68:1404–1409.

Brewster, W. 1886. Bird migration. Part 1. Observations on nocturnal bird flights at the light-house at Point Lepreaux, Bay of Fundy, New Brunswick. *Memoirs of the Nuttall Ornithological Club* 1:5–10.

Brown, L. H., E. K. Urban, and K. Newman. 1982. *The birds of Africa*. Volume I. Academic Press, London.

Bryant, S. L. 1994. *Influences of* Larus *gulls and nocturnal environmental conditions on Leach's storm-petrel activity patterns at the breeding colony*. M.S. thesis, Memorial University of Newfoundland, St. John's.

Burke, C. M., G. K. Davoren, W. A. Montevecchi, and F. K. Wiese. 2005. Seasonal and spatial trends of marine birds along offshore support vessel transects and at oil platforms on the Grand Banks. Pages 587–614 in S. L. Armsworthy, P. J. Cranford, and K. Lee (eds.), *Offshore oil and gas environmental effects monitoring: approaches and technologies*. Battelle Press, Columbus, Ohio.

Carlisle, J. G. Jr., C. H. Turner, and E. E. Ebert. 1964. Artificial habitat in the marine environment. *California Department of Fish and Game, Fish Bulletin* 124:1–93.

Carter, H. R., D. L. Whitworth, J. Y. Takekawa, T. W. Keeney, and P. R. Kelly. 2000. At-sea threats to Xantus' murrelets (*Synthliboramphus hypoleucus*) in the Southern California Bight. Pages 435–447 in D. R. Browne, K. L. Mitchell, and H. W. Chaney (eds.), *Proceedings of the Fifth California Islands Symposium, 29 March to 1 April 1999*. U.S. Minerals Management Service, Camarillo, California.

Chaffey, H. 2003. *Integrating scientific knowledge and local ecological knowledge (LEK) about common eiders (*Somateria mollissima*) in southern Labrador*. M.S. thesis, Memorial University of Newfoundland, St. John's.

Cherel, Y., and V. Ridoux. 1992. Prey species and nutritive value of food fed during summer to king penguin *Aptenodytes patagonica* chicks at Possession Island, Crozet Archipelago. *Ibis* 134:118–127.

Clark, R., and P. Leppert-Slack. 1994. Cumulative effects assessment under the National Environmental Policy Act in the United States. Pages 37–44 in A. J. Kennedy (ed.), *Cumulative effects assessment in Canada: from concept to practice*. Alberta Association of Professional Biologists, Edmonton.

Crawford, R. L. 1981. Bird kills at a lighted man-made structure: often on nights close to a full moon. *American Birds* 35:913–914.

Day, R. H., B. A. Cooper, and T. C. Telfer. 2003. Decline of Townsend's (Newell's) shearwaters (*Puffinus auricularis newelli*) on Kauai, Hawaii. *Auk* 120:669–679.

Dick, M. H., and W. Davidson. 1978. Fishing vessel endangered by crested auklet landings. *Condor* 80:235–236.

Duffy, M. 1975. From rigs to reefs. *Louisiana Conservationist* 27:18–21.

Duinker, P. N. 1994. Cumulative effects assessment: what's the big deal? Pages 11–24 in A. J. Kennedy (ed.), *Cumulative effects assessment in Canada: from concept to practice*. Alberta Association of Professional Biologists, Edmonton.

Dunn, E. H., and E. Nol. 1980. Age-related migratory behavior of warblers. *Journal of Field Ornithology* 53:254–269.

Evans, G. 1968. Lighthouse report 1968. *Bardsey Observatory Report* 16:49–53.

Farner, D. S. 1964. The photoperiodic control of reproductive cycles in birds. *American Scientist* 52:137–156.

Groot, S. J. de. 1996. Quantitative assessment of the development of the offshore oil and gas industry in the North Sea. *ICES Journal of Marine Science* 53:1045–1050.

Harris, M. P. 1966. Breeding biology of the Manx shearwater *Puffinus puffinus*. *Ibis* 108:17–33.

Hope-Jones, P. 1980. The effect on birds of a North Sea gas flare. *British Birds* 73:547–555.

Hughes, L. 2000. Biological consequences of global warming: is the signal already apparent? *Trends in Ecology and Evolution* 15:56–61.

Huntington, C. E., R. G. Butler, and R. A. Mauck. 1996. Leach's storm-petrel, *Oceanodroma leucorhoa*. Pages 1–28 in A. Poole and F. Gill (eds.), *The birds of North America*, No. 233. The Academy of Natural Sciences, Philadelphia, Pennsylvania, and the American Ornithologists' Union, Washington, D.C.

Imber, M. J. 1975. Behaviour of petrels in relation to the moon and artificial lights. *Notornis* 22:302–306.

Johnston, D. W., and T. P. Haines. 1957. Analysis of mass bird mortality in October, 1954. *Auk* 74:447–458.

Keitt, B. S. 1998. *Ecology and conservation biology of the black-vented shearwater (*Puffinus opisthomelas*) on Natividad Island, Vizcaino Biosphere Reserve, Baja California Sur, México*. M.S. thesis, University of California, Santa Cruz.

Kemper, C. A. 1964. A tower for TV: 30,000 dead birds. *Audubon Magazine* 66:86–90.

Kumlien, L. 1888. Observations on bird migration at Milwaukee. *Auk* 5:325–328.

Le Corre, M., A. Ollivier, S. Ribes, and P. Jouventin. 2002. Light-induced mortality of petrels: a 4-year study from Réunion Island (Indian Ocean). *Biological Conservation* 105:93–102.

Lipinski, M. R., and S. Jackson. 1989. Surface-feeding on cephalopods by procellariiform seabirds in the southern Benguela region, South Africa. *Journal of Zoology, London* 218:549–563.

Maillard, J. 1898. Notes on the nesting of the fork-tailed petrel (*Oceanodroma furcata*). *Auk* 15:230–233.

McNeil, R., P. Drapeau, and R. Pierotti. 1993. Nocturnality in colonial waterbirds: occurrence, special adaptations, and suspected benefits. *Current Ornithology* 10:187–246.

Melvin, E. F., and J. K. Parrish (eds.). 2001. *Seabird bycatch: trends, roadblocks, and solutions*. University of Alaska Sea Grant, AK-SG-01-01, Fairbanks.

Montevecchi, W. A. 1993. Seabird indication of squid stock conditions. *Journal of Cephalopod Biology* 2:57–63.

Montevecchi, W. A., F. K. Wiese, G. K. Davoren, A. W. Diamond, F. Huettmann, and J. Linke. 1999. *Seabird attraction to offshore platforms and seabird monitoring from offshore support vessels and other ships: literature review and monitoring designs*. Report for Canadian Association of Petroleum Producers (CAPP), Calgary, Alberta.

Morse, D. H., and C. W. Buchheister. 1977. Age and survival of breeding Leach's storm-petrels in Maine. *Bird-Banding* 48:341–349.

Mougeot, F., and V. Bretagnolle. 2000. Predation risk and moonlight avoidance in nocturnal seabirds. *Journal of Avian Biology* 31:376–386.

Muirhead, K., and A. P. Cracknell. 1984. Identification of gas flares in the North Sea using satellite data. *International Journal of Remote Sensing* 5:199–212.

Murie, O. J. 1959. Fauna of the Aleutian Islands and Alaska Peninsula. *U.S. Fish and Wildlife Service North American Fauna* 61:1–364.

Murphy, R. C. 1936. *Oceanic birds of South America*. Macmillan, New York.

Newman, R. J. 1960. Spring migration: central southern region. *Audubon Field Notes* 14:392–397.

Ortego, B. 1978. Blue-faced boobies at an oil production platform. *Auk* 95: 762–763.

Pacific Seabird Group. 2002. *Petition to the U.S. Fish and Wildlife Service/California Department of Fish and Game to list the Xantus's murrelet under the United States/California Endangered Species Act*. Pacific Seabird Group, La Jolla, California.

Pauly, D., V. Christensen, J. Dalsgaard, R. Froese, and F. Torres Jr. 1998. Fishing down marine food webs. *Science* 279:860–863.

Pipeline and Gas Journal. 2005. Dramatic increase in offshore spending predicted. *Pipeline and Gas Journal* 232(4):59.

Reed, J. R., J. L. Sincock, and J. P. Hailman. 1985. Light attraction in endangered procellariiform birds: reduction by shielding upward radiation. *Auk* 102:377–383.

Roberts, P. 1982. Birds at Bardsey Lighthouse 1982. *Bardsey Observatory Report* 26:33–34.

Rodhouse, P. G., M. R. Clarke, and A. W. A. Murray. 1987. Cephalopod prey of the wandering albatross *Diomedea exulans*. *Marine Biology* 96:1–10.

Rodhouse, P. G., C. D. Elvidge, and P. N. Trathan. 2001. Remote sensing of the global light-fishing fleet: an analysis of interactions with oceanography, other fisheries and predators. *Advances in Marine Biology* 39:261–303.

Sage, B. 1979. Flare up over North Sea birds. *New Scientist* 81:464–466.

Sillett, T. S., and R. T. Holmes. 2002. Variation in survivorship of a migratory songbird throughout its annual cycle. *Journal of Animal Ecology* 71:296–308.

Slotterback, J. W. 2002. Band-rumped storm-petrel (*Oceanodroma castro*) and Tristram's storm-petrel (*Oceanodroma tristrami*). Pages 1–28 in A. Poole and F. Gill (eds.), *The birds of North America*, No. 673. The Birds of North America, Inc., Philadelphia, Pennsylvania.

Sonnier, F., J. Teerling, and H. D. Hoese. 1976. Observations on the offshore reef and platform fish fauna of Louisiana. *Copeia* 1976:105–111.

Stehn, R. A., K. S. Rivera, S. Fitzgerald, and K. Wohl. 2001. Incidental catch of seabirds by longline fisheries in Alaska. Pages 61–77 in E. F. Melvin and J. K. Parrish (eds.), *Seabird bycatch: trends, roadblocks, and solutions*. University of Alaska Sea Grant, AK-SG-01-01, Fairbanks.

Stenhouse, I. J., G. J. Robertson, and W. A. Montevecchi. 2000. Herring gull *Larus argentatus* predation on Leach's storm-petrels *Oceanodroma leucorhoa* breeding on Great Island, Newfoundland. *Atlantic Seabirds* 2:35–44.

Tasker, M. L., P. Hope-Jones, B. F. Blake, T. Dixon, and A. W. Wallis. 1986. Seabirds associated with oil production platforms in the North Sea. *Ringing and Migration* 7:7–14.

Telfer, T. C., J. L. Sincock, G. V. Byrd, and J. R. Reed. 1987. Attraction of Hawaiian seabirds to lights: conservation efforts and effects of moon phase. *Wildlife Society Bulletin* 15:406–413.

Terres, J. K. 1956. Death in the night. *Audubon Magazine* 58:18–20.

Verheijen, F. J. 1980. The moon: a neglected factor in studies on collisions of nocturnal migrant birds with tall lighted structures and with aircraft. *Die Vogelwarte* 30:305–320.

Verheijen, F. J. 1981. Bird kills at tall lighted structures in the USA in the period 1935–1973 and kills at a Dutch lighthouse in the period 1924–1928 show similar lunar periodicity. *Ardea* 69:199–203.

Vitousek, P. M., H. A. Mooney, J. Lubchenco, and J. M. Melillo. 1997. Human domination of Earth's ecosystems. *Science* 277:494–499.

Vojkovich, M. 1998. The California fishery for market squid (*Loligo opalescens*). *California Cooperative Oceanic Fisheries Investigations Reports* 39:55–60.

Wallis, A. 1981. North Sea gas flares. *British Birds* 74:536–537.

Warham, J. 1960. Some aspects of breeding behaviour in the short-tailed shearwater. *Emu* 60:75–87.

Watanuki, Y. 1986. Moonlight avoidance behavior in Leach's storm-petrels as a defense against slaty-backed gulls. *Auk* 103:14–22.

Watanuki, Y. 2002. Moonlight and activity of breeders and non-breeders of Leach's storm-petrels. *Journal of the Yamashina Institute of Ornithology* 34:245–249.

Weimerskirch, H., D. Capdeville, and G. Duhamel. 2000. Factors affecting the number and mortality of seabirds attending trawlers and long-liners in the Kerguelen area. *Polar Biology* 23:236–249.

Weimerskirch, H., P. Jouventin, and J. C. Stahl. 1986. Comparative ecology of the six albatross species breeding on the Crozet Islands. *Ibis* 128:195–213.

Weir, R. D. 1976. *Annotated bibliography of bird kills at man-made obstacles: a review of the state of the art and solutions*. Department of Fisheries and the Environment, Environmental Management Service, Canadian Wildlife Service, Ontario Region, Ottawa.

Whittow, G. C. 1997. Wedge-tailed shearwater (*Puffinus pacificus*). Pages 1–24 in A. Poole and F. Gill (eds.), *The birds of North America*, No. 305. The Academy of Natural Sciences, Philadelphia, Pennsylvania, and the American Ornithologists' Union, Washington, D.C.

Wiese, F. K., W. A. Montevecchi, G. K. Davoren, F. Huettmann, A. W. Diamond, and J. Linke. 2001. Seabirds at risk around offshore oil platforms in the North-west Atlantic. *Marine Pollution Bulletin* 42:1285–1290.

Wolfson, A., G. Van Blaricom, N. Davis, and G. S. Lewbel. 1979. The marine life of an offshore oil platform. *Marine Ecology: Progress Series* 1:81–89.

Wood, D. 1999. Hibernia. *Air Canada en Route* 2:48–57.

Chapter 6

Road Lighting and Grassland Birds: Local Influence of Road Lighting on a Black-Tailed Godwit Population

Johannes G. de Molenaar,
Maria E. Sanders, and Dick A. Jonkers

Public awareness of the ecological effects of outdoor lighting has increased. In the Netherlands public attention caused the subject to be placed on the political agenda in 1995. As a result, the Road and Hydrologic Engineering Division of the Dutch Ministry of Transport, Public Works and Water Management commissioned the research institute Alterra to investigate effects of road lighting on nature. This resulted in the experiment presented here, which concentrates on the effects of roadway lighting on a breeding bird in an open grassland habitat. Although much research has illustrated the effects of artificial lights on birds during migration, few studies have investigated the effects of lights in and near breeding habitats.

In this chapter we first review the effects of artificial night lighting on the physiology and spatial behavior of birds. We then present the design and results of an experiment to investigate the effect of roadway lighting on the breeding ecology of black-tailed godwits (*Limosa l. limosa*) in a wet grassland in the Netherlands. Finally we discuss the outcome of this experiment and report the actions taken by the Dutch government to reduce the ecological effects of roadway lighting.

Influence of Artificial Lighting on the Physiology and Spatial Behavior of Birds

The effects of light and lighting on the physiology of birds are known primarily from laboratory experiments, especially those aiming to increase the production and quality of poultry (de Molenaar et al. 1997). These experiments concentrate on the effect of the relative length of daily light and dark periods (Elliot and Edwards 1991, Lewis and Perry 1990a, 1990b, May et al. 1990, Newcombe et al. 1991, Scheideler 1990, Siopes and Pyrzak 1990, Zimmermann and Nam 1989, and recent volumes of *Poultry Science*). Because of their scope and specific design, extrapolating these experiments to the effects of outdoor lighting is difficult. Nevertheless, combined with the results of the laboratory experiments with wild birds, they indicate that for birds in temperate regions, night length initiates growth, metabolism, skeletal development, sexual development, courtship and mating, reproductive cycles, migration, and molting (see de Molenaar et al. 1997). Knowledge from the field ranges from incidental observation to comparison of different situations over time. Experiments are few and are limited to after–control–impact studies (de Molenaar et al. 1997).

Artificial Lighting and Biorhythms

The effect of artificial lighting on circadian rhythms is known from many anecdotal reports of songbirds starting to sing early under the influence of artificial lighting in the morning or at night, such as blackbird (*Turdus merula*; Mitchell 1967), song thrush (*Turdus philomelos*; Van Lynden 1978), European robin (*Erithacus rubecula*; King 1966, Hollom 1966, Labberté 1978), chiffchaff (*Phylloscopus collybita*; Labberté 1978), and dunnock (*Prunella modularis*; Van Lynden 1978, Labberté 1978). Other birds that have been reported in the literature as singing at night under the influence of artificial lighting include bluethroat (*Luscinia svecica*), reed bunting (*Emberiza schoeniclus*), and nightingale (*Luscinia megarhynchos*; Outen 2002). Birds begin to sing within a minute of a critical light level being reached (Rawson 1923). Artificial lighting may extend the daily feeding time, to the bird's benefit (Freeman 1981, King 1966, Taapken, personal communication in de Molenaar et al. 1997), but it also may increase the risk of exposure to intraspecific competition (e.g., between diurnal raptors and nocturnal raptors) and predation (de Molenaar et al. 1997).

For circannual rhythms, artificially increasing daylength induces hormonal, physiological, morphological, and behavioral changes initiating breeding (Rowan 1928, 1929, 1937, Farner 1964, 1966, Wolfson 1959, Follett 1980, Wingfield et al. 1992, Silverin 1994). Lofts and Merton (1968) found that of about 60 wild bird species, all were brought into premature breeding condition by experimental exposure to artificially short nights in winter. Songbirds kept in a constant daylight period (light:dark mostly 12:12) showed a shortening of their annual cycle by about one to four months (Berthold 1980). Mitchell (1967) reports blackbirds singing in early January 2.5 hours after sunset in a small, lighted garden in Berlin. Havlin (1964) and Lack (1965) noted that European robins and blackbirds began laying one or two weeks earlier in urban environments than in woods; Lack also reported that their breeding extended later into the season. This could result partly from the influence of artificial lighting but partly also from the effects of food supply (Havlin 1964) and higher temperatures in the winter in urban areas. Conversely, artificially short daylengths are found to retard sexual development (Miller 1955, Farner 1959, Wolfson 1959). The photoperiodic induction of the secretion of testosterone and the development of the gonads in spring stimulates the development of secondary male sex characteristics, which by visual stimuli play a role in synchronizing the endocrinology and behavior of female birds (Witschi 1961, Wingfield and Farner 1978, de Molenaar et al. 1997).

Artificially increased daylength may also affect the timing of migration by influencing hormonal and physiological processes and by allowing a longer daily feeding time. Thus Bewick's swans (*Cygnus columbianus bewickii*) feeding under illumination at Slimbridge lay down fat more rapidly and therefore can reach spring migration condition much more quickly (Rees 1982). Early departure could result in birds arriving too early in their arctic breeding habitat, with damaging consequences (Rees 1982). Conversely, autumn migration may be delayed, exposing birds to the risk of poor fat reserves because of diminishing food supply in autumn and unfavorable conditions during migration (de Molenaar 2003). Furthermore, artificially increased daylength may also affect the timing of the molt (e.g., Rautenberg 1957, Wagner 1956, Wallgren 1954, Wolfson

1942), even in some tropical birds (e.g., Brown and Rollo 1940, Farner 1959, Rollo and Domm 1943, Wolfson 1959, Gwinner 1986).

Artificial Lighting and Spatial Behavior

Spatial behavior of animals is influenced by luminance—the visibility and intensity of the surface brilliance of objects (such as lamps and lighted windows)—and by illuminance through changes in orientation. Many articles, especially in North America, report attraction of birds to bright lights and brightly lit structures (see bibliographies by Avery et al. 1980 and Trapp 1998 and reviews by Evans Ogden 1996 and in Chapters 4 and 5, this volume). Estimates of resulting annual numbers of victims in the United States range from 4 to 50 million birds (Malakoff 2001).

Birds migrating at night, especially with an overcast sky, become disoriented by continuous orange to red light (Gochfeld 1973, Wiltschko and Wiltschko 1995, Wiltschko et al. 1993, Cochran and Graber 1958). The effect is stronger in young birds than in adults (Baldaccini and Bezzi 1989) and absent in light with green to blue wavelengths (Wiltschko and Wiltschko 1995, Wiltschko et al. 1993). Seabirds, waterbirds, and marsh birds may mistake lamplight-reflecting surfaces such as wet roads, greenhouses, and even damp grass fields for water surfaces and land on them (Kraft 1999). Having trouble taking off again, they become exposed to predation and exhaustion (de Molenaar et al. 1997, R. Podolsky, personal communication, 2002).

Effects of Roadway Lighting on Breeding of a Grassland Bird

Having found no experiments or specific observations of the effect of road lighting in the literature, my research team set up a field experiment on the breeding birds of open landscapes, where the effects of road lighting, if any, were expected to be more distinct than in woodland habitats. The experiment focused on the following question: does road lighting have an effect on establishment of breeding birds (i.e., the selection of nest site), timing of the breeding period, and egg predation by, for example, crows, gulls, and foxes because of expansion of their active period? The experiment did not include a distinction between the possible influences of luminance and illumination because the illuminated area was negligibly small compared with the area from which the bright lights were visible. A complete description of the experiment and results is found in de Molenaar et al. (2000).

The reclamation of peat lands in the west of the Netherlands in the Middle Ages required drainage, for which ditches were dug, dividing the new farmland into parcels (Lambert 1971). This reclamation resulted in the development of wet grassland (Lambert 1971), in which a characteristic community of meadow birds developed and flourished for centuries (Beintema 1995b, Moedt 1995b). These agricultural fields were poorly fertilized and mown for hay or grazed by livestock. The core of the bird community consists of limnicolous species: black-tailed godwit (*Limosa l. limosa*), lapwing (*Vanellus vanellus*), redshank (*Tringa totanus*), ruff (*Philomachus pugnax*), snipe (*Gallinago gallinago*), curlew (*Numenius arquata*), and in a wider sense others such as garganey (*Anas querquedula*), shoveler (*Anas clypeata*), oystercatcher (*Haematopus ostralegus*), and blue-headed wagtail (*Motacilla f. flava*; Beintema 1995b, Moedt 1995b).

Black-tailed godwit, which is generally considered an indicator species for the overall birdlife of open grasslands, was selected for this study. These wet grasslands covered large parts of the Netherlands, and up to circa 1990 about half of the world population of black-tailed godwits (100,000 pairs) bred in the Netherlands. However, increasing intensification of agricultural use, in particular earlier large-scale mowing enabled by lowering of the water table, threatens the survival of eggs and chicks (Kruk et al. 1997). As a result, there has been a decrease of about 50% in the Dutch breeding population since 1990, which makes the black-tailed godwit an endangered (i.e., Red List) species. Conservation action by the Netherlands is needed for this species. Therefore, the influence of possible pressures, such as road lighting, on habitat quality in breeding areas should be identified and, if considered negative, eliminated, mitigated, or compensated.

Black-tailed godwits return from their winter quarters starting in late February. They prefer to breed in moist to wet grassland with a rough sward in which they select tussocks as sites to build their nests (Moedt 1995c, Beintema 1995a). The earliest breeders lay their first egg in late March; the median egg laying date is around April 16 (Beintema 1995a). Incubation starts about five days after the first egg is laid, when the clutch is complete. Chicks hatch after 24–25 days. The peak in hatching is about May 10 (Moedt 1995c, Beintema 1995a). Therefore experiments dealing with effects on reproduction should start before the middle of February and continue through May.

The field experiment compared one year (1998) with the road lights switched off and the next year (1999) with the lights switched on. The influence of the road itself, that is, the traffic noise (Reijnen 1995), should be the same in both years.

The study area met various necessary conditions. Its size and management guaranteed a sufficiently large and stable breeding black-tailed godwit population and a minimal risk of external disturbance. It was adjacent to a representative highway with regularly used roadway lighting that could be switched off. In addition, there was a comparable nearby area free from the road's influence, where road lights were temporarily installed that were switched on and off simultaneously with the ones along the highway.

The study site was in an open grassland area of 230 ha (568 acres) in the northwest of the country, south of the city of Alkmaar and largely owned by the private nature conservation organization Het Noordhollands Landschap. The grassland is divided by ditches into parcels or plots of varying size and form (Figure 6.1), which are drained by shallow trenches. These parcels are management units, so vegetation structure within them is mostly homogeneous. With more than 50 pairs per 100 ha (250 acres), the local black-tailed godwit population ranked among the best in the country. In contrast with conditions prevailing elsewhere in the country, the population was stable, even slightly increasing. The A9 highway runs from north to south right through the middle of this grassland area. According to the road manager, Rijkswaterstaat (Directorate-General of Public Works and Water Management), the daily traffic intensity on this road was about 90,000 vehicles in 1998 and about 93,000 vehicles in 1999. The traffic noise was considered approximately equal in both years. According to Reijnen et al. (1992) and Reijnen (1995), these traffic densities and a speed limit of 120 km/h (75 mph) influence habitat quality of breeding birds generally up to 800 m (about 0.5 mile) in open grassland, and somewhat farther for black-tailed godwits. West of the A9, the grassland was fully exposed to this influence and provided a study site of about 100 ha (250 acres) receiving lighting from the light poles along the road. East of the A9, a stretch of swamp, bushes, and trees along the motorway muffled the traffic noise to the extent that it should not influence most of the approximately 135 ha (334 acres) of this area (Reijnen et al. 1992, de Molenaar et al. 2000). In this control area, away from the road, 24 light poles were installed in the middle of the grassland (Figure 6.1).

The environmental variables that determine habitat quality in open grasslands, and hence the black-tailed godwit's choice of nest site, are soil moisture or drainage conditions of the grassland parcels, the structure of the sward, the grass height, and fertilizing with coarse stable dung (Moedt 1995a, Beintema 1995c). These environmental variables are considered

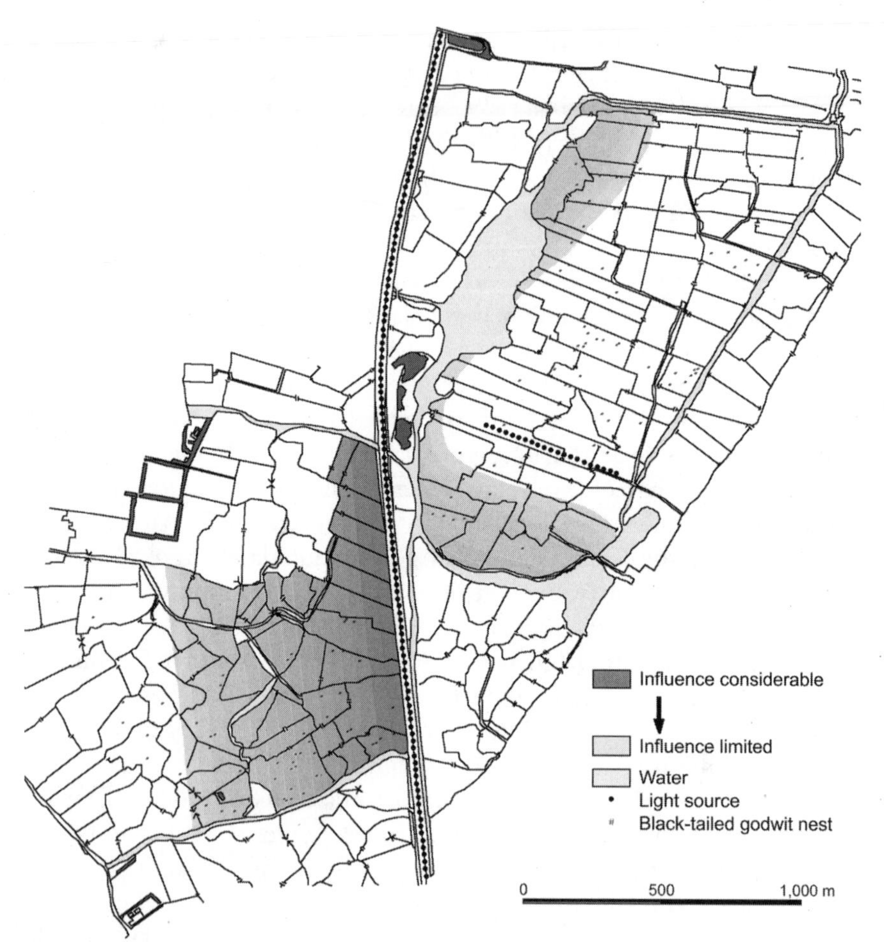

Figure 6.1. Wet meadow study area with road influence (after Reijnen et al. 1992) and location of light sources (poles) and nests in 1999.

homogeneous within a parcel but may vary widely from year to year, depending on weather conditions and the farmer's management. Consequently, the spatial distribution of nests may vary from year to year as well. Table 6.1 shows the number of nests per environmental condition class for both years. The average values expressed in classes of the environmental variables were estimated per parcel (see full results in de Molenaar et al. 2000) and digitized in a geographic information system (GIS).

Two teams inventoried black-tailed godwit nests. The teams made a weekly round in each of the two sites to find new nests and to monitor the previously found nests. It is plausible that all nests were located, given the

Table 6.1. Number of nests per class of each environmental condition in 1998 (n = 123) and 1999 (n = 140).

Height	*Structure*	*1998*				*1999*			
		Dry	*Wet*	*Very Wet*	*Extremely Wet*	*Dry*	*Wet*	*Very Wet*	*Extremely Wet*
Low	Smooth		7			2	18	3	
	Tussocky		6	3		17	35	11	12
	Very tussocky		1	6	12		8	8	18
Medium	Smooth		1	5	5		8		
	Tussocky		14	21	3				
	Very tussocky								
High	Smooth		1	22					
	Tussocky		8		8				
	Very tussocky								

experience of the surveyors, the amount of time spent in the field, and the open, easily visible character of the area. The team visited each nest up to three times, until the chicks were hatched or until the clutch appeared to be predated or otherwise abandoned. Nest location was determined by measuring the distance at right angles to the two nearest squared ditches or other fixed points of reference, then digitized in the GIS, which calculated the distance of each nest to the road and to the lights. The team started to search for nests in the second week of April, and stopped in the second week of May to avoid including renesting attempts. The team measured the incubation stage of the clutches (in days) with an incubometer, a refined variant of the common immersion test (Beintema 1995a). The result was used to establish the timing of breeding and as a check to avoid including renesting attempts.

Predation and other causes of loss were determined by the way in which the egg or eggs were damaged. When one or more eggs had disappeared the cause was left open. Clutches were considered to be abandoned if eggs were still present when the average incubation time had expired.

Regression analysis can be used to explore the relation between species and their environment. Many studies have combined GIS and regression techniques to assess the distribution of animal and plant species or their potential habitat (Austin et al. 1996, Sperduto and Congalton 1996, Narumalani et al. 1997, Bian and West 1997, Sanders 1999). Regression analysis focuses on how a particular species is related to environmental variables such as grass height and distance to light poles. The method is intended to assess which environmental variables a species responds to most strongly and which environmental variables are unimportant. Such an assessment proceeds through tests of statistical significance (Jongman et al. 1987). A standard statistical test could not be used here because nominal, quantitative, and ordinal explanatory variables had to be tested simultaneously; nests were recorded as presences, whereas absences were not recorded; and several classes of the environmental variables did not contain any examples in the field.

Quantitative and nominal variables can be tested simultaneously with logit regression, in which the statistical significance of the effect of quantitative explanatory variables is assessed by deviance tests (Jongman et al. 1987). For nominal explanatory variables the deviance test is closely related to the usual chi-square test. Regression techniques can easily cope with nominal and quantitative environmental variables but not with ordinal ones. When there are few possible values, it is better to treat an ordinal value as a nominal value (Jongman et al. 1987).

Logit regression attempts to express the probability that a species is present as a function of the explanatory variables (Hosmer and Lemeshow 1989, Jongman et al. 1987) and requires presence and absence data. The map of nest locations used in this study supplies only presences, so a map of absences is needed. An approximate solution is to add to the data a large number of random points, for each of which the environmental values have been determined from the GIS. The distribution of the random points over the classes of the environmental variables should approach the proportional cover of these classes. This approximate solution works not only for single nominal predictors but also multidimensionally because the proportional cover of combinations of classes of different variables is also estimated. The distribution of each quantitative variable is also approximated by its sample distribution. In fact, the multivariate distribution of both quantitative and qualitative variables is approximated in the sample. The more dimensions the logit model has, the larger the random dataset should be. The quantitative variables (distance to the light poles) make it impossible to determine the exact size. In this study, 10,000 random points were assumed to represent absences, and multiple logit regression was applied.

Absolute probability estimates of species occurrence are impossible to obtain because the probability estimates depend on the number of random points taken. When the number is large the absolute probability becomes unreasonably small. Therefore only a relative measure is appropriate, such as the log-odds ratio (Hosmer and Lemeshow 1989; see de Molenaar et al. 2000 for details).

We used the statistical program Genstat to perform a screening test (marginal and conditional) for generalized or multivariate linear models (Genstat 5 Committee 1987). The conditional test was applied to establish which environmental variables are significantly associated with the godwit's preference of a nest location. The test assessed the significance of each variable given the other ones. Next, we applied a logit regression analysis to the environmental variables that appeared significant ($p \leq 0.05$) from the conditional test. In the GIS, the classes of the environmental variables were substituted by their corresponding probability and were summarized per parcel. The outcome is a parcel suitability or relative probability of occupation map of black-tailed godwit nest locations.

The results of the regression analysis to predict species occurrence could not be used to validate the reliability of the predictions. The calculated variances are based on the assumption that the sites are independent. This assumption, however, is unlikely to be true because of spatial autocorrelation and the arbitrariness of the number of absences. No independent data from a different area were available to check the predictions based on regression analysis.

The statistical tests reveal which environmental variables predict the selection of nest sites (probability of occupation) (Table 6.2).

In 1998, black-tailed godwits preferred to nest in high or predominantly tussocky wet grassland. Grass height was more important than sward structure. Wet parcels were preferred above moist parcels and parcels with puddles. In 1999, the preference concentrated on predominantly tussocky, somewhat moist grassland and wet, partly inundated

Table 6.2. Significance of environmental variables predicting habitat suitability, tested separately (marginal test) and in relation to each other (conditional test).

	1998		*1999*	
	Marginal	*Conditional*	*Marginal*	*Conditional*
Grass height	***	***	*ns*	*ns*
Sward structure	*ns*	*	***	*
Drainage	***	**	***	***
Stable dung	**	*	**	*ns*
Road influence	*ns*	*ns*	*ns*	*ns*

ns, not significant; *$0.01 < p \leq 0.05$; **$0.001 < p \leq 0.01$; ***$p \leq 0.001$.

Table 6.3. Results of logit regression in 1998.

Variable	*Class*	*Estimate*	*SE*	*T(*)*
Constant		–5.471	0.295	–18.53
Structure	Tussocky	0.330	0.235	1.41
Structure	Very tussocky	0.899	0.381	2.36
Height	Medium	0.265	0.272	0.97
Height	High	1.210	0.299	4.05
Drainage	Very wet	0.729	0.219	3.33
Drainage	Extremely wet	0.257	0.297	0.87
Stable dung	Present	0.730	0.373	1.96

grassland. The preference for predominantly tussocky grassland in 1999 was consistent with the preference in 1998. Because of the weather, the overall grass height was low in 1999 when the birds started breeding and thus not an important factor in nest site selection. The statistical analysis revealed no significant road influence (Table 6.2). The environmental variables found to be significant were used in the logit regression (Table 6.3).

The results from the logit regression, the estimates, are a relative probability of occupation (log-odds). For example, the probability of occupation in medium-high grass (0.265) is about one-fifth of that in high grass (1.210). The first class is the basis at which the probability is compared and thus assimilated in the constant value. The estimates from Table 6.3 were used to calculate habitat suitability in the GIS. The estimates were linked to the corresponding classes in the GIS and summed per parcel to make a habitat suitability map (Figure 6.2).

We compared the distance of the nest locations to the road lights of the A9 and to the temporary installed lights in the field in the dark year and the lighted year. The conditional statistical test and the logit regression (Table 6.4) reveal that the lighting had a small significant influence ($p \leq 0.05$) on nest site selection along with parcel suitability (probability of occupation). In the marginal test, light was not significant, whereas parcel suitability remained highly significant ($p \leq 0.001$). There is a small but statistically significant effect of the lighting.

From the dark year 1998 to the lighted year 1999, the change in the distribution of the nests showed a division in three distinct zones: from 0 to 300 m (0–984 ft), from 300 to 500 m (984–1,640 ft), and from 500 to 1,000 m (1,640–3,281 ft) from the road lights of the motorway and the experimental road lights in the field (Table 6.5). Because the statistics predict the probability of occurrence without considering nest site tenacity,

Figure 6.2. Habitat suitability (probability of occupation) for black-tailed godwit in 1998.

Table 6.4. Results of logit regression in 1999.

Variable	*Estimate*	*SE*	*T(*)*
Constant	–0.4780	0.6890	–0.69
Probability of occupation	1.0420	0.1640	6.36
Lighting	0.0322	0.0130	2.48

Table 6.5. Relative number of located black-tailed godwit nests per hectare.

Distance from Light Poles (m)	*Number of Nests per Hectare (proportion of annual total)*		*Relative Number of Nests per 100 m*
	1998 (dark)	*1999 (lighted)*	*1998:1999*
0–300	1.37	0.90	1:0.7
300–500	1.05	1.96	1:1.9
500–1,000	0.70	0.75	1:1.1

the data in this table take into account only the parcels for which the probability of occupation did not change or changed slightly (one class) from 1998 to 1999 and disregard the part of the eastern area south of the row of experimental road lights because of its limited size and the influence of lights from two directions. The data suggest an effect distance of more than 300 m (984 ft), followed by a "bow wave" ascribed to nest site fidelity. An effect distance of 300 m (984 ft) extrapolates to 60 ha/km (239 acres/mile) of highway that is negatively influenced by lighting.

After settling the date of laying of the first egg, we traced whether environmental variables had any influence on that date and found no clear relation between the dates of laying of the first egg and parcel suitability, defined as the relative probability of occupation. Consequently, we studied the possible relation between distance to the lights and dates of laying of the first egg without correcting for grassland suitability. Although there is no statistically strong correlation between distance to the light poles and egg laying dates (Figure 6.3), the trend line indicates a later start of breeding period closer to the lighting than farther away from it.

Aerial predators, mainly carrion crows (*Corvus corone*), were regularly seen during fieldwork. Ground predators, such as red fox (*Vulpes vulpes*) and ermine (*Mustela erminea*), were observed very few times; indications of their presence were encountered occasionally. The number of observed predated clutches (six in 1998 [5%] and six in 1999 [4%]) is small in both years and well within the margin of the relative extent of predation of clutches of black-tailed godwits elsewhere in the Netherlands. Apart from evident predation, four nests in 1998 and eight nests in 1999 were abandoned for unknown reasons. Investigator effects cannot be excluded but do not help to explain the increase in the second year.

The distribution of predation over the area was wide and apparently random. The drawback to the limited extent of predation is that the number of cases was considered to be too small for statistical analysis. We did

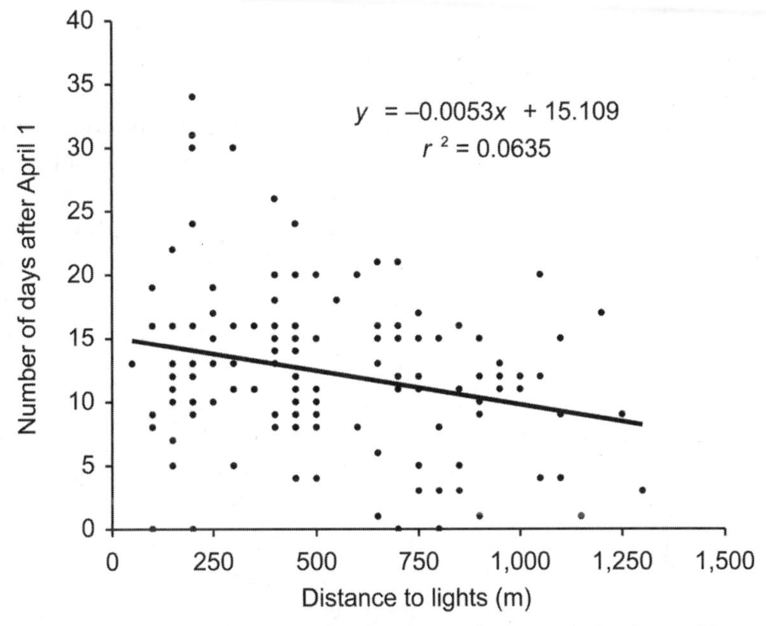

Figure 6.3. Correlation between the distance to lights and the date of laying the first egg in 1999.

observe, however, that the predated nests were situated mainly along the ditches in the verges of the parcels. This suggests a relationship with the way in which predators take their bearings and move through the field.

Apart from annual variation in grassland conditions, searching for the influence of road illumination on the nest choice of black-tailed godwits meets a serious complication: black-tailed godwits demonstrate marked breeding site tenacity. According to Groen (1993a, 1993b) half of them breed within 50 m (164 ft) of the previous year's nest site. This is a complication in the experiment that could not be avoided. Other grassland breeders such as lapwing (*Vanellus vanellus*) and redshank (*Tringa totanus*) do the same (Moedt 1995a). This breeding site tenacity, enforced by the tendency to cluster together under favorable conditions, could seriously suppress the possible effects of all sorts of environmental changes, including the effect of road lighting on nest choice and population density.

The statistical analysis revealed no significant effect for road influence, suggesting that the overall quality of the breeding habitat is so favorable that it compensates for this influence. This indicates that there is a certain buffer capacity for potentially negative external factors. Previous research revealed that motorways influence the nest choice of black-tailed godwit: the population density was low near the road and increased with increasing distance from the road (Reijnen et al. 1992, Reijnen 1995). However, Reijnen (1995) also observed that the negative influence of road traffic on the quality of the breeding habitat may be compensated by particularly favorable other factors, including management, sward structure, and grass height. The concentration of nests about 200 m (656 ft) away from the A9 motorway in 1998 presents a striking illustration of this phenomenon.

The difference in the preference for drainage conditions between 1998 and 1999 seems to be related at least in part to weather conditions, especially differences in the temporal pattern and intensity of rainfall, against the background of the overall poor drainage conditions of the parcels. It is generally assumed that black-tailed godwits prefer to nest in grass tussocks or on the rougher and somewhat drier verge of a trench (Ruitenbeek et al. 1990, Beintema 1995c). In 1998, however, grass height played a large role in nesting preference. This is attributable largely to a concentration of 13 nests about 200 m (656 ft) away from the A9 on two parcels with an early high and dense perennial ryegrass cover. This cover resulted from previous recultivation by killing the old sward, plowing, fertilizing, and resowing. The growth of the permanent, moderately to poorly drained, minimally fertilized grassland elsewhere started late and slow. In an area with a prevailing overall low sward, such early parcels with high dense grass seem for reasons of shelter to be attractive for nesting. If this supposition is true, the average date of laying of the first egg of the clutches there should be earlier than the average first date of egg laying elsewhere in the area. Indeed, an analysis of the 13 clutches on both renewed ryegrass parcels revealed that they were laid about one week (i.e., 5–9 days) earlier than average.

In 1999, the influence of the grass height was not significant. Evidently, it could not have been important in this year because at the time when the birds choose their nests, high grass plots were absent. In this year, the grass was low, and the nests of the black-tailed godwits were concentrated on tussocky parcels. Tussocks seem to become attractive when the growth of high dense grass is retarded by an overall late and slow regrowth.

The influence on nest choice apparently is caused not by the illumination as such, because the measurable illumination is near zero at a distance of 50 m (164 ft) from the road lights, but rather by the visibility of the strong light sources and perhaps also by the visibility of the illuminated space. Because no significant road influence could be observed, it is tempting to suggest that the influence of the lighting might be stronger than the influence of the road, even though it is less than the influence of the grassland quality. The design of the experiment, however, does not support that conclusion. To prove the negative effect of road lighting on the selection of nest sites, there should be control areas fully independent of the road, there must be replicates at independent test sites, and the effects should be monitored for a much longer period to study trends over time. Therefore the results of this study are only an indication that road lighting has a negative influence on the selection of nest sites.

The questions of the experiment refer to attraction or repulsion by the lights, deregulation of the day–night rhythm, or both. Deregulation of the circannual rhythm requires prolonged daily exposure in an illuminated space. In the case of migratory birds such as black-tailed godwit, this is quite unlikely, although it is possible for birds foraging under artificial illumination (Rees 1982). The measurable illumination on the ground reaches only 50 m (164 ft), and the birds arrive from their winter quarters only a few weeks before they start breeding.

It may be assumed that the first arriving birds can occupy the best positions (cf. Reijnen 1995). Therefore there should be a positive relation between the probability of occupation and the dates on which the first eggs are laid, but this relation could not be demonstrated. Apparently, the area was not yet fully occupied; both early birds and late birds seemed to be able to choose the most suitable place for their nests. Conversely, the findings suggest that early birds nested farther away from lighting than birds arriving later. The large spread in the data may be explained, among other things, by nest fidelity and the insignificant relation between the dates of laying of the first egg and the probability of occupation of the parcels. Traffic volume did not change over time from early nesters to late nesters. There was a regular daily pattern caused by intensive commuter traffic.

Conclusion

Any conclusions are preliminary because of the limited design of the experiment. Breeding site tenacity in particular, but also clustering, weather conditions, habitat quality at the beginning of the breeding season, the lack of more independent test sites, and the short duration of the study are major complications. The conclusions are as follows:

- Road lighting appeared to have a small but statistically significant negative effect on the quality of habitat for breeding by the local black-tailed

godwit population. The effect seemed to reach over several hundreds of meters. An effect distance of 300 m (984 ft) extrapolates to 60 ha/km (239 acres/mile) of highway that is negatively influenced by lighting.
- Road lighting seemed to have a negative but small effect on the timing of the breeding period.
- The observations did not permit any conclusions about a possible action of road lighting on egg predation.

It might be expected that the negative effect of lighting will increase in the long run. Breeding site tenacity covers several tens of meters, and so repulsed birds may shift their nest site every year only a short distance farther away from the lighting. It might also be expected that the negative effects of lighting will be stronger in grasslands where less optimal field conditions offer less compensation (Reijnen 1995).

A statistically significant effect of the road (i.e., road traffic noise) on breeding habitat quality near the A9 highway was absent. It seems that such an effect can be compensated by favorable conditions of environmental variables such as grass height, drainage conditions, sward structure, and fertilizing with stable dung. The statistically significant effect of road lighting on breeding habitat quality indicated that the influence of this lighting was less compensated by the site factors than was the influence of the road. This does not justify, however, the conclusion that the influence of road illumination was stronger than the influence of road traffic. A relation between the start of egg laying and the suitability of the parcels where the nests are found was not demonstrable. Consequently, it is assumed that both early and late starters could select the most suitable nest sites, but apparently they were also influenced by, for example, breeding site tenacity.

An early result of my literature study (de Molenaar et al. 1997) was the publication of guidelines for road lighting in nature areas (CROW 1997). Almost immediately thereafter, the Utrecht regional directorate of the Directorate-General of Public Works and Water Management introduced an experiment in the center of the country. Along three major motorways (A12, A27, and A28), the main road lights were switched off during the quiet hours between 11 P.M. and 6 A.M. Weak 9-W lights two-thirds up the light poles were switched on instead for orientation. A study revealed that this regime does not affect traffic safety. The experiment has now become permanent. Along the A9, where the experiment with the black-tailed godwits was carried out, the Noord-Holland regional directorate has the road lights switched off during the breeding season unless traffic and weather conditions dictate otherwise. Recently, the Ministry of

Transport, Public Works and Water Management finished a policy document on lighting of national roads, providing the basis for future policy decisions considering optimal environmental protection and energy savings as well as traffic safety. Apart from this, it should be mentioned that road signs in the country are made of highly reflective material or diffusely lit from the inside.

Acknowledgments

This study was made possible by the Road and Hydrologic Engineering Division of the Dutch Ministry of Transport, Public Works and Water Management. The Road and Hydrologic Engineering Division deserves additional gratitude for its attention to all the necessary technical provisions. It also deserves admiration for the decision to switch off the lights along the highway and appreciation for the way in which it informed road users and the public in the area and elsewhere of this measure, which at that time was felt to be risky. Furthermore, we are greatly indebted to Het Noordhollands Landschap and the local landowners for permission to enter their grounds and conduct observations and to all who contributed to the fieldwork and the technical aspects of data analysis.

Literature Cited

Austin, G. E., C. J. Thomas, D. C. Houston, and D. B. A. Thompson. 1996. Predicting the spatial distribution of buzzard *Buteo buteo* nesting areas using a geographical information system and remote sensing. *Journal of Applied Ecology* 33:1541–1550.

Avery, M. L., P. F. Springer, and N. S. Dailey. 1980. *Avian mortality at man-made structures: an annotated bibliography* (revised). U.S. Fish and Wildlife Service, Biological Services Program, FWS/OBS-80/54.

Baldaccini, N. E., and E. M. Bezzi. 1989. Orientational responses to different light stimuli by adult and young sedge warbler (*Acrocephalus schoenobaenus*) during autumn migration: a funnel technique study. *Behaviour* 110:115–124.

Beintema, A. 1995a. Eieren [Eggs]. Pages 121–142 in A. Beintema, O. Moedt, and D. Ellinger (eds.), *Ecologische atlas van de Nederlandse weidevogels*. Schuyt & Co., Haarlem, The Netherlands.

Beintema, A. 1995b. Inleiding: wat is een weidevogel? [Introduction: what is a meadow bird?]. Pages 11–18 in A. Beintema, O. Moedt, and D. Ellinger, *Ecologische atlas van de Nederlandse weidevogels*. Schuyt & Co., Haarlem, The Netherlands.

Beintema, A. 1995c. Weidevogels in Nederland [Meadow birds in the Netherlands]. Pages 172–192 in A. Beintema, O. Moedt, and D. Ellinger, *Ecologische atlas van de Nederlandse weidevogels*. Schuyt & Co., Haarlem, The Netherlands.

Berthold, P. 1980. Die endogene Steuerung der Jahresperiodik: eine kurze Übersicht

[The endogenous control of annual rhythms: a short overview]. Pages 473–478 in R. Nöhring, *Acta XVII Congressus Internationalis Ornithologici*. Verlag der Deutschen Ornithologen-Gesellschaft, Berlin.

Bian, L., and E. West. 1997. GIS modeling of elk calving habitat in a prairie environment with statistics. *Photogrammetric Engineering & Remote Sensing* 63:161–167.

Brown, F. A. Jr., and M. Rollo. 1940. Light and molt in weaver finches. *Auk* 57:485–498.

Cochran, W. W., and R. R. Graber. 1958. Attraction of nocturnal migrants by lights on a television tower. *The Wilson Bulletin* 70:378–380.

[CROW] Centrum voor Regelgeving en Onderzoek in de Grond-, Water- en Wegenbouw en de Verkeerstechniek. 1997. *Richtlijn openbare verlichting natuurgebieden* [Directive on lighting reduction in nature areas]. Publication 112. CROW, Ede, The Netherlands.

Elliot, M. A., and H. M. Edwards Jr. 1991. Studies on the effect of dihydroxycholecalciferol, cholecalciferol, and fluorescent lights on growth and skeletal development in broiler chickens. *Poultry Science* 70:S39.

Evans Ogden, L. J. 1996. *Collision course: the hazards of lighted structures and windows to migrating birds*. World Wildlife Fund Canada and the Fatal Light Awareness Program, Toronto, Canada.

Farner, D. S. 1959. Photoperiodic control of annual gonadal cycles in birds. Pages 717–750 in R. B. Withrow (ed.), *Photoperiodism and related phenomena in plants and animals*. Publication No. 55 of the American Association for the Advancement of Science, Washington, D.C.

Farner, D. S. 1964. The photoperiodic control of reproductive cycles in birds. *American Scientist* 52:137–156.

Farner, D. S. 1966. Über die photoperiodische Steuerung der Jahreszyklen bei Zugvögeln [On the photoperiodic control of yearly cycles in migratory birds]. *Biologische Rundschau* 4:228–241.

Follett, B. K. 1980. Gonadotrophin secretion in seasonally breeding birds and its control by daylength. Pages 239–244 in R. Nöhring (ed.), *Acta XVII Congressus Internationalis Ornithologici*. Verlag der Deutschen Ornithologen-Gesellschaft, Berlin.

Freeman, H. J. 1981. Alpine swifts feeding by artificial light at night. *British Birds* 74:149.

Genstat 5 Committee. 1987. *Genstat 5 reference manual*. Oxford University Press, Oxford.

Gochfeld, M. 1973. Confused nocturnal behavior of a flock of migrating yellow wagtails. *Condor* 75:252–253.

Groen, N. M. 1993a. Breeding site tenacity and natal philopatry in the black-tailed godwit *Limosa l. limosa*. *Ardea* 81:107–113.

Groen, N. M. 1993b. De broedbiologie van de Grutto in een Noordhollands weidevogelreservaat [The breeding biology of the black-tailed godwit in a North Holland grassland reserve]. *De Graspieper* 13:13–19.

Gwinner, E. 1986. *Circannual rhythms: endogenous annual clocks in the organization of seasonal processes*. Springer Verlag, Berlin.

Havlin, J. 1964. Zur Lösung der Amselfrage [The solution to the blackbird question]. *Angewandte Ornithologie* 2:9–14.

Hollom, P. A. D. 1966. Nocturnal singing and feeding by robins in winter [Notes]. *British Birds* 59:501–502.

Hosmer, D. W. Jr., and S. Lemeshow. 1989. *Applied logistic regression*. Wiley, New York.

Jongman, R. H. G., C. J. F. ter Braak, and O. F. R. van Tongeren. 1987. *Data analysis in community and landscape ecology*. Pudoc, Wageningen, The Netherlands.

King, B. 1966. Nocturnal singing and feeding by robins in winter [Notes]. *British Birds* 59:501–502.

Kraft, M. 1999. Nocturnal mass landing of migrating cranes in Hesse and Northrine–Westphalia, Germany, in November 1998. *Vogelwelt* 120:349–351.

Kruk, M., M. A. W. Noordervliet, and W. J. ter Keurs. 1997. Survival of black-tailed godwit chicks *Limosa limosa* in intensively exploited grassland areas in the Netherlands. *Biological Conservation* 80:127–133.

Labberté, K. R. 1978. Nachtelijke zang van zangvogels [Nocturnal song of songbirds]. *Vogeljaar* 26:304.

Lack, D. 1965. *The life of the robin*. Fourth edition. H. F. & G. Witherby, London.

Lambert, A. M. 1971. *The making of the Dutch landscape: an historical geography of the Netherlands*. Seminar Press, London.

Lewis, P. D., and G. C. Perry. 1990a. Response of laying hens to asymmetrical interrupted lighting regimens: physiological aspects. *British Poultry Science* 31:45–52.

Lewis, P. D., and G. C. Perry. 1990b. Response of laying hens to asymmetrical interrupted lighting regimens: reproductive performance, body weight and carcass composition. *British Poultry Science* 31:33–43.

Lofts, C., and D. Merton. 1968. Photoperiodic and physiological adaptations regulating avian breeding cycles and their ecological significance. *Journal of the Zoological Society of London* 155:327–394.

Malakoff, D. 2001. Faulty towers. *Audubon* 103(5):78–83.

May, J. D., B. D. Lott, and J. W. Deaton. 1990. The effect of light and environmental temperature on broiler digestive tract contents after feed withdrawal. *Poultry Science* 69:1681–1684.

Miller, A. H. 1955. The expression of innate reproductive rhythm under conditions of winter lighting. *Auk* 72:260–264.

Mitchell, K. D. G. 1967. Nocturnal activity of city blackbird. *British Birds* 60:373–374.

Moedt, O. 1995a. Gedrag van weidevogels [Behavior of meadow birds]. Pages 75–120 in A. Beintema, O. Moedt, and D. Ellinger (eds.), *Ecologische atlas van de Nederlandse weidevogels*. Schuyt & Co., Haarlem, The Netherlands.

Moedt, O. 1995b. Geschiedenis van de weidevogels en hun leefgebieden [History of the meadow birds and their habitats]. Pages 56–74 in A. Beintema, O. Moedt, and D. Ellinger (eds.), *Ecologische atlas van de Nederlandse weidevogels*. Schuyt & Co., Haarlem, The Netherlands.

Moedt, O. 1995c. Signalementen van weidevogels [Descriptions of meadow birds].

Pages 19–55 in A. Beintema, O. Moedt, and D. Ellinger (eds.), *Ecologische atlas van de Nederlandse weidevogels*. Schuyt & Co., Haarlem, The Netherlands.

Molenaar, J. G. de. 2003. *Lichtbelasting: overzicht van de effecten op mens en dier* [Light pollution: survey of the effects on man and animal]. Alterra-rapport 778. Alterra, Wageningen, The Netherlands.

Molenaar, J. G. de, D. A. Jonkers, and R. J. H. G. Henkens. 1997. *Wegverlichting en natuur. I. Een literatuurstudie naar de werking en effecten van licht en verlichting op de natuur* [Road illumination and nature. I. A literature review on the function and effects of light and lighting on nature]. DWW Ontsnipperingsreeks deel 34, Delft.

Molenaar, J. G. de, D. A. Jonkers, and M. E. Sanders. 2000. *Road illumination and nature. III. Local influence of road lights on a black-tailed godwit (*Limosa l. limosa*) population*. DWW Ontsnipperingsreeks deel 38A, Delft.

Narumalani, S., J. R. Jensen, S. Burkhalter, J. D. Althausen, and H. E. Mackey Jr. 1997. Aquatic macrophyte modeling using GIS and logistic multiple regression. *Photogrammetric Engineering & Remote Sensing* 63:41–49.

Newcombe, M., A. L. Cartwright, and J. M. Harter-Dennis. 1991. The effect of lighting and feed restriction on growth and abdominal fat in male broilers raised to roaster weight. *Poultry Science* 70:S87.

Outen, A. R. 2002. The ecological effects of road lighting. Pages 133–155 in B. Sherwood, D. Cutler, and J. Burton (eds.), *Wildlife and roads: the ecological impact*. Imperial College Press, London.

Rautenberg, W. 1957. Vergleichende Untersuchungen über den Energiehaushalt des Bergfinken (*Fringilla montifringilla* L.) und des Haussperlings (*Passer domesticus* L.) [Comparative investigations of the energy balance of the brambling (*Fringilla montifringilla* L.) and the house sparrow (*Passer domesticus* L.)]. *Journal für Ornithologie* 98:36–64.

Rawson, H. E. 1923. A bird's song in relation to light. *Transactions of the Hertfordshire Natural History Society and Field Club* 17:363–365.

Rees, E. C. 1982. The effect of photoperiod on the timing of spring migration in the Bewick's swan. *Wildfowl* 33:119–132.

Reijnen, M. J. S. M. 1995. *Disturbance by car traffic as a threat to breeding birds in the Netherlands*. Ph.D. thesis, Rijksuniversiteit, Leiden.

Reijnen, M. J. S. M., G. Veenbaas, and R. P. B. Foppen. 1992. *Het voorspellen van het effect van snelverkeer op broedvogelpopulaties* [Predicting the effect of highway traffic on populations of breeding birds]. Dienst Weg- en Waterbouwkunde van Rijkswaterstaat, Delft and DLO-Instituut voor Bos- en Natuuronderzoek, Wageningen, The Netherlands.

Rollo, M., and L. V. Domm. 1943. Light requirements of the weaver finch. 1. Light period and intensity. *Auk* 60:357–367.

Rowan, W. 1928. Reproductive rhythm in birds. *Nature* 122:11–12.

Rowan, W. 1929. Experiments in bird migration. I. Manipulation of the reproductive cycle: seasonal histological changes in the gonads. *Proceedings of the Boston Society of Natural History* 39:151–208.

Rowan, W. 1937. Effects of traffic disturbance and night illumination on London starlings. *Nature* 139:668–669.

Ruitenbeek, W., C. J. G. Scharringa, and P. J. Zomerdijk (eds.). 1990. *Broedvogels van Noord-Holland* [Breeding birds of North Holland]. Stichting Samenwerkende Vogelwerkgroepen Noord-Holland and Provinciaal Bestuur van Noord-Holland, Assendelft.

Sanders, M. E. 1999. *Remotely sensed hydrological isolation: a key factor predicting plant species distribution in fens*. IBN Scientific Contributions 17. DLO Institute for Forestry and Nature Research, Wageningen, The Netherlands.

Scheideler, S. E. 1990. Research note: effect of various light sources on broiler performance and efficiency of production under commercial conditions. *Poultry Science* 69:1030–1033.

Silverin, B. 1994. Photoperiodism in male great tits (*Parus major*). *Ethology Ecology & Evolution* 6:131–157.

Siopes, T. D., and R. Pyrzak. 1990. Effect of intermittent lighting on the reproductive performance of first-year and recycled turkey hens. *Poultry Science* 69:142–149.

Sperduto, M. B., and R. G. Congalton. 1996. Predicting rare orchid (small whorled pogonia) habitat using GIS. *Photogrammetric Engineering and Remote Sensing* 62:1269–1279.

Trapp, J. L. 1998. *Bird kills at towers and other human-made structures: an annotated partial bibliography (1960–1998)*. U.S. Fish and Wildlife Service, Office of Migratory Bird Management, Arlington, Virginia.

Van Lynden, A. J. H. Baron. 1978. Nachtelijke zang van zangvogels [Nocturnal song of songbirds]. *Vogeljaar* 26:187.

Wagner, H. O. 1956. Die Bedeutung von Umweltfaktoren und Geschlechtshormonen für den Jahresrhythmus der Zugvögel [The influence of environmental factors and sex hormones on the yearly rhythms of migratory birds]. *Zeitschrift für Vergleichende Physiologie* 38:355–369.

Wallgren, H. 1954. Energy metabolism of two species of the genus *Emberiza* as correlated with distribution and migration. *Acta Zoologica Fennica* 84:1–110.

Wiltschko, W., U. Munro, H. Ford, and R. Wiltschko. 1993. Red light disrupts magnetic orientation of migratory birds. *Nature* 364:525–527.

Wiltschko, W., and R. Wiltschko. 1995. Migratory orientation of European robins is affected by the wavelength of light as well as by a magnetic pulse. *Journal of Comparative Physiology A* 177:363–369.

Wingfield, J. C., and D. S. Farner. 1978. The annual cycle of plasma irLH and steroid hormones in feral populations of the white-crowned sparrow, *Zonotrichia leucophrys gambelii*. *Biology of Reproduction* 19:1046–1056.

Wingfield, J. C., T. P. Hahn, R. Levin, and P. Honey. 1992. Environmental predictability and control of gonadal cycles in birds. *Journal of Experimental Zoology* 261:214–231.

Witschi, E. 1961. Sex and secondary sexual characters. Pages 115–168 in A. J. Marshall (ed.), *Biology and comparative physiology of birds*. Volume II. Academic Press, New York.

Wolfson, A. 1942. Regulation of spring migration in juncos. *Condor* 44:237–263.

Wolfson, A. 1959. The role of light and darkness in the regulation of spring migration and reproductive cycles in birds. Pages 679–716 in R. B. Withrow (ed.), *Photoperiodism and related phenomena in plants and animals*. Publication No. 55 of the American Association for the Advancement of Science, Washington, D.C.

Zimmermann, N. G., and C. H. Nam. 1989. Temporary ahemeral lighting for increased egg size in maturing pullets. *Poultry Science* 68:1624–1630.

Part III

Reptiles *and* Amphibians

Night, Tortuguero

Because the turtles come out to nest after dark, much of my work was done at night. There was a great deal of waiting between turtles, plenty of time to sit on a driftwood log and think. In the first years of my research I was often the only one on the beach for miles. After ten or twenty minutes of sitting without using my flashlight, my eyes adapted to the dark and I could make out forms against the brown-black sand: the beach plum and coconut palm silhouettes in back, the flicker of the surf in front, sometimes even the shadowy outline of a trailing railroad vine or the scurry of a ghost crab at my feet. The air was heavy and damp with a distinctive primal smell that I can remember but not describe. The rhythmic roar of the surf a few feet away never ceased—my favorite sound. I hear it as I write in my landlocked office in New Jersey. And then, with ponderous, dramatic slowness, a giant turtle would emerge from the sea.

Usually I would see the track first, a vivid black line standing out against the lesser blackness, like the swath of a bulldozer. If I was closer, I could hear the animal's deep hiss of breath and the sounds of her undershell scraping over logs. If there was a moon, I might see the light glistening off the parabolic curve of the still wet shell. Size at night is hard to determine: even the sprightly 180-pounders, probably nesting for the first time, looked big when nearby, but the 400-pound ancients, with shells nearly four feet long, were colossal in the darkness. Then when the excavations of the body pit and egg cavity were done, if I slowly parted the hind flippers of the now-oblivious turtle, I could watch the perfect white spheres falling and falling into the flask-shaped pit scooped into the soft sand.

Falling as they have fallen for a hundred million years, with the same slow cadence, always shielded from the rain or stars by the same massive bulk with the beaked head and the same large, myopic eyes rimmed with crusts of sand washed out by tears. Minutes and hours, days and months dissolve into eons. I am on an Oligocene beach, an Eocene beach, a Cretaceous beach—the scene is the same. It is night, the turtles are coming back, always back; I hear a deep hiss of breath and catch a glint of wet shell as the continents slide and crash, the oceans form and grow. The turtles were coming here before here was here. At Tortuguero I learned the meaning of place, and began to understand how it is bound up with time.

David Ehrenfeld

From an article that originally appeared in *Orion* (888-909-6568) and was reprinted in *Beginning Again: People and Nature in the New Millennium*, Oxford University Press (1993), by David Ehrenfeld. This excerpt is published by permission of the author.

Chapter 7

Protecting Sea Turtles from Artificial Night Lighting at Florida's Oceanic Beaches

Michael Salmon

Artificial night lighting is a well-documented cause of mortality among migratory birds and hatchling sea turtles. Consequently, the plight of both groups has received significant public attention. For sea turtles, a substantial literature has been produced since McFarlane (1963) first described the effect of lighting on these animals. In response, local and state governments have expended considerable resources on efforts to ameliorate this problem.

In the United States, Florida's beaches are major rookery sites for loggerheads and northern breeding areas for increasing numbers of leatherback and green turtles. But coastal development in Florida continues unabated, increasing beach exposure directly to the lights themselves and indirectly to sky glow from lights not directly visible. Both influence female choice of nest sites and hatchling orientation. The Florida coast has of necessity become a laboratory for testing methods designed to protect turtles from "photopollution."

In this chapter I first review how, under natural conditions, females choose nesting sites and hatchlings that emerge from those nests locate the sea. I then describe how behavior of both females and hatchlings is affected by exposure to artificial night lighting. Next, I critically evaluate two approaches to protecting hatchlings at local beaches: those that prevent the turtles from responding to illumination and those that manage lighting. The second approach is preferred because it promotes habitat restoration. Finally, I review the design, philosophy, and implementation of plans to control lighting at the community, county, and state levels. Plans that concentrate efforts to reduce lighting only on beach habitats ignore the deleterious effects of lighting from adjacent and more distant areas. For this reason, conservation of marine turtles ultimately depends on local efforts but also on national and international light management policies.

Sea Turtles in Florida

The coastal waters of Florida serve as important feeding habitats for juvenile and adult marine turtles, and Florida's sandy beaches serve as important rookery sites for three species—loggerhead (*Caretta caretta*), green turtle (*Chelonia mydas*), and leatherback (*Dermochelys coriacea*). Surveys of the coastline since 1979 have established that most nesting (more than 90%) occurs on Florida's southeastern shores, that nesting numbers for some species have increased, and that most nesting is by loggerheads: well over 70,000 nests annually produced by at least 17,000 females (Meylan et al. 1995). This contribution represents about 80% of all nesting by loggerheads in western Atlantic waters, the second largest population of this species in the world.

Since the 1920s, Florida's human population has grown from about 1 million to more than 16 million residents, a rate of increase at least 2.5 times greater than that of the population of the United States (Bouvier and Weller 1992). Immigration has transformed Florida from a largely agricultural to a predominantly metropolitan state, with most of the major cities located on the coast. Once isolated and pristine beaches have become sites for resorts and high-rise condominiums, many adjacent to major ports such as those at Tampa–St. Petersburg, St. Augustine, Miami, Ft. Lauderdale, and West Palm Beach. City and suburban development along the coast has also transformed the lighting environment, although there are few quantitative data to estimate by how much. But it is clear to anyone indulging in a nocturnal beach stroll that almost everywhere in south Florida, lighting from beach dwellings, roadways, shopping centers, hotels, and office buildings reaches the beach, either directly from sources visible at the horizon or indirectly as the result of sky glow, light reflected from these sources down to the beach from the atmosphere. Put simply, photopollution, defined as the "degradation of the photic habitat by artificial light" (Verheijen 1985:2), has become another threat to sea turtle populations that, worldwide, are already seriously depleted.

Florida is a paradox. Thanks to conservation and management efforts in this country and abroad, the number of female turtles returning to nest in Florida has increased. But at the same time coastal development and artificial night lighting degrade Florida's nesting beaches as a habitat. Without a comprehensive solution to this problem the gains seen in sea turtle nesting might be offset or even reversed.

Several excellent reviews have described how artificial night lighting affects the physiology and behavior of organisms (Verheijen 1958, 1985, Outen 1998, Longcore and Rich 2004), including sea turtles (Raymond 1984, Witherington and Martin 1996, Witherington 1997, Salmon 2003). Witherington and Frazer (2003) discuss social and economic aspects of marine turtle conservation and management, including in their essay the resolution of lighting problems.

Sea Turtle Behavior in the Absence of Artificial Night Lighting

I begin with a brief review of sea turtle behavior at the nesting beach in the absence of artificial night lighting.

Nest Site Selection by Females

Sea turtles normally nest on remote beaches, shrouded in darkness. Some sites are more attractive than others, but why? It has proved impractical to do many controlled experiments with gigantic (150- to 400-kg [330- to 880-lb]) females, but we have a general idea of what processes must be involved. These may be conceived as consisting of decisions made at different spatial (geographic) scales. At the largest scale, females show preferences for particular nesting locations, manifested behaviorally by site tenacity. Tenacity is demonstrated by capturing females found near a nesting beach, then displacing them. They typically return within hours or days, depending on distance, to the capture site (Luschi et al. 1996).

Site tenacity is also manifested genetically. Females nesting at particular sites have similar mitochondrial DNA signatures. These indicate that the females are descendants of one or a few original matrilineages. It follows that each female hatchling learns and remembers the location of its natal site and returns there after many years of growth to sexual maturity. The sensory cues used for habitat imprinting are unknown. Evidence suggests that hatchlings respond to magnetic landmarks and use these cues to gauge their spatial position in the open ocean (Lohmann et al. 2001). Such a capacity may also underlie their ability as adults to return to natal beaches (Lohmann et al. 1999, Lohmann and Lohmann 2003). Experimental evidence demonstrates that juvenile turtles are also capable of navigation and that orientation is based on both visual and geomagnetic cues (Avens and Lohmann 2003, 2004).

Populations within species differ in their specificity for rookery sites. For example, loggerheads nesting in Florida consist of four genetically distinct matrilineages, each nesting in a different part of the state. Within each of these populations, females may deposit eggs in one to seven nests, each at 12- to 14-day intervals, many kilometers apart. Loggerheads in Australia, however, often nest on a single small island but not another that may be only a few hundred meters distant.

The selection of a nesting site also involves decisions on a finer spatial scale, that is, the search for an attractive site at a particular location. Attractive sites have certain ecological characteristics. For example, prime nesting beaches usually are adjacent to the nearshore oceanic currents needed for hatchling transport to nursery habitats. Nesting beaches also have a favorable underwater nearshore approach profile and contain few obstructions, such as shallow water rocks or reefs, that might injure a female attempting to reach the surf zone.

Once a female is in shallow water adjacent to the beach, and also during her crawling ascent on the beach itself, she can assess local terrestrial features such as the dune or vegetation profile behind the beach, beach slope, depth, and elevation of the beach "platform" above sea level. Sand characteristics, such as temperature and moisture content, may also be detected. For loggerheads, beach slope is an important cue (Wood and Bjorndal 2000), but other species of sea turtles may have different, and currently unknown, requirements.

Females usually nest at night, when temperatures are lower and when nests, and the females digging them, are less likely to be detected by terrestrial predators. Nesting can take an hour or more and involves digging a shallow body pit with all four flippers, then an egg chamber with the rear flippers. The egg chamber can receive a clutch of more than a hundred eggs, which is covered with sand. Extra sand is scattered to mask the location of the egg chamber. The female then returns to the sea, abandoning the unprotected nest to its fate.

If no predators locate the egg chamber, and if the nest is not flooded by storm-generated high tides or wave action, embryological development will be completed in 45–75 days depending on temperature.

Hatchling Orientation: Locating the Sea from the Nest

After extricating themselves from their eggs, the hatchlings dig their way upward en masse toward the sand surface. If the surface sand is hot, they stop and become inactive until the sand cools after sunset. This response normally results in a simultaneous nocturnal emergence of most of the hatchlings to the beach surface. Their appearance is immediately followed by a rapid (two minutes or less) crawl directly to the ocean, an orientation behavior known as seafinding.

Seafinding is mediated visually, using a perceptual filter that confines the visual field to directional cues located in a horizontally wide (180°) and vertically narrow (–10° to +30°) "cone of acceptance" (Lohmann et al. 1997). The visual cues that the hatchlings use are simple. They crawl away from tall or dark objects located against the landward horizon, characterized by the dune or vegetation behind the beach, and toward the lower, uniformly flatter beach-facing horizon. This horizon also typically reflects and emits more light from the stars or moon. In the surf zone, incoming waves lift the turtles off the sand surface and induce vigorous synchronized paddling with the foreflippers. Hatchlings are carried seaward by the retreating waves. Continued orientation offshore is then directed by swimming into wave-induced orbital currents (Lohmann et al. 1995). Because surface waves typically approach the beach parallel to shore, the hatchlings move into deeper water. The turtles swim continuously during a swimming frenzy that lasts 24–36 hours. This migratory activity is the turtle equivalent of migratory restlessness, or *Zugunruhe*, which has been so well studied in birds (Berthold 1993).

Artificial Night Lighting and Sea Turtles

Artificial night lighting disrupts the normal behavior of sea turtle females searching for appropriate nest sites and of hatchlings attempting to orient toward the ocean.

Effects of Lighting on Female Nest Site Selection

The number of nests placed on a length of beach can vary locally for reasons that are not always obvious. For example, nest densities typically are lower at beaches exposed to artificial night lighting, but lighting may not be the only or even the primary cause. Lighting is associated with coastal development, and coastal development may be correlated with a host of changes such as dune alteration, deliberate changes in beach profile, or other anthropogenic modifications. These can include increased compaction, shoreline armoring to retard sand loss, accumulated debris, and human traffic that can cause a decline in nesting by disturbing females, either as they begin their crawl ascent or as they begin to dig their nests. Thus to determine whether lighting causes a decline in nesting activity, these other potential variables must be excluded.

Witherington (1992) completed the critical experiments with loggerheads in Florida and green turtles in Costa Rica. Portable generators were used to power lights that illuminated an otherwise dark and, for the females, attractive nesting beach. When the beach was exposed to mercury vapor lighting, the number of nesting attempts, whether or not they resulted in nests, was reduced almost to zero. But when the beach was exposed to near-monochromatic yellow light from low pressure sodium vapor lamps or when the lamps were left in place but turned off, both unsuccessful and successful nesting attempts returned to normal.

Many beaches in south Florida are exposed to less intense and more diffuse lighting than in Witherington's experiment, and as a consequence lowered levels of nesting occur in these areas. At these locations, the influence of lighting often is revealed by the spatial distribution of the nests. In Boca Raton, Florida, we found that most loggerhead nests were clustered in front of tall condominiums, largely dark and unoccupied during the summer nesting season (Salmon et al. 1995a). Further study showed that clustering was unrelated to beach physical attributes such as width, elevation, or slope or differences in nearshore, underwater profile. But clustering was significantly and positively correlated with building elevation. The buildings apparently acted as light barriers, shadowing the beach from city lighting in the interior. That observation led to the hypothesis that at urban locations exposed to lighting, nesting females found shadowed patches of beach most attractive. Apparently the more that shadow extended above the horizon, the more attractive the location was as a nest site.

The correlation between relative darkness and nesting density is also evident on a larger geographic scale. Most loggerhead nesting in the

United States occurs in south Florida, but within that area the distribution of the major nesting sites is not uniform. Three species—loggerheads, green turtles, and leatherbacks—favor the same, darkest beaches (Salmon et al. 2000; Figure 7.1). This association suggests that female choice of nesting site is strongly biased by coastal development and its associated lighting.

We cannot determine whether choosing nest sites to avoid lighting affects reproductive success. Choice of nest site seems to have no effect on nest success, defined as the proportion of eggs that ultimately result in hatchlings that leave the nest. We lack information to compare the proportion of hatchlings that reach their offshore goals from the few and darkest available beaches today with that proportion before human settlement in Florida without the influence of lighting on nest site choice. We can, however, draw two conclusions. The first is that if nesting sites are currently selected by the absence of lighting, then selection based on factors that in the past did not include lighting must be weakened. The second conclusion

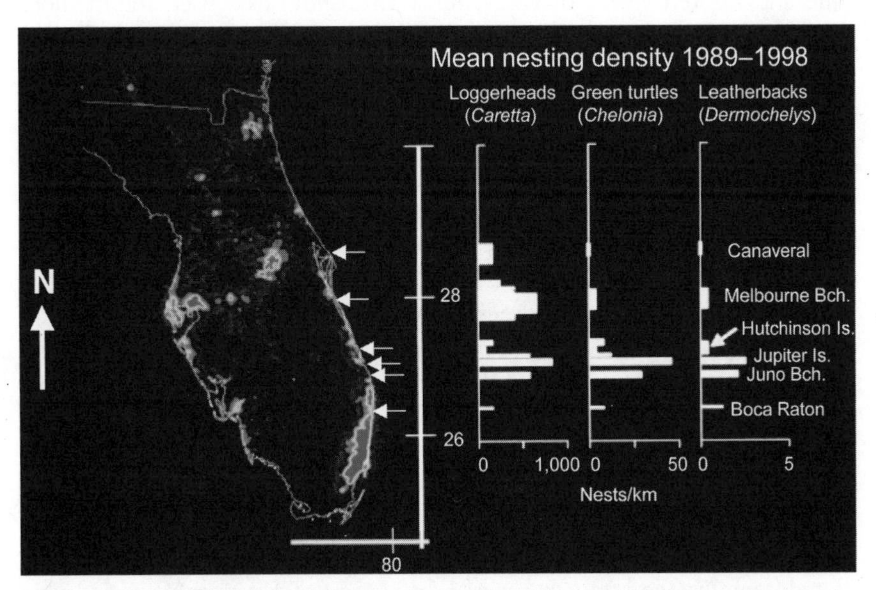

Figure 7.1. Artificial light radiating from ground level in Florida, as measured by satellite photographs. Gray areas surrounded by white patches radiate the most light; white patches are intermediate; black regions are dark. Nesting on Florida's east coast is clustered at five beaches where levels of development and lighting are relatively low. Boca Raton (a sixth site) has relatively high nesting densities for a metropolitan area. It is located in a small patch of intermediate light radiance, surrounded to the north and south by brighter regions. Redrawn from Salmon et al. (2000).

is that if current trends continue, more nests are likely to be concentrated in an ever-declining area of remaining dark sites. We already know that when nests are concentrated in space, under natural (Gyuris 1994) or artificial conditions, rates of hatchling mortality increase.

Effects of Lighting on Hatchling Orientation

Artificial lighting disrupts the normally accurate seaward orientation of hatchlings. Disruption typically is discovered through inspections conducted in daylight, when the tracks, or flipper prints, of the turtles can be seen on the sand surface. *Disoriented* hatchlings crawl in circuitous paths, as if unable to detect directional cues. *Misoriented* hatchlings crawl on straight paths, but they often lead directly toward light sources visible from the beach at night (Salmon et al. 1995b). When their orientation is disrupted, the prospects for hatchling survival diminish (Witherington and Martin 1996). Disoriented hatchlings may crawl on the beach for hours, wasting time and limited stores of yolk energy that should be used for offshore migration during the dark period. Some of the disoriented turtles may eventually locate the sea, but the fate of misoriented turtles is far worse. Those not trapped in dune vegetation may exit the beach, traverse coastal roadways where they are crushed by passing vehicles, or gather at the base of light poles. Misoriented hatchlings are also weakened by exhaustion, physiologically stressed by dehydration, taken by terrestrial predators, or killed after sunrise by exposure to lethally high temperatures. Florida hatchlings lost annually as a consequence of disrupted orientation are estimated to number in the hundreds of thousands (Witherington 1997).

Why is hatchling orientation so seriously affected by artificial lighting, whereas the orientation of their mothers is rarely affected? One possibility is that hatchlings are simply more sensitive to lighting than adults. Another is that the two life history stages respond to different visual features even though both stages show orientation. Females nesting at illuminated beaches are attracted to dark patches of beach or some correlate thereof, such as a tall, dark object behind the beach. After nesting is completed, females need only to reverse the sign of this preference (i.e., crawl away from dark patches) to locate the sea. Hatchlings, however, scan a broader (180°) length of horizon and are naturally attracted to areas reflecting more light, which is usually the seaward horizon. Luminaires, by virtue of their brightness in comparison with the remainder of the visual environment, may simply be supernormal substitutes for naturally directing stimuli (Witherington 1997).

The lights themselves do not have to be directly visible to hatchlings. At many developed sites, lighting from sky glow, or from gaps between buildings, dunes, or vegetation behind the beach, compromises hatchling orientation (Figure 7.2).

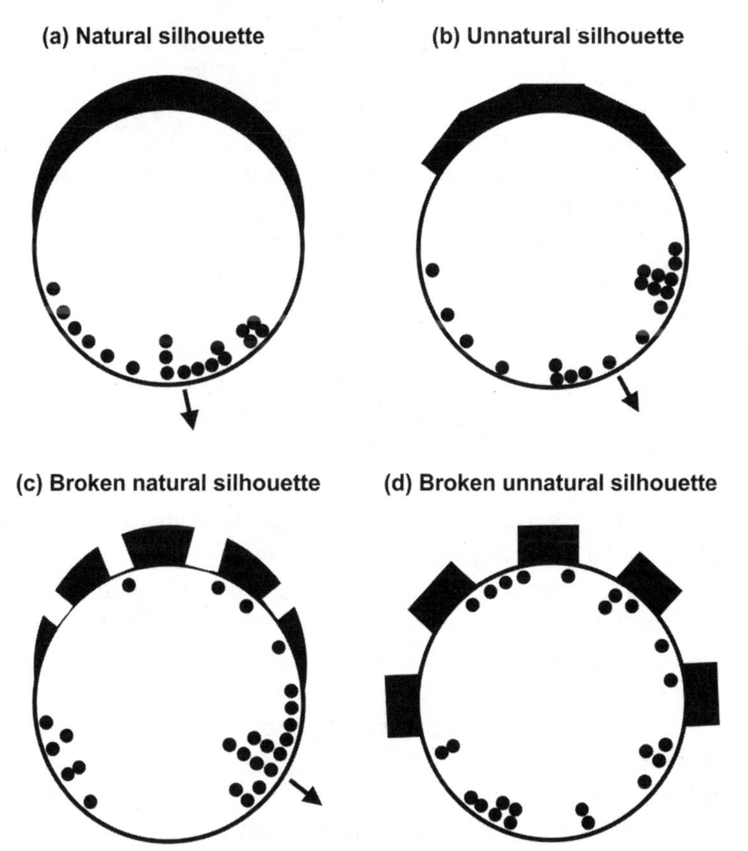

Figure 7.2. The orientation shown by loggerhead hatchlings in a laboratory arena. The turtles are presented with natural (crescent-shaped and unbroken) or artificial (odd-shaped, broken, or both) "landward" silhouettes. Solid dots within circle diagrams show the mean angle of orientation for each hatchling; arrow outside circle is the mean angle for the three groups that show statistically significant orientation. *(a)* Hatchlings crawl away from "land" when presented with a natural silhouette, but *(c)* show less accurate orientation when the silhouette is broken and allows light into the arena. *(b)* Orientation is more variable when hatchlings are exposed to a solid, unnaturally shaped silhouette. *(d)* Turtles show no significant orientation when the silhouette is broken. Thus, hatchlings depend on both horizon shape and continuity for accurate seafinding. Redrawn from Salmon et al. (1995b).

Mitigating Effects of Artificial Lighting by Manipulating Nests or Hatchlings

How can managers best protect turtles from lighting? What are the common management strategies, and what have we learned about their efficacy?

For very practical reasons, managers of nesting beaches have concentrated on protecting hatchlings from artificial lighting, ignoring any effects on nesting female turtles. In Florida, most nesting beaches are surveyed by volunteer groups, private educational and research organizations that showcase marine turtles, park and wildlife personnel, and biologists hired by local governments. The primary concern of monitoring personnel is to document nest success in terms of hatchling production and to minimize any loss of hatchlings associated with exposure to artificial lighting. When such losses occur, the signs are evident—abnormal hatchling tracks, turtles reported crossing coastal roadways, and hatchlings collected under light poles. Although most efforts are concentrated on hatchlings, artificial lighting affects both hatchlings and nesting females. It is much more difficult, however, to document effects on the females; doing so takes years of data to establish that there has been a nesting decline. Even if such a trend is documented, it is difficult to determine a causal relationship between more lighting and less nesting.

Protecting both females and their hatchlings ultimately requires a coordinated effort that involves monitoring, local code enforcement personnel, and state and federal agencies responsible for resolving lighting problems. In the last 10–15 years the need for such cooperation has increased, and successful collaborations have become more common. As a result, management practices have shifted in emphasis from protecting hatchlings to habitat restoration through large-scale planning. Such an approach obviously benefits both hatchlings and females. I return to discuss this more holistic approach to sea turtle management and recovery at the end of this chapter, after discussing methods of hatchling protection. This discussion begins with nest relocation and nest caging, two procedures intended to prevent hatchlings from being affected by artificial lighting.

Nest Relocation

In Florida some beaches are exposed to so much lighting that emerging hatchlings cannot locate the sea. Nests deposited at these beaches are relocated to hatcheries, sites where the eggs from each nest are reburied in sand cavities that mimic, in size and proportion, the egg chambers excavated by females. Some hatcheries are fenced to exclude predators and confine the turtles. Hatchlings that emerge in fenced locations are collected twice during the dark period, in late evening and early morning, then released at a dark beach. Self-releasing hatcheries are located at dark beaches and have no fences. After emergence, the turtles crawl to the sea unassisted.

Managers recognize that hatcheries are costly to operate, that relocation not done properly (e.g., within 12 hours of deposition by the female) can damage embryonic membranes and cause egg death, and that spatially concentrating nests can result in low hatching success and poor hatchling quality. Hatcheries therefore are considered a method of last resort (Mortimer 1999), to be used only when conditions virtually guarantee that nests left in place will not survive.

Despite the drawbacks, managers until recently considered hatcheries successful if sufficient care was taken to minimize egg mortality and if hatchling tracks led to the sea. These criteria, however, have been shown to be inadequate (Wyneken and Salmon 1996). For more than a decade Broward County has managed a large self-releasing hatchery where more than 1,500 nests annually were routinely placed chronologically in neat rows by deposition date (Figure 7.3). From late May through September, hatchlings from several nests deposited on the same day or within 1–2 days of each other emerged each night, crawled to the surf zone, and entered the sea. But waiting for them, often within just a few meters of

Figure 7.3. A self-releasing hatchery at Hillsborough Beach, Broward County, Florida. Each stake marks the position of a nest.

shore, were predatory fishes (e.g., tarpon, mangrove snapper, and sea catfishes) and squid that had learned where to find a reliable source of prey. To quantify predation rates, observers in kayaks followed hatchlings at a distance as they swam offshore. Predators took about 29% of the turtles within 15 minutes after they entered the sea.

After this discovery, an alternative hatchery system was explored for the next two seasons. Instead of a single large hatchery, three hatcheries separated by several hundred meters were used. Nests were transferred to a single hatchery for no more than two weeks; thereafter, they were transferred to another hatchery. As a result, hatchlings entered the ocean at any one site for only brief (two-week) periods. Presumably, this schedule reduced the time that predators had to learn where prey were available. Predation rates were assayed again by observers following hatchlings offshore. Rates averaged 2.5% at control sites between the three hatcheries and 17% in front of the hatcheries. Thus spreading hatchling risk, both spatially and temporally, resulted in lower turtle mortality levels than those at a single large hatchery. But even smaller, separate hatchery sites resulted in an average predation rate seven times higher than those at the control, nonhatchery sites.

Nest Caging

At some rookery sites, sea turtle nests are covered with wire mesh cages, open at the bottom and anchored in the sand. Cages are the analogs of hatcheries, reduced in size to protect single nests. Self-release cages are constructed so that the mesh on the ocean-facing panel permits the hatchlings to escape from the cage, then crawl unassisted to the sea. These cages typically are used to protect nests from natural predators such as raccoons, foxes, armadillos, and skunks. Restraining cages are used to protect hatchlings from artificial lighting (Florida Fish and Wildlife Conservation Commission 2002). They do not permit hatchling escape. Beach monitoring personnel must inspect these cages twice each night for turtles, then release them at a dark site.

At our study site in Boca Raton, Florida, self-releasing cages commonly were used at locations where raccoon predators were a serious threat or where light levels were believed to be too low to seriously affect hatchling orientation. At sites where light levels were low, cages were supposed to prevent the hatchlings from immediately crawling toward the lights until they could adapt to local conditions. But was this protection effective? Because some of the turtles left tracks that led from the cage directly to the ocean, the initial assumption was that the method was successful. But observations and experiments led to different conclusions (Adamany et al. 1997).

Hatchling emergences were staged inside cages placed on dark beaches and inside cages at sites where they were exposed to low lighting. Caging did not alter hatchling orientation at dark sites, but at illuminated sites many turtles crawled against the landward-facing wall. They remained there for part and sometimes all of the dark period.

At most of the illuminated sites, artificial lighting diminished after midnight, and many turtles eventually escaped from the cage. But some escapees crawled only a short distance before their orientation was again disrupted. At other sites where lighting levels were higher, the hatchlings remained trapped inside the cage until dawn and only then crawled to the ocean. They left behind a record of tracks that were spatially normal but temporally inappropriate because in the absence of darkness the turtles were vulnerable to visual predators on land and in the sea. Finally, all caged hatchlings, whether they escaped from the cages before dawn or at dawn, spent time and energy crawling within a cage. That energy should have been used for offshore migration.

At Boca Raton, caging also provided inadequate protection against predators. Raccoons learned to use the cages to locate nests (Mroziak et al. 2000).

Mitigating Effects of Artificial Lighting by Controlling Light Emissions

Adverse effects of lights on sea turtles can be reduced at the source through the design of lighting systems used in coastal environments. Such approaches include the use of streetlight filters on existing or new lamps and nontraditional lights embedded in roadways instead of mounted on light poles.

Streetlight Filters

Hatchlings vary in their response to light of different pure wavelengths. Turtles are strongly attracted to the shorter, violet to green wavelengths but are either indifferent to or, uniquely in the case of loggerheads, repelled by longer amber wavelengths (Lohmann et al. 1997). These results led to the development of dyed acrylic filters designed for streetlights and other luminaires that permit the transmission of only the longer wavelengths. The assumption underlying filter development was that the responses elicited by a range of single light wavelengths could be

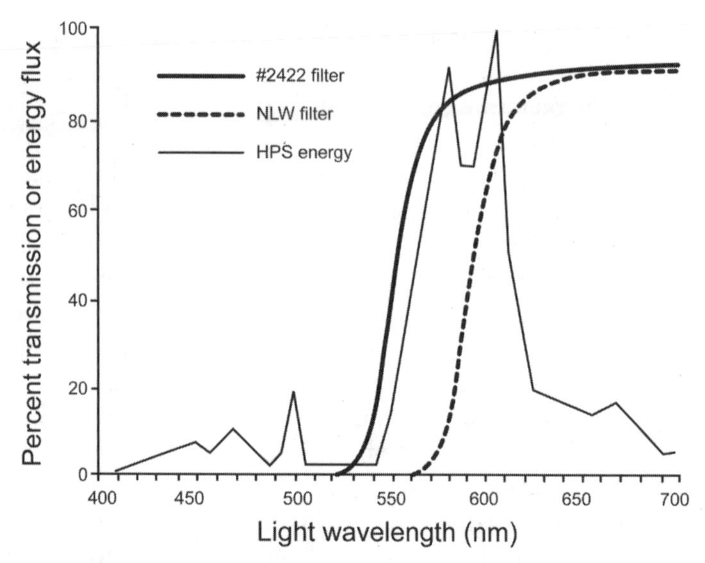

Figure 7.4. Transmission characteristics of the two General Electric Lighting Systems (#2422 and NLW) streetlight filters, in relation to the spectral output of a high pressure sodium vapor (HPS) luminaire. Both filters exclude the shorter wavelengths that are especially attractive to, and disrupt the orientation of, sea turtle hatchlings.

used to predict the responses elicited by a spectrum that included the same light wavelengths.

Two filters were designed by General Electric Lighting Systems, Inc. (GELS) to exclude transmission of wavelengths less than 530 nm (#2422 filter) or 570 nm (NLW filter; Figure 7.4). The Florida Power and Light Company installed #2422 filters in poled streetlights along coastal roadways throughout south Florida. These streetlights were equipped with high pressure sodium vapor (HPS) luminaires that transmitted wavelengths known to attract hatchlings. If completely effective, the filters would render these lights unattractive to the turtles; if partly effective, they might substantially reduce orientation disruption. Unfortunately, GELS produced fliers advertising the filters as "turtle friendly" before they were adequately field tested.

Such filters have several potential advantages. First, they immediately modify the spectral output of HPS streetlights at a modest cost. Second, they reduce the amount of light energy transmitted to the environment and therefore make the luminaires less likely to affect hatchling orientation. Third, they can be removed after turtle nesting season, if desired. The critical question, then, is whether filtered lighting is effective, either by not attracting hatchlings or by being less attractive than unfiltered HPS lighting.

Response of Females to Filtered Lighting

Pennell (2000) monitored loggerhead nesting attempts over an entire summer at a dark beach in Palm Beach County, Florida. Highway A1A, which is illuminated by numerous streetlights, runs parallel to and just behind the beach. The site was divided into three 440-m (quarter-mile) sections: north and south control sites, where the streetlights were turned off, and a central experimental zone where three filtered streetlights on poles were alternately turned on and off for one-week periods.

When the lights were on there was no evidence that nesting attempts, both successful and unsuccessful, or their ratios were affected because there were no statistical differences in nesting density between the control and experimental sections. There were also no differences in nesting attempts in the experimental section when the streetlights were turned on or off (Figure 7.5). Nest densities recorded during that summer fell within

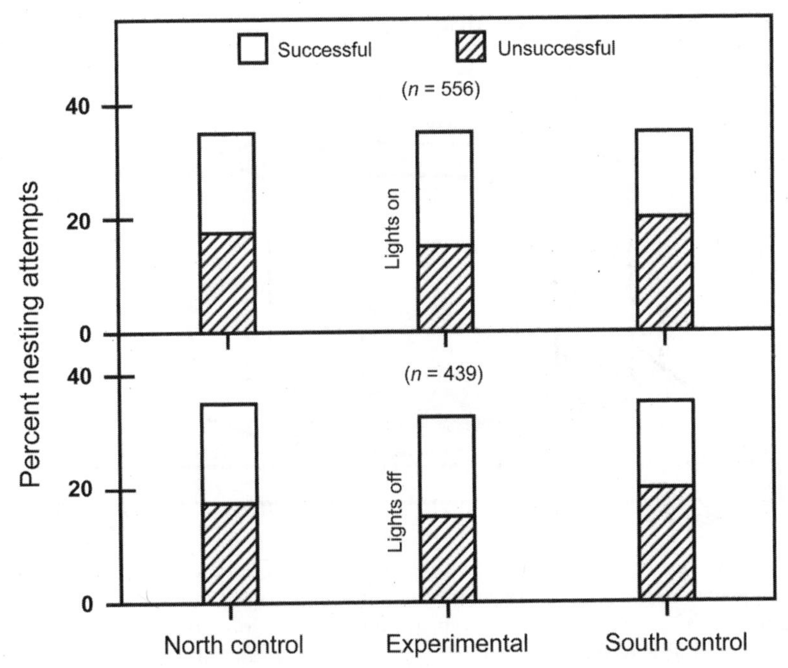

Figure 7.5. Nesting attempts by loggerheads at Carlin Park, Florida. The beach was divided into north and south control sections, where streetlights were turned off, and a central section exposed to filtered (#2422) street lighting. There were no statistical differences in nesting attempts between the control and experimental sections. Nesting attempts in the experimental section did not differ when the streetlights were on or off. n = the total number of nesting attempts. Modified from Pennell (2000).

the range of those recorded in the previous 12 years, when all of the streetlights were turned off. But because females often nest where lighting can affect their offspring, these studies were followed by experiments with hatchlings. Unfortunately, those results were less encouraging.

Response of Hatchlings to Filtered Lighting

Tuxbury (2001) performed laboratory experiments in a circular arena, located inside a windowless room. Hatchlings were tethered by a short line to the center of the arena but could crawl in any direction. Half of the arena presented the turtles with a flat, unobstructed "seaward" horizon. The opposite "landward" half had two book lights placed upright against the arena wall, 90° apart (Figure 7.6). These lights simulated in position two streetlight poles about 33 m (108 ft) apart, which is a typical spacing of these luminaires along some coastal roadways. The bulbs used in the book lights emitted a broader spectrum of wavelengths than HPS luminaires. When fitted with filters, however, emissions were confined to the expected amber wavelengths.

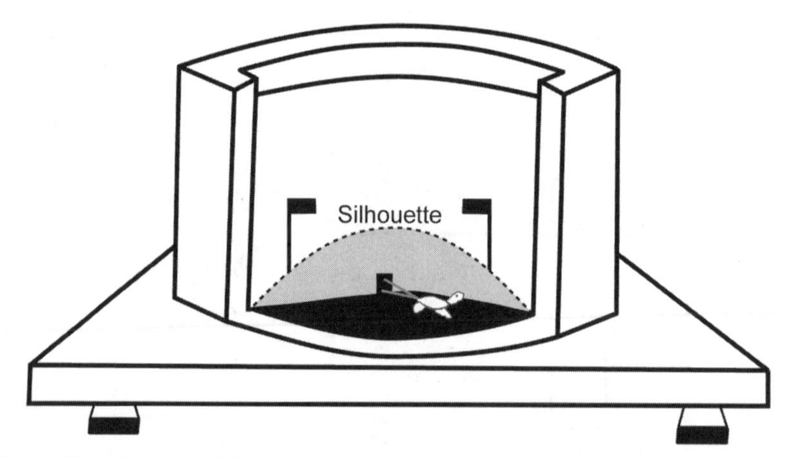

Figure 7.6. Arena used by Tuxbury (2001) to investigate hatchling orientation during exposure to filtered street lighting and silhouettes. Hatchlings are tethered inside the arena but are free to move in any direction. Their average orientation direction is noted. Two surrogate streetlights (miniature book lights whose openings are covered with a General Electric Lighting Systems filter) are placed against the "landward" half of the visual field. A dark, crescent-shaped silhouette mimics the presence of a vegetated dune behind the beach. The "seaward" half of the visual field is dark and flat. Responses of the turtles to light transmitted through each of two filters (#2422, NLW) were measured in the presence or absence of the silhouette. Redrawn from Bertolotti and Salmon (in press).

Orientation was examined under four treatment conditions: when light passed through a #2422 or NLW filter, in the presence or absence of a silhouette, 15° high, measured at center. Green turtle and loggerhead hatchlings, about to emerge from their nests, were collected during the afternoon and served as the experimental subjects that evening.

In the absence of silhouettes, hatchlings of both species crawled toward the lamps, regardless of which filter was used. In the presence of a silhouette, however, some of the turtles crawled "seaward," or away from the lamps. The response of each species to the "light with silhouette" treatment differed. Loggerheads, but not green turtles, crawled away from #2422 filtered lighting, whereas green turtle hatchlings, but not loggerheads, crawled away from NLW filtered lighting.

The results led to three conclusions. First, filtered lighting was not "turtle friendly," because it attracted the turtles. Second, attraction could be reversed by a stronger natural cue, a high silhouette, resulting in seaward orientation even when the turtles were exposed to filtered lighting. Third, the two species differed in their response to lighting in the presence of identical silhouettes, depending on small differences in the transmitted light spectra. Thus, species-typical differences in light perception must be considered in management decisions. At many rookery sites, nesting by two or more marine turtle species is common; a single kind of filtered light may reduce orientation disruption in one hatchling species, but not in others.

Is Filtered Lighting Less Attractive to Hatchlings Than Unfiltered High Pressure Sodium Lighting?

Although filtered lighting attracted loggerhead and green turtle hatchlings, it might be less attractive than unfiltered lighting. If so, then filtering might be an effective management tool under certain conditions. For example, if filtered lighting were only weakly attractive to hatchlings, normal orientation might be restored by coupling the use of filters with lower-wattage luminaires, by moving lights farther away from nesting sites, or by making the silhouette behind the beach taller or more complete.

Nelson (2002) conducted experiments in the laboratory in which loggerhead hatchlings were exposed to filtered and unfiltered HPS lighting. A T-maze apparatus was used to determine how the hatchlings responded. Turtles initially crawled down a runway and then, at the T-intersection where lighting was visible, turned either to the left or right. In one set of tests a single light (HPS or filtered HPS) was presented for turtles to choose between an illuminated and dark side of the maze. In a second set of experiments, two different lights (filtered and unfiltered) were viewed simultaneously, one light from each side. Intensities of the lights were matched to those measured at a beach when light poles were 60 m (about 200 ft) distant.

Almost all (more than 96%) of the turtles turned toward the HPS light when it was presented alone. When filtered light was presented alone, attraction was weaker; 68% of the hatchlings turned toward the #2422, whereas 85% turned toward the NLW. When filtered and unfiltered lights were presented simultaneously, the HPS light attracted more than 90% of the turtles. Reducing HPS brightness by one or two log units, thereby rendering it dimmer than the filtered light with which it was paired, made the two lights equally attractive; that is, about half of the hatchlings turn toward each luminaire. A significant attraction to the filtered light occurred only when the HPS light was reduced in intensity by three log units. These results indicate that filtering an HPS luminaire does make its light less attractive to the turtles. They also show that attraction depends on both the intensity and the spectral composition of these lights.

Embedded Lighting on Coastal Roadways

If street lighting on poles is visible at the beach, filtering may reduce orientation disruption, but it does not eliminate the light stimulus. A better alternative is to confine roadway lighting to the street surface. The Florida Department of Transportation sponsored an embedded lighting project in Boca Raton. Light-emitting diodes were installed along a 1-km (0.6-mile) section of coastal roadway (Highway A1A; Figure 7.7). Streetlights with filters, which were already present along the roadway, were left in place.

The project site was located at a park bordering the nesting beach. Because few other lights were present, this location was ideal for experiments designed to compare hatchling orientation under three conditions: when only the filtered streetlights were on, when only the embedded lights were on, and when both lighting systems were switched off. Bertolotti and Salmon (in press) used beach arena assays to measure turtle orientation under each of these conditions. Hatchlings were captured in the afternoon of the day they would naturally emerge and then taken to the beach that evening. They were released in the center of a 4-m (13-ft) diameter circle drawn on the beach surface. Hatchlings showing normal orientation all crawled east, toward the ocean. Hatchlings whose ori-

Figure 7.7. The coastal roadway used for the embedded lighting project in Boca Raton, Florida. The nesting beach is located to the right (not visible). Above, view at night with the traditional poled streetlights turned on; below, view with the streetlights turned off and the embedded lights turned on. The streetlights, but not the embedded lights, were visible from the beach.

entation was disturbed by lighting either crawled toward the lights or failed as a group to show any significant orientation preference.

At a control site, vegetation between the beach and the roadway acted as a light barrier, and the hatchlings under all treatment conditions crawled toward the sea. At two experimental sites the streetlights were visible; orientation was disrupted when the streetlights were turned on but not when the embedded lights were on or when both the streetlights and embedded lights were switched off.

Embedded lighting was also an effective lighting alternative for people. Pedestrians, cyclists, and motorists all responded favorably to the lighting modification.

Comprehensive Plans to Reduce Artificial Light at Sea Turtle Nesting Beaches

Restoration of natural levels of darkness on sea turtle nesting beaches will require large-scale plans. When implemented fully, such plans can dramatically reduce artificial light experienced at beaches. Following are three examples of comprehensive light management plans from Florida.

A Light Management Plan for Broward County

The beaches of Broward County in southeastern Florida receive about 2,500 sea turtle nests each year. Most (about 70%) are deposited on beaches exposed to so much lighting that they must be relocated. The cost of this effort to the county is substantial (e.g., $95,000 in 2001). As discussed earlier in this chapter, concentrating nests in hatcheries increases predation rates on hatchlings, but that is not the only problem. Relocated eggs may be damaged in transport, increasing probabilities of egg death or sublethal effects during development that could reduce hatchling vigor. Additionally, concentrating nests in one location year after year increases the risk that local perturbations, such as a storm, a raid by terrestrial predators, or accumulating sand pathogens, may destroy large numbers of eggs.

An alternative approach is light management, eliminating the need to cage or relocate nests. Although also initially expensive, the benefits of habitat restoration are long-term and obvious. Restoration is encouraged by both the state (Florida Fish and Wildlife Conservation Commission) and federal (U.S. Fish and Wildlife Service) agencies responsible for coordinating sea turtle recovery efforts. The issue then becomes one of devising a plan that promotes habitat restoration most efficiently and on the largest scale possible. Such a plan (Ernest and Martin 1997), developed for Broward County by an environmental consulting firm, is being implemented.

The county spans about 23 miles (37 km) of coastline that includes eight jurisdictional boundaries. Some beaches in the county are relatively dark and undeveloped, such as in front of single-family residences or parks, whereas others are brightly illuminated by buildings constructed just behind the beach (e.g., Fort Lauderdale and Hollywood). Nest density was inversely correlated with beachfront development. Given this variation, and considering that resources to implement management plans are always limited, the challenge of the plan is to restore the natural light regime of the habitat, thereby reducing the need to relocate nests.

The plan consisted of an initial assessment phase followed by an implementation phase. The assessment phase began with a lighting inspection and a review of the most recent (1994–1995) nest density data. This information was used to rank the beaches into management areas. Ranks were based on the sum of scores for nest densities, number of nests needing relocation, extent of coastal development, and proximity to other management areas, specifically the potential for lighting in one area to cause problems in an adjacent, darker area. Sites with low scores were those that were least developed and contained many nests. A second element of the plan was public awareness. Coastal property owners were informed about the plan's objectives and its benefits for sea turtles, the environment, and coastal residents. Residents were also provided with guidelines for voluntary compliance. Finally, a single simplified lighting ordinance stating the rules and regulations for protecting nesting turtles and their hatchlings was created for the county, designed to establish uniform criteria to identify lighting problems and to enforce compliance.

The implementation phase began in 2001; more time will be needed to evaluate the plan's strengths and weaknesses. But its basic elements seem appropriate and workable. Initial light management efforts will be directed to the sites with low scores, that is, sites where the largest numbers of nests are located and where light control can be most easily achieved. To ensure that plan goals are continuously achieved, lighting inspections will be continued, and property owners will be notified of any lighting infractions. They will also be provided with assistance in the effort to resolve them. Finally, once changes are made they will be evaluated to ensure that they are effective.

Overall, the plan represents an approach toward, and provides a framework for, the management of problem lighting at any coastal habitat. What is essential for its success is a firm commitment by the community and its regulators to achieve those goals.

Patrick Air Force Base and Cape Canaveral

Brevard County receives more than 40% of Florida's sea turtle nests and, for this reason, is an area of special concern. It is also home to the 45th Space Wing of the U.S. Air Force, which maintains two facilities on the coast, the Patrick Air Force Base (PAFB) and the Cape Canaveral Air Force Station (CCAFS). On average, about 1,800 loggerhead nests are deposited annually on the 7-km (4.3-mile) beach in front of the PAFB, and an average of 3,500 nests of the same species are placed on the 23-km (14.3-mile) beach in front of the CCAFS.

In 1988, meetings were initiated between the U.S. Fish and Wildlife Service and the U.S. Air Force to resolve lighting issues that had caused serious hatchling misorientation and disorientation problems at both sites. A lighting plan was developed for the CCAFS in 1988 and for the PAFB in 1995. Light reduction was complicated by the necessity at both sites to maintain lighting essential for human safety and national security. Nevertheless, both sites are now impressively dark, thanks to changes that collectively involved more than 1,000 luminaires. Modifications included replacing high pressure sodium with low pressure sodium luminaires, reducing wattage, shielding and recessing lights, installing motion detector controls to turn on lights only when they were needed, and eliminating unnecessary lights at both facilities. The affected areas included roadways on the facility and between the facility and the beach, parking lots, family housing units, hangars, runways, launch pads, and sports fields. Lighting curfews were imposed during turtle nesting season for all outdoor sporting and social activities. Once the project was completed, the transformation was remarkable. The coastal roadway between the beach and the base is extremely dark.

The Air Force took responsibility for ensuring compliance at existing facilities and for reviewing lighting plans for all new construction. The Air Force also agreed to annually monitor and record sea turtle nesting activity and hatchling behavior, to support beach dune enhancement by planting native dune vegetation, and to add light screens at sites where hatchling orientation problems persisted. Finally, the Air Force agreed to support monitoring efforts, to report the annual "take" of turtles (primarily losses of hatchlings caused by lighting problems) to the Fish and Wildlife Service, and to limit take to 2% or less of all hatchlings from all nests.

This example illustrates the successful enforcement of the Endangered Species Act, where the military in cooperation with the Fish and Wildlife Service achieved conservation goals without sacrificing readiness. It also shows that a structurally complex coastal community consisting of residential, specialized industrial, service, and recreational components can function effectively while having a minimal adverse effect on sea turtles.

Lighting Plans for Coastal Roadways

Disruption of hatchling orientation is especially common at coastal roadway sites in Florida. In recognition of this problem, a technical working group met to formulate a *Coastal Roadway Lighting Manual*. The working group consisted of representatives from industry, state and federal government, and technical experts. The manual (Ernest and Martin 1998) presents a step-by-step approach to the diagnosis of roadway lighting problems at sea turtle nesting beaches and their resolution though effective light management. It is intended for a wide audience of regulators, traffic planners and engineers, utility company personnel, conservationists, and environmental planners. Tables in the manual list the efficacy of lighting alternatives as a function of local conditions. An appendix provides technical specifications, costs, and sources of standard roadway luminaires.

Lighting problems are identified by nighttime surveys, by hatchling disorientation reports submitted by permit holders to the Florida Fish and Wildlife Conservation Commission, or by beach arena assays. Specific solutions appropriate to each site vary, but a standard approach is advocated that applies to any location. It involves three elements: keep lighting off the beach by repositioning or shielding the light; reduce luminance by turning lights off, installing fewer lights, or lowering wattage; and minimize the disruptive wavelengths by using light filters or low pressure sodium luminaires. Finally, the manual stresses the importance of incorporating new technology as it becomes available.

Conclusion

Forty years have passed since McFarlane (1963) published the first report of sea turtle hatchling disorientation by artificial roadway lights in Florida. Since then, other studies have stressed the effects of artificial lighting on all wildlife (Verheijen 1985, Outen 1998, Longcore and Rich 2004) and on sea turtle nesting beaches. Some (Raymond 1984, Witherington and Martin 1996) have also described methods that work best to achieve light control. Thanks to the efforts of permit holders, concerned citizens, municipal and county environmental regulators, the federal government, and private environmental organizations (especially the Center for Marine Conservation), many historically dark beaches remain dark, and others previously exposed to stray lighting have been partially, and in a few cases completely, restored. There has also been an increase in public awareness of the sea turtle lighting problem. That awareness has resulted in the adoption of strict local lighting regulations. For example, it is now impossible to obtain a building permit for a coastal structure without having an approved lighting plan.

But as this review indicates, some strategies to manage and protect marine turtles have been more successful than others. Those least successful have sought to remove the turtles from areas of problem lighting or prevent the turtles from responding to the lights by caging. These strategies fail for two reasons. First, they create new problems for the turtles. Second, they fail to deal with causes, in this case habitat degradation by lighting, and for this reason have been criticized as "halfway technology" (Frazer 1992). The alternative approach advocates habitat restoration through light management to reduce the need to manipulate either sea turtle nests or hatchlings. The scale of light management has varied from small patches of beach to entire communities or municipalities. Obviously, small-scale modification will be effective where there are few, easily modified sources of artificial lighting. But large-scale plans are needed at locations where development is more extensive and where there are many kinds and sources of artificial lighting.

In the last few years there has been a gradual increase in the number of green turtle and leatherback nests at Florida's beaches, whereas loggerhead nesting has slightly declined, for reasons that remain unknown. For green turtles and leatherbacks, these changes may be a consequence of a widespread international effort to protect marine turtles at their nesting beaches, feeding grounds, and nursery habitats and along their migratory routes. But despite that positive trend, continued vigilance will be required; this is particularly true when it comes to the artificial lighting problem.

In Florida, there are two elements of continued concern. The first is the absence of a plan to limit population growth in the state. The current population is dangerously near the state's carrying capacity (Bouvier and Weller 1992). Continued development places an excessive burden on infrastructure such as roads, schools, water, sewage, police and fire protection, and family and other social services. It also portends dire consequences for the preservation of natural areas and wildlife, including sea turtles.

The second concern is that current efforts to manage artificial lighting at nesting beaches stress a highly regional near-coastal approach. A regional approach will reduce lighting directly visible from the beach but will not reduce inland sources that produce ever-increasing sky glow. In fact, one might predict that if regional approaches succeed and beach habitats become darker, the problem of sky glow from interior locations will become even more serious. This prediction arises because the influence of artificial lighting depends on its contrast with adjacent, interior location environments.

What is needed in Florida is a statewide (or, one could argue, national) policy for artificial light management. Organizations around the world have recognized this need and are actively proposing change through public education, stressing the energy-saving, ecological, and aesthetic benefits of light management. But the task will take time, hard work, and patience. For the moment, the best we can do as conservation scientists is to act locally to protect wildlife in critical habitats. But we must also promote through our conversations with public officials, our writings, and our lectures a message that artificial lighting must be managed everywhere.

Acknowledgments

New data reported in this chapter come from the theses of students Stephanie Adamany, Lesley Bertolotti, Kristen Nelson, Jeff Pennell, and Susan Tuxbury. Their diligence and dedication are appreciated. The Florida Power and Light Company, the Florida Department of Transportation, and the National Save-the-Sea-Turtle Foundation of Fort Lauderdale, Florida provided financial support. Sandra MacPherson (U.S. Fish and Wildlife Service) and Jane Provancha (Dynamac, U.S. Air Force, and National Aeronautics and Space Administration) documented the history of the lighting plans for the CCAFS and the PAFB; Don George gave me a tour of both facilities. Chris Elvidge (National Oceanic and Atmospheric Administration) graciously provided the satellite photos of Florida. The manuscript was improved by comments from Robert Ernest (Ecological Associates, Inc.), Blair Witherington (Florida Fish and Wildlife Conservation Commission), and Jeanette Wyneken (Center for Sea Turtle Research, Florida Atlantic University).

Literature Cited

Adamany, S. L., M. Salmon, and B. E. Witherington. 1997. Behavior of sea turtles at an urban beach. III. Costs and benefits of nest caging as a management strategy. *Florida Scientist* 60:239–253.

Avens, L., and K. J. Lohmann. 2003. Use of multiple orientation cues by juvenile loggerhead sea turtles *Caretta caretta*. *Journal of Experimental Biology* 206:4317–4325.

Avens, L., and K. J. Lohmann. 2004. Navigation and seasonal migratory orientation in juvenile sea turtles. *Journal of Experimental Biology* 207:1771–1778.

Berthold, P. 1993. *Bird migration: a general survey*. Oxford University Press, Oxford.

Bertolotti, L., and M. Salmon. In press. Do embedded roadway lights protect sea turtles? *Environmental Management*.

Bouvier, L. F., and B. Weller. 1992. *Florida in the 21st century: the challenge of population growth*. Center for Immigration Studies, Washington, D.C.

Ernest, R. G., and R. E. Martin. 1997. *Beach lighting management plan: a comprehensive approach for addressing coastal lighting impacts on sea turtles in Broward County, Florida*. Ecological Associates, Inc., Jensen Beach, Florida.

Ernest, R. G., and R. E. Martin. 1998. *Coastal roadway lighting manual: a handbook of practical guidelines for managing street lighting to minimize impacts to sea turtles*. Ecological Associates, Inc., Jensen Beach, Florida.

Florida Fish and Wildlife Conservation Commission. 2002. *Sea turtle conservation guidelines*. Revised edition. Florida Fish and Wildlife Conservation Commission, Tallahassee, Florida.

Frazer, N. B. 1992. Sea turtle conservation and halfway technology. *Conservation Biology* 6:179–184.

Gyuris, E. 1994. The rate of predation by fishes on hatchlings of the green turtle (*Chelonia mydas*). *Coral Reefs* 13:137–144.

Lohmann, K. J., S. D. Cain, S. A. Dodge, and C. M. F. Lohmann. 2001. Regional magnetic fields as navigational markers for sea turtles. *Science* 294:364–366.

Lohmann, K. J., J. T. Hester, and C. M. F. Lohmann. 1999. Long-distance navigation in sea turtles. *Ethology Ecology & Evolution* 11:1–23.

Lohmann, K. J., and C. M. F. Lohmann. 2003. Orientation mechanisms of hatchling loggerheads. Pages 44–62 in A. B. Bolten and B. E. Witherington (eds.), *Loggerhead sea turtles*. Smithsonian Books, Washington, D.C.

Lohmann, K. J., A. W. Swartz, and C. M. F. Lohmann. 1995. Perception of ocean wave direction by sea turtles. *Journal of Experimental Biology* 198:1079–1085.

Lohmann, K. J., B. E. Witherington, C. M. F. Lohmann, and M. Salmon. 1997. Orientation, navigation, and natal beach homing in sea turtles. Pages 107–135 in P. L. Lutz and J. A. Musick (eds.), *The biology of sea turtles*. Volume I. CRC Press, Boca Raton, Florida.

Longcore, T., and C. Rich. 2004. Ecological light pollution. *Frontiers in Ecology and the Environment* 2:191–198.

Luschi, P., F. Papi, H. C. Liew, E. H. Chan, and F. Bonadonna. 1996. Long-distance migration and homing after displacement in the green turtle (*Chelonia mydas*): a satellite tracking study. *Journal of Comparative Physiology A* 178:447–452.

McFarlane, R. W. 1963. Disorientation of loggerhead hatchlings by artificial road lighting. *Copeia* 1963:153.

Meylan, A., B. Schroeder, and A. Mosier. 1995. Sea turtle nesting activity in the state of Florida 1979–1992. *Florida Marine Research Publications* 52:1–51.

Mortimer, J. A. 1999. Reducing threats to eggs and hatchlings: hatcheries. Pages 175–178 in K. L. Eckert, K. A. Bjorndal, F. A. Abreu-Grobois, and M. Donnelly (eds.), *Research and management techniques for the conservation of sea turtles*. IUCN/SSC Marine Turtle Specialist Group Publication Number 4.

Mroziak, M. L., M. Salmon, and K. Rusenko. 2000. Do wire cages protect sea turtles from foot traffic and mammalian predators? *Chelonian Conservation and Biology* 3:693–698.

Nelson, K. A. 2002. *The effect of filtered high pressure sodium lighting on hatchling loggerhead (*Caretta caretta *L.) and green turtle (*Chelonia mydas *L.) hatchlings.* M.S. thesis, Florida Atlantic University, Boca Raton.

Outen, A. R. 1998. *The possible ecological implications of artificial lighting*. Hertfordshire Biological Records Centre, Hertfordshire, UK.

Pennell, J. P. 2000. *The effect of filtered roadway lighting on nesting by loggerhead sea turtles (*Caretta caretta *L.)*. M.S. thesis, Florida Atlantic University, Boca Raton.

Raymond, P. W. 1984. *Sea turtle hatchling disorientation and artificial beachfront lighting: a review of the problem and potential solutions*. Center for Environmental Education, Washington, D.C.

Salmon, M. 2003. Artificial night lighting and turtles. *Biologist* 50:163–168.

Salmon, M., R. Reiners, C. Lavin, and J. Wyneken. 1995a. Behavior of loggerhead sea turtles on an urban beach. I. Correlates of nest placement. *Journal of Herpetology* 29:560–567.

Salmon, M., M. G. Tolbert, D. P. Painter, M. Goff, and R. Reiners. 1995b. Behavior of loggerhead sea turtles on an urban beach. II. Hatchling orientation. *Journal of Herpetology* 29:568–576.

Salmon, M., B. E. Witherington, and C. D. Elvidge. 2000. Artificial lighting and the recovery of sea turtles. Pages 25–34 in N. Pilcher and G. Ismail (eds.), *Sea turtles of the Indo-Pacific: research, management and conservation*. Asean Academic Press, London.

Tuxbury, S. M. 2001. *Seafinding orientation of hatchlings exposed to filtered lighting: effects of varying beach conditions*. M.S. thesis, Florida Atlantic University, Boca Raton.

Verheijen, F. J. 1958. The mechanisms of the trapping effect of artificial light sources upon animals. *Archives Néerlandaises de Zoologie* 13:1–107.

Verheijen, F. J. 1985. Photopollution: artificial light optic spatial control systems fail to cope with. Incidents, causations, remedies. *Experimental Biology* 44:1–18.

Witherington, B. E. 1992. Behavioral responses of nesting sea turtles to artificial lighting. *Herpetologica* 48:31–39.

Witherington, B. E. 1997. The problem of photopollution for sea turtles and other nocturnal animals. Pages 303–328 in J. R. Clemmons and R. Buchholz (eds.), *Behavioral approaches to conservation in the wild*. Cambridge University Press, Cambridge.

Witherington, B. E., and N. B. Frazer. 2003. Social and economic aspects of sea turtle conservation. Pages 355–384 in P. L. Lutz, J. A. Musick, and J. Wyneken (eds.), *The biology of sea turtles*. Volume II. CRC Press, Boca Raton, Florida.

Witherington, B. E., and R. E. Martin. 1996. Understanding, assessing, and resolving light-pollution problems on sea turtle nesting beaches. *Florida Marine Research Institute Technical Report* TR-2:1–73.

Wood, D. W., and K. A. Bjorndal. 2000. Relation of temperature, moisture, salinity, and slope to nest site selection in loggerhead sea turtles. *Copeia* 2000:119–128.

Wyneken, J., and M. Salmon. 1996. *Aquatic predation, fish densities, and potential threats to sea turtle hatchlings from open-beach hatcheries: final report.* Broward County, Department of Natural Resource Protection, Technical Report Number 96-04.

Chapter 8

Night Lights and Reptiles: Observed and Potential Effects

Gad Perry and Robert N. Fisher

Reptiles are amazing creatures. They are found in most types of habitats, sometimes in great numbers; the greatest density of any terrestrial vertebrate was measured in *Sphaerodactylus macrolepis*, a tiny West Indian gecko (Rodda et al. 2001). They range in size from the tiniest known vertebrates (another *Sphaerodactylus*; Hedges and Thomas 2001) to large "man-eaters" more than seven meters long and weighing over one ton (Pooley et al. 1989). Despite this, relatively little is known about most species. Summarizing our knowledge of the better-studied West Indian taxa, Schwartz and Henderson (1991:2) stated, "It is surprising how little is known about the natural history of about 95% of the herpetofauna." Perry and Garland (2002) likewise complained that even basic ecological data, such as home range size, were unavailable for a majority of reptile species. Gibbons et al. (2000) pointed out that reptile species are disappearing at a rate that is at least comparable to that raising great alarm in amphibians, yet little attention is being paid to this decline. It is perhaps not surprising, therefore, that information on the effects of night lighting on reptiles, with the notable exception of the sea turtles (see Chapter 7, this volume), is similarly sketchy. For example, the only book to date to focus exclusively on the conservation of the Amphibia and Reptilia (Corbett 1989) does not list night lighting among the threats facing European species. Similarly, Klauber's (1956[1997]) massive monograph on rattlesnakes (two volumes, more than 1,500 pages) only briefly discusses light, under "bodily functions," while describing the function of the eye. The early review of artificial light effects by Verheijen (1958) did not discuss any effects on reptiles either.

In this chapter we bring together the few published reports on direct effects of night lighting on reptiles and add what unpublished information we have been able to locate. We begin by surveying broad taxonomic patterns in activity time, then discuss documented effects on both diurnal and nocturnal species. We then review work that might bear on this issue, such as studies on the effects of lunar lighting on activity levels. Finally, we assess the data for apparent patterns and gaps in our knowledge and make recommendations for future work.

In this chapter we focus on taxa commonly thought of as reptiles. Birds, although nested in the clade Reptilia, are covered separately in this volume (see Chapters 4–6), as are sea turtles (see Chapter 7).

The term "night light" has been used in several contexts. Night-lighting is a common technique used to search at night for taxa that possess a reflective layer called the tapetum lucidum in their retinas. The presence of this structure causes light from searchers to be reflected, allowing target animals such as crocodilians to be located (e.g., Rice et al. 1999). Here we focus entirely on the effects of ambient light pollution, a byproduct of human outdoor illumination (see, e.g., Dawson 1984).

Nocturnality in Reptiles

Nocturnal activity is widespread in reptiles, but the distribution is not taxonomically random. Rather, some clades, such as geckos, show strong tendencies toward nocturnal activity, whereas others, such as lacertid and iguanid lizards, are strongly diurnal (Figure 8.1). As with many generalizations, there are exceptions to this rule. For example, the most speciose gecko genus, *Sphaerodactylus*, contains primarily diurnal forms. Nonetheless, Figure 8.1 is useful for identifying broad taxonomic groups that may be affected by the presence of artificial lights.

Both nocturnal and diurnal species can be attracted to artificial lights (Verheijen 1958), but the effects on them are potentially quite different. For primarily nocturnal species, greater lighting levels increase visibility.

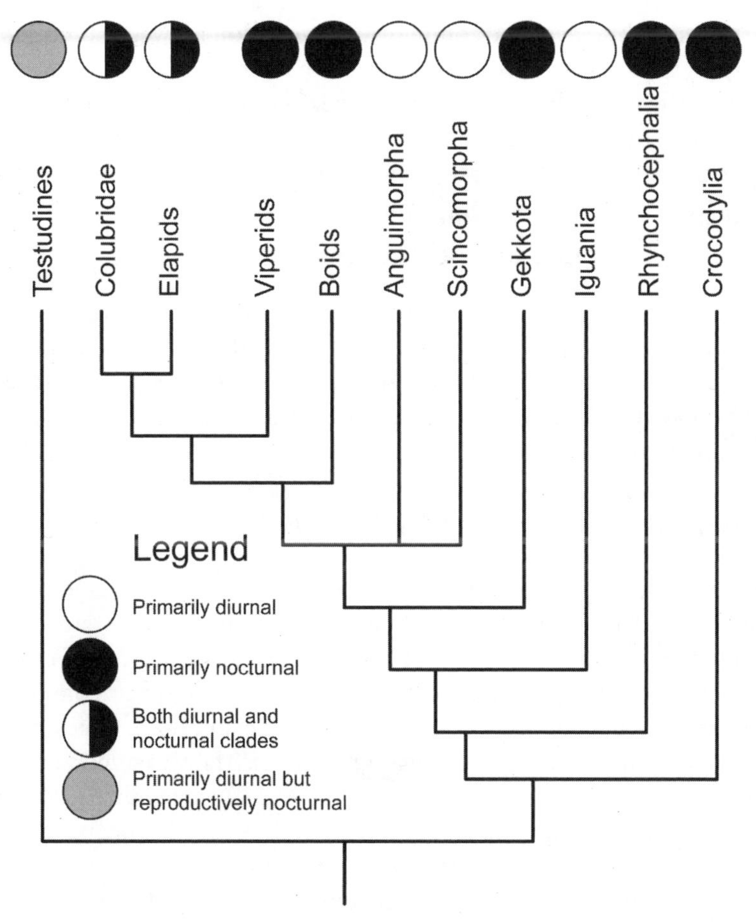

Figure 8.1. Prevalence of nocturnality in nonavian Reptilia. Note that symbols represent general trends only, and some exceptions occur in many of the clades depicted.

This might be useful to a predator, whose ability to locate prey may be improved. It might increase the chance of prey to identify a predator from a distance, however, and thus cause predators to reduce their activity levels. By increasing the perceived sense of risk experienced by prey species, artificial lights may also cause prey species to reduce activity levels. Studies relevant to these issues are reviewed in this chapter. For primarily diurnal species, the presence of lights during naturally dark periods may allow activity periods to be extended. The combined effects on an ecosystem level are hard to predict, even when effects on component species are known, because the interaction can be complex. For example, the presence of artificial light affects the competitive interaction between some geckos. Figure 8.1 therefore broadly identifies the types of effects one might expect but provides little predictive power about individual species or the interactions of multiple taxa.

"Positive" Effects: Enhanced or Extended Foraging at Night Lights

Some species of reptiles prey on invertebrates attracted to night lights. In a simplistic sense this is a positive effect because both diurnal and nocturnal species can forage successfully under these conditions. This apparent benefit may have costs, however, such as exposure to increased predation or competitive interactions.

Nocturnal Species Active at Night Lights

The light trap is one of the most common tools used by entomologists to sample nocturnal invertebrates, many of which are strongly attracted by artificial lights over various wavelengths (e.g., Verheijen 1958, Southwood 1978). Not surprisingly, nocturnal edificarian reptiles (species often found in or around human structures), primarily geckos, were quick to take advantage of the enhanced foraging opportunity. Many examples exist, and we will review only some here. Buden (2000:249) states that in the Sapwuahfik Atoll, *Gehyra mutilata* "is common in areas of human habitation [and] . . . encountered only rarely in vegetation" (these data are not based on numerical measures of population density, however). Two other geckos, *Gehyra oceanica* and *Lepidodactylus lugubris*, are also locally common around houses (Buden 2000). In Costa Rica, introduced *Lepidodactylus lugubris* "often feeds on insects attracted to lights inside buildings" (Savage 2002:486). Several authors describe edificarian habits in *Thecadactylus rapicauda* (Lazell 1995, Vitt and Zani 1997, Howard et al. 2001), also observed in buildings in Costa Rica (G. Perry, unpublished data). It has also been observed using mercury vapor lamps for foraging in Trinidad (Kaiser and Diaz 2001). Three species of *Ptyodactylus* are known to use houses in Israel (Werner 1995, Johnson and Bouskila, unpublished) and the United Arab Emirates (P. Cunningham, personal communication, 2004), occasionally feeding near lights. Interestingly, the density of *Ptyodactylus guttatus* on buildings is about 200 times higher than in nearby natural habitats, presumably because of the additional food afforded by the artificial lights (Johnson and Bouskila, unpublished). *Bunopus tuberculatus* also feeds at lights in the UAE (P. Cunningham, personal communication, 2004). Cunningham (personal communication, 2004) also recorded two species of *Pachydactylus* (*P. bibronii* and *P. turneri*) at lights in Namibia. Fisher (unpublished observation) has recorded *Nactus pelagicus* foraging, rarely, at night lights in the South Pacific.

The gecko genus *Hemidactylus* is probably the best-known example of nocturnal edificarian taxa using lights. *Hemidactylus mabouia* "is frequently associated with human establishments and is well known for utilizing areas around artificial light sources as hunting grounds," but "the availability of lights had no apparent effect on the distribution of these geckos" on buildings on Anguilla (Howard et al. 2001:287). On Guana Island (British Virgin Islands), *Hemidactylus mabouia* is commonly seen in and around buildings, but Rodda et al. (2001) found none in four forest plots that were thoroughly sampled. In urban settings in Israel, *Hemidactylus turcicus* "concentrate near sources of illumination . . . and feed on moths and other insects attracted to the light" (Werner 1966:11), and this is also true in the UAE (P. Cunningham, personal communication, 2004). Bowersox et al. (1994) report an adult *Hemidactylus haitianus* attempting to feed on a juvenile of the same species at a hotel in the Dominican Republic. Cunningham (personal communication, 2004) reports both *Hemidactylus flaviviridis* and *Hemidactylus persicus* foraging around lights in the UAE. Several species of *Hemidactylus* are known by the common name "house gecko," and although the use of night light edificarian habitats is not obligatory, they appear to depend greatly on the presence of such human environments, especially outside their native ranges. They and other edificarian species, such as *Lepidodactylus lugubris*, tend to be especially adept at human-aided dispersal, often establishing introduced populations or even becoming invasive (Case et al. 1994). Two species (*Hemidactylus frenatus* and *Hemidactylus garnotii*) have been introduced into Costa Rica, where they are locally commonly edificarian and often found at lights (Savage 2002). In the Pacific, *Hemidactylus frenatus*, *Lepidodactylus lugubris*, and *Cosymbotus platyurus* are more abundant on buildings with external lights than on unlit buildings (Case et al. 1994). *Hemidactylus mabouia* is introduced in Brazil, where it is found mostly in human dwellings (Vitt and Zani 1997) and on streetlights (Y. L. Werner, unpublished observation from Manaus). This is by no means a comprehensive list. These species have been introduced elsewhere, and similar species are also known to have been introduced. The phenomenon is common enough that Canyon and Hii (1997) studied the use of geckos as biological control agents for mosquitoes around human habitation.

We have been able to locate only a single published report of a nocturnal snake, the African brown house-snake (*Lamprophis fuliginosus*), foraging under lights (Cunningham 2002). The snake was observed capturing a gecko, *Pachydactylus capensis*, which was itself feeding at a porch light (P. Cunningham, personal communication, 2004). An additional record for snakes is the brown tree snake (*Boiga irregularis*), in its native range, observed foraging downward for geckos from the roof of a lit building toward the night light (one observation each from Papua New Guinea and Solomon Islands; R. Fisher, unpublished). Similar observations are not uncommon on Guam, where *Boiga irregularis* is an invasive species (G. Perry, unpublished). Snakes are generally more wary of humans than are other reptiles and rarely reach high population density. It is therefore likely that other species occasionally take advantage of artificial lights but that such events are not commonly observed. We have not located reports of turtles or crocodilians taking advantage of night lights, but they tend to be even more secretive than snakes, and the same caveat applies.

Diurnal Species Active at Night

Over the years, a number of anecdotal reports have appeared that detail normally diurnal reptiles extending their activity into the night near lights. Garber (1978) has called this new habitat the "'night-light' niche."

Henderson and Powell (2001) surveyed the responses of West Indian reptiles to human presence and found that 20% of lizards (67 species) and 11% of snakes (18 species) known from the region are associated, at least some of the time, with human habitations. Of those, nine species of diurnal lizard and one species of diurnal snake have been documented to extend their activity into the night near artificial lights (Henderson and Powell 2001). With a recent report of a diurnal snake lying in ambush for diurnal lizards at night lights (Perry and Lazell 2000), it appears that artificial lights may allow whole feeding webs to extend their activity times.

All known cases of use of the night-light niche are listed in Table 8.1. Interestingly, a preponderance of these observations is of West Indian anoles; nearly all are for tropical island-dwellers. Does this reflect a real biological difference between these and other reptiles? West Indian anoles have received disproportionate scientific attention (e.g., Perry and Garland 2002), suggesting a possible bias. All of the species listed in Table 8.1, however, not only the tropical anoles, are both locally abundant and common in and around human habitations. This suggests that an edificar-

Table 8.1. Use of the night-light niche by typically diurnal reptile species.

Species	*Location*	*References*
LIZARDS		
Geckos (Gekkonidae)		
Gonatodes humeralis	Peru	Dixon and Soini 1975
Gonatodes vittatus	Trinidad	Quesnel et al. 2002
Phelsuma laticauda	Hawaii	Fisher, unpublished
	Hawaii	Werner et al., unpublished
Sphaerodactylus macrolepis	Guana Island, BVI	Perry and Lazell 2000
Sphaerodactylus sputator	Anguilla	Howard et al. 2001
Anoles (Iguanidae)		
Anolis bimaculatus	Antigua	Schwartz and Henderson 1991
Anolis brevirostris	Hispaniola	Bowersox et al. 1994
Anolis carolinensis	Texas	McCoid and Hensley 1993
	Hawaii	Fisher, unpublished
Anolis cristatellus	Puerto Rico	Garber 1978
	Guana Island, BVI	Perry and Lazell 2000
	Puerto Rico	Henderson and Powell 2001
Anolis cybotes	Hispaniola	Henderson and Powell 2001
Anolis gingivinus	St. Maarten	Powell and Henderson 1992
Anolis marmoratus	Guadeloupe	Schwartz and Henderson 1991
Anolis richardii	St. George's, Grenada	Henderson, personal communication, 2003
Anolis sabanus	Saba	Schwartz and Henderson 1991
Anolis sagrei	Bahamas	Schwartz and Henderson 1991
	Florida	Carmichael and Williams 1991
Other Iguanids (Iguanidae)		
Tropidurus plica [=*Plica plica*]	Trinidad	Werner and Werner 2001
Skinks (Scincidae)		
Cryptoblepharus poecilopleurus	Cocos Island, Guam	McCoid and Hensley 1993
Lamprolepis smaragdina	Pohnpei	Perry and Buden 1999
SNAKES		
Racers (Colubridae)		
Alsophis portoricensis	Guana Island, BVI	Perry and Lazell 2000

ian lifestyle is an important contributor to a secondary use of artificial lights. We predict that additional species showing these two traits will gradually be recorded using the night-light niche. For example, the skink *Emoia cyanura*, which is common "in sparse vegetation in the vicinity of human habitation" (Buden 2000:251), may be able to take advantage of artificial lighting. Klauber (1939) lists eight common diurnal desert lizard species he observed active at night, and such species may be able to take advantage of artificial lighting near rural towns in the California deserts to extend their diel activity.

Although actual observations of diurnal species active at night are limited, there are indications that other species may also be able to use the night-light niche. Auffenberg (1988) described in detail the biology of Gray's monitor lizard (*Varanus olivaceus*). In one study he attached light-sensitive transmitters to lizards and characterized their activity pattern in terms of illumination. The results indicate that ambient light levels help determine both onset and cessation of activity: "On overcast days haul-out time is almost one hour later than on sunny mornings." However, "light is not the only factor determining haul-out time" (Auffenberg 1988:106). In addition, Auffenberg's study was conducted in a largely human-free area, and artificial lights were not a factor he examined. It is therefore not possible to use these data to predict what effect strong artificial lights might have on this species. Moreover, many monitor lizards are highly sensitive to human presence (e.g., Perry and Dmi'el 1995) and may not remain near human habitation long enough to be affected by lights.

"Negative" Effects: Increased Predation Risk and Decreased Foraging Success

Direct information on negative consequences of artificial lights in free-ranging reptiles (other than sea turtles) is not readily available. Several studies suggest, however, that such adverse effects may indeed exist. As stated above, entomologists use light traps to sample nocturnal invertebrates, and nocturnal edificarian reptiles take advantage of the enhanced foraging opportunities available at artificial lights. Entomologists have also long known that the apparent activity of many invertebrates increases during dark nights (e.g., Bowden and Church 1973), perhaps because of a reduction in predation risk. Conversely, the activity of some predators increases when a full moon offers additional light (e.g., Mills 1986, for goatsuckers), and predation success has been shown to vary as a consequence (Nelson 1989). It follows that the presence of additional lighting could affect reptiles either directly, by altering their own behaviors, or indirectly, by altering the behavior of their prey or their predators.

The first explicit observations of moonlight avoidance by snakes were made by Klauber (1939:50), who stated that the consensus among collectors of snakes at night in the desert is that "the best results are secured in the dark of the moon." With two exceptions, the studies that followed support this observation. One exception is a study of *Leptotyphlops humilis* (Brattstrom and Schwenkmeyer 1951), which found that these fossorial snakes prefer moonlit nights for their aboveground activity. In the other, *Echis coloratus* did not change their ambush sites as a function of moonlight intensity at two Israeli sites (Bouskila 1989, Tsairi and Bouskila 2004). In contrast, *Crotalus cerastes* do change their ambush sites in response to moon phase (Bouskila 2001). In addition, Bouskila (1995) reported that rattlesnake predation on free-ranging kangaroo rats was highest during the relative dark of the new moon, consistent with the snakes reducing their predation risk from owls or large mammals. Snakes may also be maximizing their foraging. The general activity patterns documented for heteromyid rodents increase during the dark of the moon, and tracking the lunar cycle is presumed to be a mechanism for reducing predation (e.g., Lockard and Owings 1974). A similar study documented the same pattern in a radically different environment. Madsen and Osterkamp (1982) found that for *Lycodonomorphus bicolor*, which is strictly nocturnal, activity and predation were both tied to the new moon. Similar patterns, with activity reduction on moonlit nights, were reported in the snakes *Acrochordus arafurae* (Houston and Shine 1994) and *Corallus grenadensis* (Henderson 2002). Studies of the snake *Phyllorhynchus decurtatus* found a negative effect of moonlight on hatchling activity (Lotz in press) but no effect on subadults or adults (Brattstrom 1953). Finally, although young *Sistrurus miliarius* were more likely to use caudal luring under relatively dark conditions, the difference was not statistically significant (Rabatsky and Farrell 1996). Increased light availability from artificial night lighting similarly could reduce foraging success of a broad range of reptilian predators. Secondarily, it could drastically increase the predation risk to reptiles from other reptile, amphibian (e.g., *Rana catesbeiana*), mammalian, or bird predators.

Two studies further suggest that different life stages might be affected in different ways. Clarke et al. (1996) studied the effects of light intensity on the foraging behavior of prairie rattlesnakes (*Crotalus viridis*). The study was conducted in captivity, under artificial lights simulating conditions ranging from new to full moon. The authors found that adult but not juvenile rattlesnakes showed strong reductions in activity levels when light intensity was high. They concluded that the difference between the two age groups might have been related to diet; juveniles feed primarily on prey that are inactive at night and would not be better able to recognize the predator, whereas the nocturnal rodent prey of adults would benefit from increased light (Clarke et al. 1996). Pacheco (1996) studied the effects of multiple environmental factors on activity in black caimans (*Melanosuchus niger*) in Bolivia. He found that moon phase had no effect on the activity of adults but that hatchlings were more active on moonlit nights. Y. Bogin (unpublished data) found the same pattern in *Stenodactylus doriae*.

Extensive work in urbanized southern California by Case and Fisher (Fisher and Case 2000, Case and Fisher 2001, unpublished data) suggests that declines are occurring in populations of many local reptiles. Many causes are operating, but for a subset of declining species (i.e., *Arizona elegans* and *Rhinocheilus lecontei*) light pollution, through both direct and indirect pathways, is the leading hypothesis (Fisher and Case, unpublished). In a preliminary analysis of capture data, Fisher and Case (unpublished) showed that capture frequencies exhibit a peak of activity around the new moon in *Rhinocheilus lecontei* (a nocturnal species) but not in the diurnal *Pituophis catenifer* (Figure 8.2). This study was conducted in a region that is highly influenced by light pollution (Figure 8.3), and some species' declines (range reductions) are consistent with the gradient of light pollution, based on comparisons with their historic distributions (Fisher and Case, unpublished). Specifically, *Arizona elegans* and *Rhinocheilus lecontei* appear to suffer, with *Arizona elegans* showing the greatest

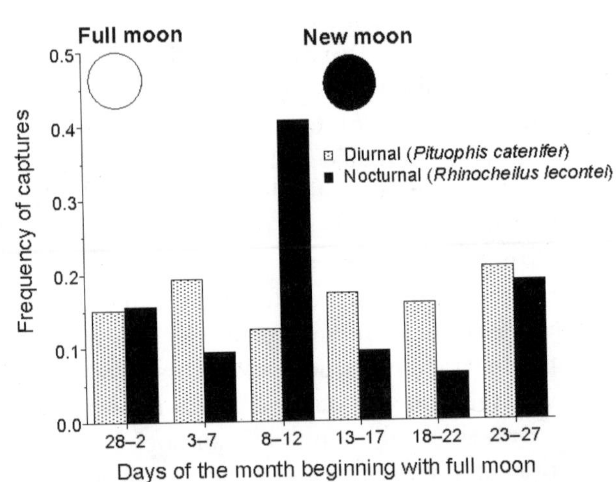

Figure 8.2. Capture frequencies of two snake species from coastal southern California relative to moon phase. Data are derived from Fisher and Case (2000), Case and Fisher (2001), and Fisher and Case (unpublished) and were collected between 1995 and 1998. The frequency of captures shows a pattern of increased activity during the new moon phase for the nocturnal snake but not for the diurnal snake.

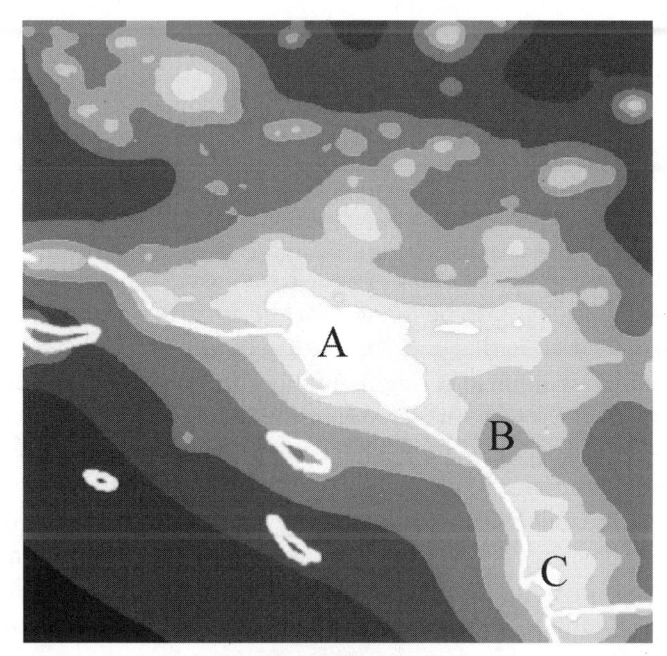

Figure 8.3. Southern California light pollution map from data provided by P. Cinzano (see Cinzano et al. 2001 for details). White and light gray shades are the brightly illuminated urban areas (A, Los Angeles; C, San Diego). Medium gray shades are less illuminated rural areas (B, southern Santa Ana Mountains). Dark gray and black areas are relatively free of artificial illumination. Recent field studies by Fisher, Case, and others have found declines in some nocturnal reptile species in this region. Specifically, *Arizona elegans* is absent from the white and lightest gray areas, and *Rhinocheilus lecontei* and *Coleonyx variegatus* are generally absent from white areas. Other species have declined in the same areas, but these declines probably are related to other causes (e.g., fragmentation, habitat destruction) and not light pollution.

pattern of decline in any reptile in coastal southern California (Fisher and Case, unpublished). Declines in these species have not been recorded in similar rural habitats in northern California, where light pollution is not yet an issue (Sullivan 2000). *Arizona elegans* preys on small mammals, specifically *Perognathus longimembris*, to a greater extent than does *Rhinocheilus lecontei* (Rodríguez-Robles et al. 1999). Extreme declines have been observed in *Perognathus longimembris* in coastal southern California (U.S. Fish and Wildlife Service 1998), indicating spatially concurrent range reductions in predator and prey, both possibly linked to light pollution.

Frankenberg and Werner (1979) studied responses to presence of moonlight in three allopatric subspecies (now considered three congeneric species) of *Ptyodactylus hasselquistii* that exhibit diurnal, diurno-nocturnal, and nocturnal activity patterns. The diurno-nocturnal form was more active with moonlight, even in a lightless chamber. The authors attributed this apparently innate response to a mechanism extending the time lizards could visually forage (Frankenberg and Werner 1979). *Stenodactylus doriae*, an insectivorous, terrestrial, nocturnal desert gecko, decreases locomotor activity during lunar eclipses and is more active on moonlit nights (Bouskila et al. 1992, Reichmann 1998). *Stenodactylus doriae* is also more likely to be found near the cover of a tree on full moon nights than on dark nights and is more likely to use the shade side of a bush on full moon nights (Y. Bogin, unpublished data). In addition, juveniles are more likely to be active during the full moon (Y. Bogin, unpublished data). These behavioral changes appear to be in response to changes in the activity levels of both the prey and the predators of the lizards (Reichmann 1998). Evidence also indicates that light intensity affects the behavior of potential prey. For example, changes in moon phase affected foraging decisions and habitat use in rodents (Kotler et al. 1991, 1994, Bouskila 1995) and scorpions (Skutelsky 1996), both of which are reptile prey. In this manner, increased light intensity can directly affect the level of risk perceived by many species, causing them to change their behavior. Indirectly, changes in prey behavior are likely to affect predator success.

Other Possible Ecological and Behavioral Consequences

In this section we summarize published studies that document the effects of any kind of light on how reptile species interact. Few of these studies address the effects of artificial lights, so the applicability of many of the conclusions listed in this chapter to interactions under night light conditions remains to be demonstrated.

Effects of Photoperiod Length on Behavior

Exposure to different artificial lighting regimes under laboratory conditions is commonly used to study behavior and physiology in (primarily diurnal) reptiles (e.g., Bertolucci et al. 2000). Such studies show that a lighting regime can affect, for example, the interaction between temperature and hormone levels (Firth et al. 1999) and response to externally administered hormones (Hyde and Underwood 2000). There is therefore good reason to suspect that chronic exposure to artificial lights could affect the biology of many reptile species under free-ranging conditions. We have not been able to locate studies addressing this possibility.

Effects of Light Intensity on Behavior

Studies on the habitat choices of single species suggest the presence of generalized preferences for particular lighting regimes. For example, Tiebout and Anderson (2001) showed that *Sceloporus woodi*, a diurnal iguanid lizard, prefers dark experimental arenas (45% of ambient sunlight) to ones with full ambient radiation. Their study was intended to test the effects of changes in ambient light as the result of logging and does not assess the effects of additional artificial lighting. Such innate preferences, however, suggest that changes in illumination can influence the way in which nocturnal animals use habitats near human settlements.

Another reason for changes in habitat use might be light dependency of specific behaviors. For example, Tang et al. (2001:252) showed that *Gekko gecko* advertisement calls are made primarily at dusk, and "the changing natural light leads tokay geckos to display vocal behavior." If these lizards perceive artificial lighting as an extension of the day period or as somehow overriding natural changes in daylength, then opportunities for calling, and thus attracting mates, may be curtailed. Our experience with other edificarian geckos is that they readily call under artificial lights, but the extent to which calls made under such conditions resemble ones made under natural lighting remains to be studied.

Klauber (1939) suggested that snakes are more likely to be encountered during the dark of the moon. Whitaker and Shine (1999) studied the effect of natural light intensity on the chance of encountering the diurnal elapid snake *Pseudonaja textilis*. Although encounter rates were similar under both cloudy and sunlit conditions, snakes were less likely to be encountered by researchers wearing light clothes under cloudy conditions and by those wearing dark clothes under well-lit conditions. This suggests that the ability of these snakes to identify possible dangers depends on the interaction between lighting conditions and predator characteristics. It is reasonable to expect the ability of nocturnal predators and prey similarly to be affected by light intensity. Andreadis (1997) studied nocturnal foraging activity of northern water snakes (*Nerodia sipedon*) over the entire lunar cycle. He found significant differences in activity levels during different lunar phases, with activity being highest during the dark and waning moon phases. Andreadis (1997) hypothesized that foraging success, predation risk, or both were responsible for this pattern.

Indeed, light intensity has been shown to affect feeding success. Zhou et al. (1998) studied feeding and growth rates in juvenile Chinese soft-shelled turtles (*Trionyx sinensis*) under four lighting regimes. They found a linear and negative relationship between light intensity and both daily food consumption and growth rate. The highest rates of intakes and growth were recorded at the lowest lighting intensity, 10 lux. In a similar design, Clarke et al. (1996) found that *Crotalus viridis viridis* activity levels changed as a function of simulated light intensity during the nocturnal activity phase. The application of such captive studies to wild individuals is always problematic, and published data are generally lacking. Werner and colleagues (personal communication, 2004) have found, however, that the phase of the moon affects foraging behavior in the Japanese gecko *Goniurosaurus kuroiwae*. These studies suggest that artificial enhancement of light intensity could affect feeding, growth, and perhaps even survival (often strongly correlated with growth) in at least some reptile species.

Because of its catastrophic ecological and economic effects on Guam and potential to cause similar effects elsewhere, the brown tree snake (*Boiga irregularis*) is one of the most studied reptiles (Rodda et al. 1999). Two studies of this nocturnal pest have involved the effects of light. Shivik et al. (2000) found that the presence of artificial illumination did not affect the way in which snakes oriented toward live mouse lures, compared with their behavior in complete darkness. The behavior was different under the two conditions, however, with strikes at the lure being typical of lighted conditions only (Shivik et al. 2000). Caprette (1997) conducted a series of studies on the effects of artificial illumination on brown tree snake behavior and management. He tested the hypothesis that bright lights would be deterrents to this large-eyed nocturnal species by combining laboratory testing with observational studies at several fenced sites. Trap success was the response variable in both approaches and generally did not vary as a direct function of light intensity. Although capture rates in one experiment were lowest under the greatest lighting intensity (ranging from 63 to 246 lux, compared with 4–10 lux for the next brightest treatment), lighting for that treatment was confounded with heat and lack of humidity because the trap was placed directly underneath the lighting source (Caprette 1997).

Species Interactions

Fleishman et al. (1997) and Leal and Fleishman (2002) examined whether light intensity is a factor in habitat partitioning among tropical anoles.

Both studies found that Caribbean *Anolis* subdivide their usual activity habitats according to light intensity and spectral quality. Leal and Fleishman (2002) also showed that the reflective qualities of the dewlaps of two of these species were adapted to be highly effective communication devices in those environments. This suggests a long-term, evolutionary division of habitat in diurnal species based on light intensity and may be relevant in assessing the effects of artificial lighting in two ways. First, it suggests that nocturnal species may be sensitive to light quality and intensity, not only for locating food and avoiding predation but also for intraspecific communication purposes. Second, it strengthens the idea that light may be important not only intraspecifically but also, at least over the long term, in determining the outcome of interspecific interactions. To date, all nocturnal observations of anoles (Table 8.1) have involved feeding behaviors, and nocturnal taxa normally do not use visual displays in communication (for a possible exception see Parker 1939). Nonetheless, species adapted to diurnal activity in full shade may be better able to adapt to the night-light niche, and nocturnal species adapted to deep darkness may be especially susceptible to negative effects from increased lighting. Addition or removal of species from an ecological community is likely to affect other members, although the precise nature of the effect is difficult to predict.

A highly publicized series of studies has focused on interspecific competitive interactions between introduced geckos in Hawaii. *Hemidactylus frenatus* has adversely affected populations of the smaller *Lepidodactylus lugubris* throughout the Pacific, although the two species appear to coexist well in native habitats (Case et al. 1994) and on many lighted structures (G. Perry, unpublished observations). Case, Petren, and their colleagues (Petren et al. 1993, Petren and Case 1996) focused on aggression, avoidance behaviors, and food competition, and McCoid and Hensley (1993) suggested that predation is also an important element in those interactions. Most relevant in the current context is the work of Petren et al. (1993), who studied the effects of artificial lights on interactions in edificarian habitats. They found that the presence of artificial lights was an important factor in the interaction between *Hemidactylus frenatus* and *Lepidodactylus lugubris*, to the extent that "asymmetric competition occurred only in the presence of light, which attracts a dense concentration of food" (Petren et al. 1993:354). Odd species communities develop at times under these conditions, where lizard families that might not normally interact or compete are sharing this niche. We have observed this in Hawaii, where *Phelsuma laticauda*, *Hemidactylus frenatus*, and *Anolis carolinensis* were observed foraging together at the same light source.

Conclusion

Artificial lights are well documented to have negative consequences for sea turtles (see Chapter 7, this volume). Considerable information now exists to support the contention that artificial lighting affects the activity of some other reptiles, but the nature of the effects is species specific and hard to predict.

Although they are often characterized as either diurnal or nocturnal, few species fit either definition perfectly. For example, Vitt and Zani (1997) describe some diurnal activity in the normally nocturnal *Thecadactylus rapicauda* in Brazil, and we have observed similar behavior in that species in Costa Rica and commonly for the South Pacific *Gehyra oceanica*. In a complementary fashion, Hoogmoed and Avila-Pires (1989) documented six diurnal lizard species active at night, under the bright illumination of a full moon. Klauber (1939) earlier described a similar situation in North America, and Frankenberg (1978) showed that many species can shift their behavior time as a function of season. This flexibility may form a part of a coevolved suite of traits (a "behavioral syndrome"; Sih et al. 2004) that allows nocturnal species, especially edificarian geckos, to take advantage of unusual lighting opportunities to extend their foraging periods. The addition of artificial lights simply allows this natural flexibility to come into play on a more regular basis in both nocturnal and diurnal species (Table 8.1). These behavioral changes appear relatively harmless. Inasmuch as the food intake of these species is increased, the overall effect even seems positive. Moreover, there is a potential benefit to humans in that geckos eat many mosquitoes and other undesirable invertebrates at lights (Canyon and Hii 1997).

What about negative effects? Our knowledge of those is spotty at best. Here we identify some possible negative effects, but these remain little more than speculation that requires testing. Having evolved with a light–dark cycle that varies in predictable ways, many organisms depend on light characteristics for synchronizing important physiological processes (e.g., Kumar 1997). At least some reptiles possess extraretinal light receptors (Underwood and Menaker 1970, Gianluca and Avery 1996) and may be able to respond physiologically to light even when their eyes are closed or removed. If reptiles perceive artificial lights as indistinguishable from natural illumination, then it is possible for extensive physiological effects to follow. A large body of work, very briefly touched on in this chapter, focuses on the effects of photoperiod in captive animals, but practically no research asks whether these findings are relevant to free-ranging individuals exposed to artificial lights. The ongoing use of night lights by geckos suggests that at least some species can take advantage of the extended photoperiod with no negative effects to them. Of possible concern would be nonedificarian taxa exposed to lights away from a direct light source. To what extent are species outside of towns affected by artificial night lighting? What are the effects on nonedificarian species that might exist in an urban matrix, such as box turtles in Lubbock, Texas, snakes in California, or alligators in Florida? We do not have the information to answer these important questions.

Another issue to consider, especially in the context of the geckos discussed in this chapter, is the negative effect of introduced species (reviewed in Pimentel 2002). Edificarian species often are excellent dispersers with humans, and geckos have become established in many localities. Work cited in this chapter shows that introduced species can negatively affect similar native taxa. It is also likely that the increased predation affects invertebrate populations, some of which may be of conservation or economic concern. For example, many pollinators are drawn to lights. If the conclusions of Canyon and Hii (1997) can be generalized, then geckos may already be having a large negative effect. Almost nothing is known about the effect of "benign" introductions on native invertebrates. To the extent that artificial lights provide a preferred habitat and enhance the ability of invaders to establish and proliferate, the overall effect must tentatively be considered negative. As with other introduced species, better biosanitary regulations are a desperately needed prophylactic measure for much of the world. Reduction in the use of lights, especially at ports of entry where initial establishments are likely to occur, might also reduce possible colonizations.

As should be clear from the paucity of detailed, specific data in this chapter, we know relatively little about the effects of artificial lights on reptiles other than sea turtles. Even for the latter, research has focused on the effects on emerging hatchlings and female nest site choice, and little is known about effects at other life stages. Thus much work remains to be done on the effects of lights on reptiles. We identify the following as the three questions of highest priority.

- Are there negative consequences of diffuse illumination for reptiles, especially in urban and suburban contexts?
- What, if any, are the effects of "benign" introductions, such as house geckos, on other vertebrate and invertebrate taxa?
- To what extent do artificial lights increase the ability of introduced species to establish and become invasive?

We are encouraged by the recent increased interest in urban ecology (e.g., Pickett et al. 2001) and hope that the ecological consequences of artificial night lighting, hitherto mostly ignored, will soon receive the attention they deserve. The data at hand for reptiles are sufficient to generate some alarming hypotheses, but careful tests are now needed.

Acknowledgments

We thank Amos Bouskila, Carlton Rochester, Russell Scarpino, Shane de Solla, Laurie Vitt, and especially Robert Henderson and Yehudah Werner for discussions of night light effects and suggestions of additional sources of information. Yael Bogin, Peter Cunningham, Robert Henderson, and Yehudah Werner kindly allowed us to cite unpublished observations. RNF thanks Ted Case for the many years of working in lighted urban settings, both in the Pacific and in southern California. This chapter is manuscript T-9-980 of the College of Agricultural Sciences and Natural Resources, Texas Tech University.

Literature Cited

Andreadis, P. T. 1997. A lunar rhythm in the foraging activity of northern water snakes (Reptilia: Colubridae). Pages 13–23 in A. F. Scott, S. W. Hamilton, E. W. Chester, and D. S. White (eds.), *Proceedings of the Seventh Symposium on the Natural History of Lower Tennessee and Cumberland River Valleys*. The Center for Field Biology, Austin Peay State University, Clarksville, Tennessee.

Auffenberg, W. 1988. *Gray's monitor lizard*. University of Florida Press, Gainesville.

Bertolucci, C., V. A. Sovrano, M. C. Magnone, and A. Foà. 2000. Role of suprachiasmatic nuclei in circadian and light-entrained behavioral rhythms of lizards. *American Journal of Physiology: Regulatory, Integrative and Comparative Physiology* 279:R2121–R2131.

Bouskila, A. 1989. Activity of snakes during different moon phases. Page 37 in T. Halliday, J. Baker, and L. Hosie (compilers), *Abstracts, First World Congress of Herpetology*. University of Kent at Canterbury, UK.

Bouskila, A. 1995. Interactions between predation risk and competition: a field study of kangaroo rats and snakes. *Ecology* 76:165–178.

Bouskila, A. 2001. A habitat selection game of interactions between rodents and their predators. *Annales Zoologici Fennici* 38:55–70.

Bouskila, A., D. Ehrlich, Y. Gershman, I. Lampl, U. Motro, E. Shani, U. Werner, and Y. L. Werner. 1992. Activity of a nocturnal lizard (*Stenodactylus doriae*) during a lunar eclipse at Hazeva (Israel). *Acta Zoologica Lilloana* 41:271–275.

Bowden, J., and B. M. Church. 1973. The influence of moonlight on catches of insects in light-traps in Africa. Part II. The effects of moon phase on light-trap catches. *Bulletin of Entomological Research* 63:129–142.

Bowersox, S. R., S. Calderón, G. Cisper, R. S. Garcia, C. Huntington, A. Lathrop, L. Lenart, J. S. Parmerlee Jr., R. Powell, A. Queral, D. D. Smith, S. P. Sowell, and K. C. Zippel. 1994. Miscellaneous natural history notes on amphibians and reptiles from the Dominican Republic. *Bulletin of the Chicago Herpetological Society* 29:54–55.

Brattstrom, B. H. 1953. Notes on a population of leaf-nosed snakes *Phyllorynchus decurtatus perkinsi*. *Herpetologica* 9:57–64.

Brattstrom, B. H., and R. C. Schwenkmeyer. 1951. Notes on the natural history of the worm snake, *Leptotyphlops humilis*. *Herpetologica* 7:193–196.

Buden, D. W. 2000. The reptiles of Sapwuahfik Atoll, Federated States of Micronesia. *Micronesica* 32:245–256.

Canyon, D. V., and J. L. K. Hii. 1997. The gecko: an environmentally friendly biological agent for mosquito control. *Medical and Veterinary Entomology* 11:319–323.

Caprette, C. 1997. *Visual perception in brown tree snakes and potential controls on dispersal*. Unpublished annual report, on file with Ohio Cooperative Fish and Wildlife Research Unit, Columbus.

Carmichael, P., and W. Williams. 1991. *Florida's fabulous reptiles and amphibians*. World Publications, Tampa, Florida.

Case, T. J., D. T. Bolger, and K. Petren. 1994. Invasions and competitive displacement among house geckos in the tropical Pacific. *Ecology* 75:464–477.

Case, T. J., and R. N. Fisher. 2001. Measuring and predicting species presence: coastal sage scrub case study. Pages 47–71 in C. T. Hunsaker, M. F. Goodchild, M. A. Friedl, and T. J. Case (eds.), *Spatial uncertainty in ecology: implications for remote sensing and GIS applications*. Springer-Verlag, New York.

Cinzano, P., F. Falchi, and C. D. Elvidge. 2001. The first world atlas of the artificial night sky brightness. *Monthly Notices of the Royal Astronomical Society* 328:689–707.

Clarke, J. A., J. T. Chopko, and S. P. Mackessy. 1996. The effect of moonlight on activity patterns of adult and juvenile prairie rattlesnakes (*Crotalus viridis viridis*). *Journal of Herpetology* 30:192–197.

Corbett, K. (ed.). 1989. *The conservation of European reptiles and amphibians*. Christopher Helm, London.

Cunningham, P. L. 2002. Colubridae: *Lamprophis fuliginosus*, brown house-snake. Foraging. *African Herp News* 34:28–29.

Dawson, D. W. 1984. Light pollution and its measurement. Pages 30–53 in R. C. Wolpert and R. M. Genet (eds.), *Advances in photoelectric photometry*. Volume 2. Fairborn Observatory, Fairborn, Ohio.

Dixon, J. R., and P. Soini. 1975. The reptiles of the upper Amazon basin, Iquitos region, Peru. I. Lizards and amphisbaenians. *Milwaukee Public Museum Contributions in Biology and Geology* 4:1–58.

Firth, B. T., I. Belan, D. J. Kennaway, and R. W. Moyer. 1999. Thermocyclic entrainment of lizard blood plasma melatonin rhythms in constant and cyclic photic environments. *American Journal of Physiology: Regulatory, Integrative and Comparative Physiology* 277:R1620–R1626.

Fisher, R. N., and T. J. Case. 2000. Distribution of the herpetofauna of coastal southern California with reference to elevation effects. Pages 137–143 in J. E. Keeley, M. Baer-Keeley, and C. J. Fotheringham (eds.), *2nd interface between ecology and land development in California*. U.S. Geological Survey Open-File Report 00-62, Sacramento, California.

Fleishman, L. J., M. Bowman, D. Saunders, W. E. Miller, M. J. Rury, and E. R. Loew. 1997. The visual ecology of Puerto Rican anoline lizards: habitat light and spectral sensitivity. *Journal of Comparative Physiology A* 181:446–460.

Frankenberg, E. 1978. Interspecific and seasonal variation of daily activity times in gekkonid lizards (Reptilia, Lacertilia). *Journal of Herpetology* 12:505–519.

Frankenberg, E., and Y. L. Werner. 1979. Effect of lunar cycle on daily activity rhythm in a gekkonid lizard, *Ptyodactylus*. *Israel Journal of Zoology* 28:224–228.

Garber, S. D. 1978. Opportunistic feeding behavior of *Anolis cristatellus* (Iguanidae: Reptilia) in Puerto Rico. *Transactions of the Kansas Academy of Science* 81:79–80.

Gianluca, T., and R. A. Avery. 1996. Dermal photoreceptors regulate basking behavior in the lizard *Podarcis muralis*. *Physiology & Behavior* 59:195–198.

Gibbons, J. W., D. E. Scott, T. J. Ryan, K. A. Buhlmann, T. D. Tuberville, B. S. Metts, J. L. Greene, T. Mills, Y. Leiden, S. Poppy, and C. T. Winne. 2000. The global decline of reptiles, déjà vu amphibians. *BioScience* 50:653–666.

Hedges, S. B., and R. Thomas. 2001. At the lower size limit in amniote vertebrates: a new diminutive lizard from the West Indies. *Caribbean Journal of Science* 37:168–173.

Henderson, R. W. 2002. *Neotropical treeboas: natural history of the* Corallus hortulanus *complex*. Krieger Publishing Company, Malabar, Florida.

Henderson, R. W., and R. Powell. 2001. Responses by the West Indian herpetofauna to human-influenced resources. *Caribbean Journal of Science* 37:41–54.

Hoogmoed, M. S., and T. C. S. de Avila-Pires. 1989. Observations on the nocturnal activity of lizards in a marshy area in Serra do Navio, Brazil. *Tropical Zoology* 2:165–173.

Houston, D., and R. Shine. 1994. Movements and activity patterns of arafura filesnakes (Serpentes: Acrochordidae) in tropical Australia. *Herpetologica* 50:349–357.

Howard, K. G., J. S. Parmerlee Jr., and R. Powell. 2001. Natural history of the edificarian geckos *Hemidactylus mabouia*, *Thecadactylus rapicauda*, and *Sphaerodactylus sputator* on Anguilla. *Caribbean Journal of Science* 37:285–288.

Hyde, L. L., and H. Underwood. 2000. Effects of melatonin administration on the circadian activity rhythm of the lizard *Anolis carolinensis*. *Physiology & Behavior* 71:183–192.

Kaiser, H., and R. E. Diaz. 2001. *Thecadactylus rapicauda* (turnip-tail gecko). Behavior. *Herpetological Review* 32:259.

Klauber, L. M. 1939. Studies of reptile life in the arid southwest. Part I. Night collecting on the desert with ecological statistics. *Bulletins of the Zoological Society of San Diego* 14:7–64.

Klauber, L. M. 1956[1997]. *Rattlesnakes: their habits, life histories, and influence on mankind*. Volume 1. Second edition. University of California Press, Berkeley.

Kotler, B. P., Y. Ayal, and A. Subach. 1994. Effects of predatory risk and resource renewal on the timing of foraging activity in a gerbil community. *Oecologia* 100:391–396.

Kotler, B. P., J. S. Brown, and O. Hasson. 1991. Factors affecting gerbil foraging behavior and rates of owl predation. *Ecology* 72:2249–2260.

Kumar, V. 1997. Photoperiodism in higher vertebrates: an adaptive strategy in temporal environment. *Indian Journal of Experimental Biology* 35:427–437.

Lazell, J. 1995. Natural Necker. *The Conservation Agency Occasional Paper* 2:1–28.

Leal, M., and L. J. Fleishman. 2002. Evidence for habitat partitioning based on adaptation to environmental light in a pair of sympatric lizard species. *Proceedings of the Royal Society of London. Series B, Biological Sciences* 269:351–359.

Lockard, R. B., and D. H. Owings. 1974. Moon-related surface activity of bannertail (*Dipodomys spectabilis*) and Fresno (*D. nitratoides*) kangaroo rats. *Animal Behaviour* 22:262–273.

Lotz, A. In press. Influence of environmental factors on nocturnal activity of snakes in the Colorado desert. *Journal of Herpetology*.

Madsen, T., and M. Osterkamp. 1982. Notes on the biology of the fish-eating snake *Lycodonomorphus bicolor* in Lake Tanganyika. *Journal of Herpetology* 16:185–188.

McCoid, M. J., and R. A. Hensley. 1993. Shifts in activity patterns in lizards. *Herpetological Review* 24:87–88.

Mills, A. M. 1986. The influence of moonlight on the behavior of goatsuckers (Caprimulgidae). *Auk* 103:370–378.

Nelson, D. A. 1989. Gull predation on Cassin's auklet varies with the lunar cycle. *Auk* 106:495–497.

Pacheco, L. F. 1996. Effects of environmental variables on black caiman counts in Bolivia. *Wildlife Society Bulletin* 24:44–49.

Parker, H. W. 1939. Luminous organs in lizards. *Journal of the Linnean Society of London, Zoology* 40:658–660.

Perry, G., and D. W. Buden. 1999. Ecology, behavior and color variation of the green tree skink, *Lamprolepis smaragdina* (Lacertilia: Scincidae), in Micronesia. *Micronesica* 31:263–273.

Perry, G., and R. Dmi'el. 1995. Urbanization and sand dunes in Israel: direct and indirect effects. *Israel Journal of Zoology* 41:33–41.

Perry, G., and T. Garland Jr. 2002. Lizard home ranges revisited: effects of sex, body size, diet, habitat, and phylogeny. *Ecology* 83:1870–1885.

Perry, G., and J. Lazell. 2000. *Liophis portoricensis* anegadae. Night-light niche. *Herpetological Review* 31:247.

Petren, K., D. T. Bolger, and T. J. Case. 1993. Mechanisms in the competitive success of an invading sexual gecko over an asexual native. *Science* 259:354–358.

Petren, K., and T. J. Case. 1996. An experimental demonstration of exploitation competition in an ongoing invasion. *Ecology* 77:118–132.

Pickett, S. T. A., M. L. Cadenasso, J. M. Grove, C. H. Nilon, R. V. Pouyat,

W. C. Zipperer, and R. Costanza. 2001. Urban ecological systems: linking terrestrial ecological, physical, and socioeconomic components of metropolitan areas. *Annual Review of Ecology and Systematics* 32:127–157.

Pimentel, D. (ed.). 2002. *Biological invasions: economic and environmental costs of alien plant, animal, and microbe species*. CRC Press, Boca Raton, Florida.

Pooley, A. C., T. Hines, and J. Shield. 1989. Attacks on humans. Pages 172–187 in C. A. Ross (ed.), *Crocodiles and alligators*. Facts on File, New York.

Powell, R., and R. W. Henderson. 1992. *Anolis gingivinus*. Nocturnal activity. *Herpetological Review* 23:117.

Quesnel, V. C., T. Seifan, N. Werner, and Y. L. Werner. 2002. Field and captivity observations of the lizard *Gonatodes vittatus* (Gekkonomorpha: Sphaerodactylini) in Trinidad and Tobago. *Living World: Journal of the Trinidad and Tobago Field Naturalists' Club* 2002:8–18.

Rabatsky, A. M., and T. M. Farrell. 1996. The effects of age and light level on foraging posture and frequency of caudal luring in the rattlesnake, *Sistrurus miliarius barbouri*. *Journal of Herpetology* 30:558–561.

Reichmann, A. 1998. *The effect of predation and moonlight on the behavior and foraging mode of* Stenodactylus doriae. M.S. thesis, Ben Gurion University, Israel.

Rice, K. G., H. F. Percival, A. R. Woodward, and M. L. Jennings. 1999. Effects of egg and hatchling harvest on American alligators in Florida. *Journal of Wildlife Management* 63:1193–1200.

Rodda, G. H., G. Perry, R. J. Rondeau, and J. Lazell. 2001. The densest terrestrial vertebrate. *Journal of Tropical Ecology* 17:331–338.

Rodda, G. H., Y. Sawai, D. Chiszar, and H. Tanaka (eds.). 1999. *Problem snake management: the habu and the brown treesnake*. Cornell University Press, Ithaca, New York.

Rodríguez-Robles, J. A., C. J. Bell, and H. W. Greene. 1999. Food habits of the glossy snake, *Arizona elegans*, with comparisons to the diet of sympatric long-nosed snakes, *Rhinocheilus lecontei*. *Journal of Herpetology* 33:87–92.

Savage, J. M. 2002. *The amphibians and reptiles of Costa Rica: a herpetofauna between two continents, between two seas*. University of Chicago Press, Chicago.

Schwartz, A., and R. W. Henderson. 1991. *Amphibians and reptiles of the West Indies: descriptions, distributions, and natural history*. University of Florida Press, Gainesville.

Shivik, J. A., J. Bourassa, and S. N. Donnigan. 2000. Elicitation of brown treesnake predatory behavior using polymodal stimuli. *Journal of Wildlife Management* 64:969–975.

Sih, A., A. M. Bell, J. C. Johnson, and R. E. Ziemba. 2004. Behavioral syndromes: an integrative overview. *Quarterly Review of Biology* 79:241–277.

Skutelsky, O. 1996. Predation risk and state-dependent foraging in scorpions: effects of moonlight on foraging in the scorpion *Buthus occitanus*. *Animal Behaviour* 52:49–57.

Southwood, T. R. E. 1978. *Ecological methods: with particular reference to the study of insect populations*. Chapman and Hall, London.

Sullivan, B. K. 2000. Long-term shifts in snake populations: a California site revisited. *Biological Conservation* 94:321–325.

Tang, Y.-Z., L.-Z. Zhuang, and Z.-W. Wang. 2001. Advertisement calls and their relation to reproductive cycles in *Gekko gecko* (Reptilia, Lacertilia). *Copeia* 2001:248–253.

Tiebout, H. M. III, and R. A. Anderson. 2001. Mesocosm experiments on habitat choice by an endemic lizard: implications for timber management. *Journal of Herpetology* 35:173–185.

Tsairi, H., and A. Bouskila. 2004. Ambush site selection of a desert snake (*Echis coloratus*) at an oasis. *Herpetologica* 60:13–23.

Underwood, H., and M. Menaker. 1970. Extraretinal light perception: entrainment of the biological clock controlling lizard locomotor activity. *Science* 170:190–193.

U.S. Fish and Wildlife Service. 1998. Pacific pocket mouse (*Perognathus longimembris pacificus*) recovery plan. USFWS, Portland, Oregon.

Verheijen, F. J. 1958. The mechanisms of the trapping effect of artificial light sources upon animals. *Archives Néerlandaises de Zoologie* 13:1–107.

Vitt, L. J., and P. A. Zani. 1997. Ecology of the nocturnal lizard *Thecadactylus rapicauda* (Sauria: Gekkonidae) in the Amazon region. *Herpetologica* 53:165–179.

Werner, Y. L. 1966. *The reptiles of Israel* [in Hebrew]. The Hebrew University of Jerusalem, Israel.

Werner, Y. L. 1995. *A guide to the reptiles and amphibians of Israel* [in Hebrew]. Nature Reserve Authority, Jerusalem, Israel.

Werner, Y. L., and N. Werner. 2001. An iguanid lizard shamming a house gecko. *Herpetological Bulletin* 77:28–30.

Whitaker, P. B., and R. Shine. 1999. When, where and why do people encounter Australian brownsnakes (*Pseudonaja textilis*: Elapidae)? *Wildlife Research* 26:675–688.

Zhou X.-Q., Niu C.-J., Li Q.-F., and Ma H.-F. 1998. The effects of light intensity on daily food consumption and specific growth rate of the juvenile soft-shelled turtle, *Trionyx sinensis* [in Chinese with English abstract]. *Acta Zoologica Sinica* 44:157–161.

Chapter 9

Observed and Potential Effects of Artificial Night Lighting on Anuran Amphibians

Bryant W. Buchanan

Anuran amphibians (frogs and toads) are experiencing global declines in population size and diversity (Alford and Richards 1999, Stuart et al. 2004). Researchers studying declining amphibians have identified several anthropogenic factors that are likely to contribute to such declines, including habitat destruction and disruption; acid precipitation; ultraviolet radiation damage caused by ozone depletion; environmental toxicants such as pesticides, herbicides, and industrial waste; changes in predator, prey, parasite, or competitor abundance; and the introduction of nonindigenous predators, competitors, or parasites (Alford and Richards 1999, Stuart et al. 2004). As more data become available, more factors probably will be identified that are contributing to amphibian declines.

Light pollution, the introduction of artificial lighting into areas where it changes the illumination or spectral composition of natural lighting, may be one such factor. In recent years, the amount of artificial light entering amphibian habitats has increased radically in conjunction with increases in human population growth, industrialization, and urban and suburban sprawl (Cinzano et al. 2001). Almost the entire eastern United

States, most of Europe, and Japan are considered to have measured nocturnal sky brightnesses that qualify as polluted, as do most other industrialized regions of the world (Cinzano et al. 2001).

In this chapter, I argue that anurans may be particularly susceptible to adverse effects from light pollution. One should expect artificial night lighting to affect frogs as much as or more than other taxa for a number of reasons:

- Most species of frogs and toads are partly or completely nocturnal, putting them at risk for exposure to light pollution.
- Nocturnal frogs and toads are predators and prey of other nocturnal animals, and artificial night lighting may cause changes in prey or predator density or behavior.
- Frogs and toads are widely distributed in all but the driest deserts and in arctic and antarctic regions, exposing different species to diverse environmental conditions.
- As a group, anurans display a great diversity of reproductive modes, exposing different species within a given habitat to a wide range of environmental conditions.
- Anurans may be less mobile than many other animals because of their modes of locomotion and their dependence on sources of moisture, making them less able to compensate for changes in night lighting by moving.
- Many anuran species have complex life cycles (Wilbur 1980), with different life stages of the same species occupying different microhabitats and exposing individuals to diverse environmental conditions over their lifespans.

Although anecdotal reports of the effects of artificial lights are common in the literature on frog natural history (e.g., Goin 1958, Goin and Goin 1957, Wright and Wright 1949:118, 167, 169, 188, 314, 347), there have been few direct experimental studies of the effects of artificial night lighting on anurans. The few studies reported in the literature demonstrate that anurans are sensitive and responsive to artificial night lighting. More common are laboratory studies in which lighting was manipulated for reasons other than the study of light pollution. I do not attempt a complete review of research related to frog photobiology because the purpose of this chapter is to stimulate interest and further research on the effects of light pollution on anurans. With the notable exception of birds (see Chapters 4–6, this volume) and sea turtles (see Chapter 7, this volume), the general study of light pollution and its effects on wildlife is a relatively new discipline (e.g., Verheijen 1985, Frank 1988), and therefore most pertinent papers have been published with other theoretical purposes in mind and do not specifically address light pollution or artificial lighting, making it difficult to conduct comprehensive literature reviews on the subject. I present some key examples of ways in which the existing literature can be used to generate hypotheses and predictions and to draw conclusions concerning the effects of light pollution on natural populations of frogs.

For persons who are not familiar with basic frog biology, I provide a review of frog natural history for species that exhibit more typical natural histories. Readers should consider this chapter only a starting point for the investigation of the effects of artificial night lighting on taxa of special concern because I hesitate to generalize when it comes to a group as diverse as anurans. Broad generalizations are difficult, and the effects of artificial night lighting may differ for each taxon. Those conducting environmental impact analysis must investigate the details of the natural history of target species because each taxon may respond differently to particular environmental perturbations.

After introducing anuran biology, I discuss the relation between illumination and foraging patterns. Effects of artificial night lighting are considered in two sections, the first evaluating the effects of chronic increases in illumination and its effects and the second considering the effects of dynamic shifts in illumination. I then discuss the role of light spectrum in anuran responses to light and offer conclusions about the mitigation of the effects of artificial night lighting on anuran amphibians.

Natural History of Anurans

Duellman and Trueb (1994), McDiarmid and Altig (1999), and Stebbins and Cohen (1997) are excellent sources of information on the natural history of anurans, as are regional field guides (e.g., Conant and Collins 1998, Stebbins 1985, Wright and Wright 1949). About 5,000 species of frogs occupy terrestrial, aquatic, semiaquatic, fossorial, and arboreal habitats and are found on every major land mass in the world except Antarctica, Greenland, and Iceland. Most taxa have complex life cycles in which eggs develop into free-swimming larvae before metamorphosing and adopting the terrestrial, arboreal, or semiterrestrial adult form. Offspring of many species develop directly from egg to small adult, bypassing the typical tadpole stage. Most species, however, have aquatic eggs and larvae that metamorphose into adults in the absence of parental care. Almost all anuran species have external fertilization, with males and females cooperating to shed eggs and sperm in a process requiring several hours. Many species exhibit parental care of eggs, larvae, or both. Frogs are not usually social, but most may be found in dense seasonal breeding aggregations at oviposition sites such as ponds, where males advertise for mates, primarily acoustically. Males often compete with each other acoustically and physically for display sites or directly for females (Gerhardt and Huber 2002, Ryan 2001, Wells 1977). Visual information also may be used in mate choice (Hödl and Amézquita 2001). Females often choose mates based on the quality of their territories or displays (Gerhardt and Huber 2002, Ryan 2001). Oviposition sites are usually in shallow areas of temporary or permanent bodies of water. The creation of ditches and other artificial ponds during the construction of roads, businesses, and residences has created many breeding habitats that are exposed to artificial night lighting from automobiles, streetlights, and other lights.

Whereas many anuran larvae are herbivorous (scraping periphyton off of surfaces or filtering plankton from the water column), others are carnivorous or cannibalistic (McDiarmid and Altig 1999). After a larval period of one to many months, larvae metamorphose into juvenile versions of the adults. Adult anurans typically feed on small, moving invertebrates, but individuals of most species will eat almost any moving prey they can swallow, including other vertebrates. Some species specialize on particular prey such as ants or other species of frogs.

Most foraging in frogs is visually mediated. The few species of frogs studied so far maximize the capture of light necessary to form visual images at low illuminations by having large retinal surfaces with more photoreceptors and through spatial summation, with multiple photoreceptors stimulating a single neuron, and/or temporal summation, with photoreceptors collecting multiple photons before stimulating a neuron (Warrant 1999). Anurans are thought to have color vision (Chapman 1966, Hailman and Jaeger 1974), although it is unclear whether color vision functions at low nocturnal illuminations. The evolution of, and dependence on, excellent nocturnal visual sensitivity put frogs at risk of being affected by changes to the lighting of their habitats.

Most species of frogs have adaptations for accommodating seasonal changes in temperature, photoperiod, or the availability of water. Most are more active under warm, wet conditions than under cool, dry conditions. In seasonal areas such as at temperate latitudes or in seasonally dry tropical habitats, frogs must obtain and store sufficient energy and water under favorable (warm, wet) conditions to allow them to hibernate or estivate and survive suboptimal conditions (cold, dry). Thus, most species exhibit distinct seasonality with regard to feeding, activity, and reproduction. In temperate regions, such seasonality of behavior may be controlled hormonally (Herman 1992) and appears to be triggered by changes in photoperiod. Likewise, in most habitats anurans must acquire sufficient food and water during nocturnal activity to allow them to survive drier diurnal conditions until the next suitable nocturnal activity period. Changes in the normal daily cycle of illumination (i.e., the loss or disruption of natural daylength cues) therefore may affect many aspects of anuran biology. Although these are generalizations regarding a diverse group, they illustrate general trends that can be used to make predictions about the ways in which photopollution may affect frogs in illuminated habitats.

Nocturnal Illumination and Activity Patterns

Illumination and irradiance change over the course of a single night because of changes in solar position, lunar position and phase, and cloud cover (Figure 9.1). As the sun sets in the evening, illuminations plummet several orders of magnitude. Animals active at this time (Figure 9.1, e.g., 2100–2200 h) must possess mechanisms of dark adaptation if they are to use vision in foraging or other activities (Warrant 1999). Once direct solar input is eliminated, variability in nocturnal illumination is determined largely by the amount of sunlight reflected off the moon (moon phase and position) and the amount of cloud cover that is blocking, reflecting, and refracting the moonlight and starlight (Dusenbery 1992). For example, illuminations were higher before the setting of the moon (about 2400 h) than after moonset and cloud cover increased ambient illumination by refracting light into the forest (clouds were present until about 0100 h; Figure 9.1). Cloud cover diffuses light in a way that reduces the formation of shadows in the forest that are typical on clear nights with strong lunar input. Heavy cloud cover may also block starlight on moonless nights, lowering ambient illuminations below 10^{-6} lux, even lower than the levels shown in Figure 9.1. Animals that are active at low illuminations either have visual adaptations that allow them to collect sufficient light to form a visual image (Warrant 1999) or use other modalities in place of vision. At sunrise (0500–0600 h), illuminations increase by several orders of magnitude before stabilizing at diurnal levels (Figure 9.1).

Relatively few studies have investigated the activity patterns of anurans in nature in relation to illumination (Table 9.1), but those few have demonstrated that frogs are as diverse in timing and periodicity of activity as they are in most other aspects of their biology. Adult and larval frogs can be roughly categorized as being diurnal (active at daytime illuminations), crepuscular (active at intermediate illuminations associated with dusk and dawn), nocturnal (active at lower illuminations associated with

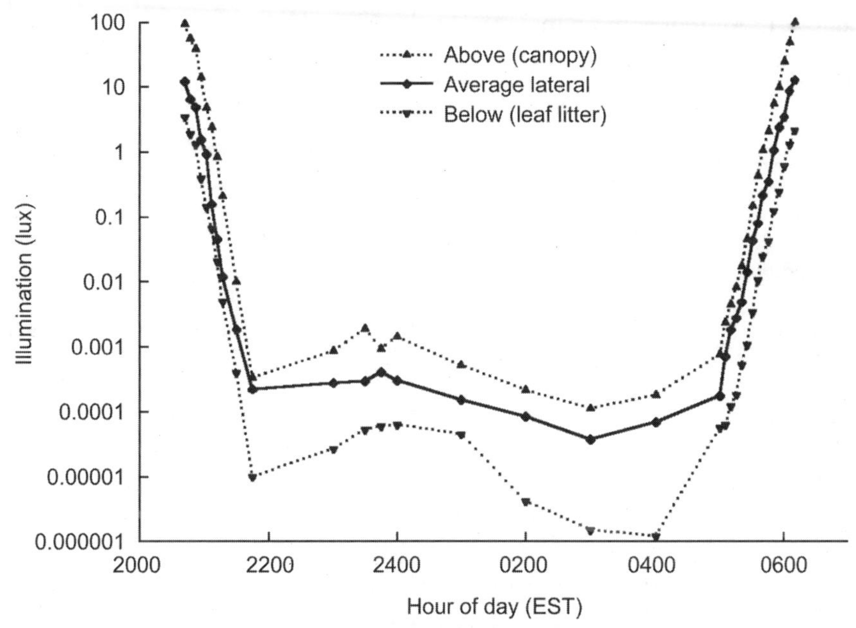

Figure 9.1. A representative illumination record for the night of June 28–29, 1993 at Mountain Lake Biological Station in Giles County, Virginia. Illuminations were recorded using an International Light IL1700 radiometer fitted with a silicon detector (SHD033) and a scotopic filter. The illuminations presented are averages ($n = 4$) of lateral illuminations taken at each time period in each cardinal compass direction at 20 cm (8 in) above the leaf litter under the forest canopy in a temperate, deciduous forest. Also presented are the minimum (toward leaf litter) and maximum (toward forest canopy) illuminations recorded at those positions.

night), or circumdiel (active at all illuminations). Most species of adult anurans are completely or partly nocturnal, seeking refugia during the day and conducting most of their activities such as foraging, finding mates, and avoiding predators under different levels of what we, as diurnal humans, would perceive as darkness. Activity of individuals within a species may also vary based on individual differences, sex, life stage, or season (Duellman and Trueb 1994, Higginbotham 1939). Obviously, nocturnally active anurans are at greatest risk of encountering and being affected by artificial night lighting, although artificial light may also affect diurnal and crepuscular frogs during their inactive phases unless they occupy light-safe refugia during those times.

In terms of changes that might be ecologically relevant to anuran amphibians, the most commonly altered characteristics of night lighting

Table 9.1. Anuran activity in natural habitats under diurnal, crepuscular, and nocturnal conditions.

Taxon	*Diurnal*	*Crepuscular*	*Nocturnal*	*References*
Ascaphus truei	No*	No	<0.0001 lux	Hailman 1982
Bufo boreas	No	No	0.1–0.00001 lux	Hailman 1984
Bufo marinus	No	20–60 lux	<0.01 lux	Jaeger and Hailman 1981
Bufo typhonius	8.5–110 lux	No	No	Jaeger and Hailman 1981
Colostethus nubicola	<50 lux	0.01–50 lux	No	Jaeger et al. 1976
Dendrobates auratus	<21 lux	1–21 lux	No	Jaeger and Hailman 1981
Eleutherodactylus coqui	No	No	Yes**	Woolbright 1985
Hyla squirella	No	No	<0.003 lux	Buchanan 1992
Leptodactylus pentadactylus	No	<1 lux	<0.01 lux	Jaeger and Hailman 1981
Physalaemus pustulosus	No	<150 lux	<0.01 lux	Jaeger and Hailman 1981

*No activity reported under these conditions.
**Activity reported but no illuminations provided.

are intensity (brightness) and spectral (color) characteristics. In turn, these characteristics can be altered further through aberrant spatial or temporal distributions of the light in anuran habitats. The following discussion of the potential effects of artificial night lighting on frogs is organized based on the type of change to the nocturnal lighting environment that frogs are likely to experience. Within each section, I begin by describing demonstrated effects of artificial lighting on frogs and end by making predictions concerning the effects that anurans may experience when subjected to such lighting changes.

Effects of Chronic Increases in Ambient Illumination

Chronic increases in nocturnal illumination are experienced wherever long-term sources of light are positioned in or near anuran habitat or where sky glow (light from distant sources such as cities, industrial complexes, or sports complexes reflected off moisture or particles in the atmosphere back to the ground) artificially increases illumination. In addition to sky glow, likely sources of this type of artificial lighting include convenience and security lighting near human habitations, industrial complexes and businesses with permanent outdoor lighting, sports complexes and stadiums, illuminated roadside signs, and especially roadway lighting, including streetlights and illuminated highway interchanges. In Utica Marsh (Oneida County, New York) near roadway lighting at a highway interchange, nocturnal, ambient illuminations as high as 1 lux were recorded in anuran breeding habitat (Buchanan and Wise, unpublished data). Because these lights are on all night, every night of the year, nocturnal illuminations in these areas of the marsh are consistently 100,000 to 1 million times greater than normal nocturnal ambient illuminations. With the recent encroachment of urban, industrial, and residential areas on historic amphibian habitat (Alford and Richards 1999), anurans are increasingly exposed to greater than predevelopment nocturnal illuminations. Chronic increases in illumination are likely to have a variety of effects. The following is a description of studies that illustrate the observed and potential effects of chronic increases to illumination in anuran habitats.

Aggregation at Lights and Phototaxis

Juvenile toads (*Bufo bufo*) tend to aggregate under streetlights, where they presumably capture insects attracted to the lights (Baker 1990). Although Baker (1990) did not include a statistical analysis in his article, I performed an analysis of his data and found that significantly more toads were found in brighter areas under streetlights than in same-sized, darker areas between streetlights, even with a sample size of only six lighted and six unlighted areas (median [interquartile range] for lighted, 10.5 toads [4–35.25]/10.8 m^2; and unlighted, 0.5 toads [0–1]/10.8 m^2 areas; Mann–Whitney U test, $U = 0$, $p = 0.004$, two-tailed, $\alpha = 0.05$). Baker (1990) concluded that aggregation under streetlights exposed juvenile toads to greater concentrations of prey (see Frank 1988; see also Chapters 12 and 13, this volume) but also put them at greater risk of being struck by automobiles (see Fahrig et al. 1995, Hels and Buchwald 2001).

Other frogs known to forage at permanent light sources include *Hyla squirella* and *Hyla cinerea* (Goin 1958, Goin and Goin 1957), *Bufo compactilis* (Wright and Wright 1949), *Bufo marinus* (Henderson and Powell 2001, Wright and Wright 1949), *Osteopilus septentrionalis* (Henderson and Powell 2001), and *Eleutherodactylus coqui* (Henderson and Powell 2001). Other references undoubtedly will be found as individuals search the natural history literature for descriptions of the effects of artificial lighting on nocturnal frog behavior.

Some of the earliest studies of the effect of lighting on frogs were laboratory studies of photopositive and photonegative behavior (e.g., Pearse

1910, Riley 1913). In an early study (without appropriate controls), Pearse (1910) provided limited evidence of positive phototaxis in *Rana clamitans* and *Rana sylvatica*. Riley (1913, in a pseudoreplicated study that also lacked appropriate controls) provided weak evidence that young toads (*Bufo americanus*) are positively phototactic to lights of intermediate brightness (about 1 lux) but respond photonegatively to very bright lights (about 1,000 lux). Later, in carefully controlled studies of more than 120 species (Jaeger and Hailman 1973, 1976a, 1976b), about 87% of the species tested were monotonically photopositive at the illuminations used in the study (0.043–89.9 lux). About 8% of the species tested were monotonically photonegative, avoiding artificial light in every treatment, apparently preferring illuminations below 0.043 lux. About 5% of the species tested exhibited modal preferences, that is, they preferred illuminations somewhere between the two extremes (0.043–89.9 lux). Thus, some species of frogs may avoid artificial lights whereas most others may be attracted to them, depending on the illuminance created by the light source and the ambient illumination at the position of the frog. If frogs consistently move toward permanent point sources of light, they could be pulled out of adjacent natural habitats and concentrated in areas around lights, a classic example of an evolutionary trap (Schlaepfer et al. 2002). Such conditions may interfere with normal interactions between frogs or with established predator–prey relationships or, as demonstrated by Baker (1990), may put the frogs at risk of being struck by automobiles.

Low-Illumination Foraging

Some nocturnal frogs have exceedingly sensitive nocturnal visual capabilities resulting from a combination of large retinal surfaces, large irises, tapeta lucida (intraocular reflective surfaces), and spatial and temporal summation, all methods of increasing the amount of light collected by the visual system (Warrant 1999). Species that have evolved under very low illuminations have adaptations that may constrain their abilities to compensate for increases in illumination or that may allow them to capitalize on slight increases in environmental illumination.

Common toads (*Bufo bufo*) are able to capture prey using vision at very low environmental illuminations (comparable to dim starlight, 10^{-6} to 10^{-5} lux; Aho et al. 1988, Larsen and Pedersen 1982). Squirrel tree frogs (*Hyla squirella*) also have excellent low-illumination visual capabilities, and some individuals attempt to capture prey using vision alone at illuminations at least as low as 10^{-5} lux (Buchanan 1998). As illuminations were increased, more tree frogs attempted to capture prey such that at illuminations above 10^{-3} lux, all frogs tested attempted to capture prey (Buchanan 1998). Slight increases in illumination caused by nearby lights, bright distant lights, or sky glow may be sufficient to allow foraging by frogs at times when visually mediated foraging would not normally be possible. Aggregations of frogs near lights could result from differential availability of prey attracted to the lights (Frank 1988; see Chapters 12 and 13, this volume), improved visual capabilities at higher illuminations, or a combination of the two effects.

Chronic illumination from point sources of light is likely to affect frogs not only because of the constant increase in ambient illumination but also because of the way the light enters the habitat. Only on clear nights when the moon is reflecting sunlight to the ground does the majority of light available to nocturnally foraging frogs come from a point source, namely the moon. When the moon is not prominent or when clouds obscure the moon, the light illuminating frog habitat comes from a variety of sources including starlight and moonlight refracted through the atmosphere, including any clouds. The more diffuse the light when it reaches the ground, the more even the illumination of objects in the habitat. Such diffuse or even lighting tends to eliminate shadows and make the illumination of various objects more uniform, reducing the need for radical dark or light adaptation as a frog moves through the environment.

Frogs foraging near bright point sources of light move constantly between brightly illuminated areas and shadows cast by objects such as vegetation between the light source and the frog. Near light sources, frogs must be able to dark adapt and light adapt quickly to changes in illumination, limit their motility by remaining in one area of constant illumination, or move slowly between areas of differing illumination to allow adaptation. Because dark adaptation takes longer after greater increases in illumination than after lesser increases in illumination (Cornell and Hailman 1984, Fain et al. 2001), frogs nearest point sources of light in complex environments with many shadows would need to spend more time dark adapting than frogs farther from light sources or in areas that are structurally simpler, potentially delaying foraging. Because frogs' visual fields in dim areas in the habitat are likely to contain areas of brighter illumination, it may be impossible for frogs' eyes to dark adapt completely to the lower illuminations in dim areas (Fain 1976, Fain et al. 2001), forcing the frogs to forage only in brightly illuminated areas. Such conditions are not completely novel to frogs because moonlight may create similar conditions on a regular, monthly basis. However, anthropogenic sources of light may be brighter and less diffuse than lunar lighting and therefore may create greater differences between maximum and minimum illuminations in a habitat at the position of a frog. Chronic exposure to anthropogenic sources of light could lead to long-term effects that may not result from intermittent exposure to moonlight. Such effects would not be limited to foraging but rather would be expected whenever frogs use vision in complex habitats near point sources of light, in situations such as visually mediated mate choice, territorial interactions, or detection or avoidance of predators.

Changes in Reproductive Behavior

Artificial lighting can alter the mate choice behavior of female frogs. Female Túngara frogs (*Physalaemus pustulosus*) were more likely to choose mates and were more discriminating of mates under darker (less than 0.01 μE) conditions than under brighter (0.04–0.05 μE) conditions (Rand et al. 1997) even when both conditions were much brighter than normal nocturnal illuminations. Although radiant flux ($\mu E/m^2/s$) cannot be converted to illuminance (lux) without knowledge of the power spectrum of the light source, the dimmer condition in that study may be estimated to have been approximately 0.7 lux, and the brighter condition was about 3 lux. The authors interpreted the change in female behavior under brighter conditions as a response to greater perceived threat of predation because females presumably would be more visible to visually oriented predators under the brighter conditions (Rand et al. 1997). Some nocturnal frogs use visual cues during mate choice (Taylor 2004) and male–male competition (for discussion of visual cues see Buchanan 1994, Hödl and Amézquita 2001). Increases in lighting may make frogs' intersexual or intrasexual displays more visible under higher illuminations. Permanent increases in illumination in mating areas probably affect mate choice decisions, interindividual spacing, antipredator behavior, or the relative reliance on different modalities (e.g., visual or auditory) in affected species.

Females of *Physalaemus pustulosus* altered their oviposition site choice in illuminated and completely dark trials (Tárano 1998). In that study, female frogs hid their foam nests in vegetation at higher light levels (0.003 lux from distant streetlight) but placed their nests randomly in total darkness (nesting area covered by opaque box). Unfortunately there was no treatment in this study that approximated natural illuminations, so the normal (relative to treatments) placement of nests is unknown. The

0.003-lux illumination provided by the streetlamp was similar to the brightest nocturnal illuminations encountered on nights with a full moon (Figure 9.1). Tárano (1998) suggested that nest site choice at different illuminations was a response to potential nest predation, with females hiding their nests at higher illuminations. It seems possible then that females might further alter their nest site choice at even higher artificial illuminations. Equally likely, however, is the idea that females normally hide their nests but do not do so under the abnormal, total darkness treatment because they are unable to see suitable nest sites.

Chronic illumination may also affect reproductive behavior of individuals, leading to chorus-wide effects. Although I have not quantified the effect, I have noted consistently that the tree frogs *Hyla squirella* (Louisiana, United States) and *Hyla leucophyllata* (Madre de Díos, Perú) are less likely to form or maintain chorus activity, particularly after human disturbance, on brightly moonlit nights or, in the case of *Hyla squirella*, when artificial lights illuminate their breeding areas. At brighter illuminations, frogs of both species seem to call from sites deeper in the vegetation and are more likely to terminate calling when disturbed. In *Hyla squirella*, frogs in an outdoor artificial breeding enclosure would not form stable breeding choruses on nights when sky glow created higher than normal illuminations unless a black plastic curtain surrounded the enclosure to block glare from distant light sources (Buchanan 1993b). I have, however, noted individuals of *Hyla squirella*, *Hyla cinerea*, and *Pseudacris crucifer* calling in permanently lighted areas, demonstrating that this effect is not universal.

Changes to Antipredator Behavior

As discussed above, female Túngara frogs (*Physalaemus pustulosus*) adjusted their nest site choice based on the amount of light available, possibly as an antipredator tactic (Tárano 1998). Likewise, Rand et al. (1997) suggested that the change in mate choice behavior of females in lighted conditions was a response to the risk of visually mediated predation. Although conducted under natural illuminations, da Silva Nunes's (1988) study of *Smilisca sila* demonstrates that male frogs adjust their calling behavior in response to illumination-dependent risk of predation. Males were more likely to call from exposed positions at higher illuminations where they may have been better able to see predatory frog-eating bats. Under darker conditions, males used more concealed positions from which to call, and they called less. Thus, in this situation, it may be the frogs' ability to see the predator and not the predator's ability to see the frog that affects frog behavior. Based on da Silva Nunes's study, one might predict more calling near artificial light sources in species such as *Smilisca sila*. These results illustrate the importance of knowing as much as possible about the biology of the species being studied. Although changes in lighting are likely to affect most nocturnal frogs, such changes are also likely to affect them in very different ways.

Effects on Tadpoles and Tadpole Behavior

Changes to the normal daily cycle of light and dark destroy normal rhythms of color change in tadpoles of *Xenopus laevis* (Binkley et al. 1988), probably through changes in melatonin production (Vanecek 1998). Tadpoles maintained in constant light (no photoperiod) never experienced normal melanophore function (Binkley et al. 1988). Melanophores become punctate at night and have dispersed melanosomes during the day. That work demonstrates that there is no intrinsic circadian rhythm that guides color change in these larvae; color change is simply affected by ambient illumination.

Toad tadpoles (*Bufo americanus*) use light cues to direct their daily movements within the pond (Beiswenger 1977). Under normal conditions, tadpoles were more likely to spend nights in deeper water that stayed warmer than in shallow water that cooled more quickly. During the day, tadpoles were found in shallower water that warmed more quickly. Tadpoles seemed to anticipate daily increases in shallow water temperature by using illumination cues. If toad tadpoles were to lose such natural lighting cues, they may lose the ability to thermoregulate and optimize their growth by adjusting their diel migrations in the pond in a manner that constantly keeps them in the warmest water throughout daily heating and cooling cycles. Young *Xenopus laevis* tadpoles use illumination cues to locate areas of shadow that may be important in attachment, predator avoidance, or thermoregulation (Jamieson and Roberts 2000). Reductions in illumination cause a pineal gland–dependent effect that results in larvae changing their swimming direction from horizontal to vertical, with movement toward the surface in dim light. This effect is induced by changes in ambient illumination in the absence of shadow; thus, it appears that absolute illumination may regulate the behavior. If such an effect is induced by some absolute level of illumination, vertical movement of larvae could be affected in brightly lit areas.

Disruption of Nocturnal Activity and Photoperiodic Behavior

Individuals of *Ascaphus truei* are active, presumably foraging, during the darkest periods of the night (about 10^{-5} lux; Hailman 1982). If these frogs emerge and forage only at exceedingly low illuminations, even slight increases in illumination in their stream environment will disrupt the timing of their emergence and subsequent foraging and activity. Likewise, other nocturnally active frogs (Table 9.1) could be affected by altered activity periods if the *Zeitgeber* that triggers their nightly emergence from refugia is not distinct or if their circadian clocks are improperly set. This becomes likely when artificial night lighting is bright enough to equal dim diurnal or twilight illuminations, as might occur under security or roadway lights, especially for forest-dwelling species that are seldom or never exposed to direct sunlight. Higginbotham (1939) studied circadian rhythms in *Bufo americanus* and *Bufo fowleri* and found wide variation in individuals' responses to illumination, whereas constant illumination was only minimally disruptive to normal activity patterns. His research suffered from small sample sizes, however, and so his conclusions are tentative. Research is needed to determine how much of a shift in illumination is needed to serve as a *Zeitgeber* to induce photoperiodic behaviors or reset the circadian clock.

Changes in the length of photophase, the light phase of the photoperiod (day), and scotophase, the dark phase (night), may affect the hormonal regulation of activity and reproduction. Bush (1963) found that artificially increasing the length of the photophase affected the foraging, growth, and induction of fat storage in *Bufo woodhousii*. Bush's design suffered from pseudoreplication, a common problem in designs of this type, because toads exposed to each lighting treatment were all exposed to the same lighting regime simultaneously (i.e., there were no independent replicates for treatments, but the data were then treated as if statistically independent). Biswas et al. (1978) presented intriguing data suggesting that *Bufo melanostictus* maintained under constant illumination suffer severe suppression of spermatogenesis coupled with increased activity by Leydig cells. Levels of sperm production were about twice as high in control toads maintained on a normal light cycle as in toads maintained in constant light. Assuming that no other factor affected spermatogenesis in the toads, this means that the lack of a photoperiod should dramatically reduce fertility in male toads living in constantly illuminated areas with no photoperiodic information. Because all toads in a treatment group were housed together and simultaneously exposed to the same treatment, this study also suffers from pseudoreplication.

Green and Besharse (1996) demonstrated that production of the polypeptide (gene product) *nocturnin* is affected by photoperiod in *Xenopus laevis*. *Nocturnin* is expressed at night, and its expression is reduced under constant illumination, although its expression retained a cyclical pattern regardless of lighting condition. Produced in the retina, *nocturnin* regulates clock component expression by degrading messenger RNA molecules in the cytoplasm of retinal cells; it is not yet known whether it targets specific messenger RNA molecules (Baggs and Green 2003). The clock develops during the first several days of embryonic development, and the amplitude of its expression, but not the underlying rhythm, depends on the existence of cyclical changes in illumination (Green et al. 1999). Rearing embryos under constant darkness while the clock was developing did not permanently alter clock function (Green et al. 1999). Steenhard and Besharse (2000) have demonstrated that *period* (another potential clock gene) is related to the circadian clock and may be a molecular link between photoperiod and the circadian clock in frogs. They found that *Per2* transcription can be phase shifted with light, probably resulting in lower melatonin production if frogs are not exposed to a distinct scotophase. Thus, although the clock itself is not altered by lack of photoperiod, clock-related gene expression is affected, resulting in altered hormone levels.

In *Rana perezi*, thyroid hormones cycle in relation to photoperiod (Gancedo et al. 1996), almost certainly as a result of melatonin cycles (Vanecek 1998). Melatonin has been implicated in the control of a variety of processes in frogs (Vanecek 1998), including color change, gonadal development, and reproduction in *Rana catesbeiana*, *Rana ridibunda*, *Xenopus laevis*, and *Rana cyanophlyctis* (Binkley et al. 1988, Camargo et al. 1999, Delgado et al. 1983, Joshi and Udaykumar 2000). Melatonin synthesis depends on the activity of serotonin *N*-acetyltransferase (NAT) with higher levels of NAT being related to greater melatonin production in *Xenopus laevis*, *Discoglossus pictus*, *Rana perezi*, and *Bufo calamita* (Alonso-Gómez et al. 1994). Similar effects have been demonstrated in *Rana esculenta* (d'Istria et al. 1994) and *Rana tigrina* (Lee et al. 1997). NAT synthesis, and thus melatonin synthesis, is greatest at night (Alonso-Gómez et al. 1994, d'Istria et al. 1994). Even one minute of exposure to light during scotophase can disrupt NAT activity, suppressing production of melatonin (Lee et al. 1997), although it is not yet clear what intensity of light is necessary to induce this effect. Disruption of melatonin is likely to have a variety of serious physiological consequences (Vanecek 1998).

Basinger and Matthes (1980) suggested that individuals of *Rana pipiens* maintained under constant illumination for 14 months experienced reversible damage to their eyes. They provided too little information to evaluate whether a suitable control condition existed; housing of frogs during the study also created pseudoreplication in their experimental design. According to the authors, frogs in constant light experienced swelling and gross reorganization of the cytoplasmic contents of the pigment epithelium cells supporting the retina, resulting in failure to shed rod cell outer segments in a normal fashion. Brief exposure to light and then darkness restored normal cell function. In this manner, constant exposure to artificial night lighting (i.e., no scotophase) disrupts normal retinal structure and recycling.

Effects on Development of Eggs and Larvae

Because larval development in species with unpigmented eggs is very sensitive to light (McDiarmid and Altig 1999), one might expect to see developmental aberrations in embryos developing in aquatic habitats that are brightly illuminated, if embryos are exposed to the greater illumination. However, it is important to remember that when oviposition occurs at night and embryos are developing under natural lighting, much of early larval development normally occurs in daylight the next day. It seems unlikely that exposure to bright illumination alone would greatly affect embryonic development unless critical gene expression is affected in the first few hours of (normally) nocturnal development or if periodic shifts in illumination are necessary for proper development. Artificial illumination of breeding habitats probably seldom approaches diurnal levels of illumination except directly under bright lamps. Artificial night lighting does, however, reach twilight levels (e.g., 1 lux measured at Utica Marsh, Oneida County, New York, near highway lighting; Buchanan and Wise, unpublished data), and therefore may change the duration of scotophase when twilight levels of illumination trigger photoperiodic events. Thus, even in species with pigmented eggs, development may be affected by changes in scotophase duration or the intensity of lighting during scotophase. For example, DNA and RNA synthesis in *Rana pipiens* are related to the timing of the onset and termination of photoperiod (Morgan and Mizell 1971a, 1971b). Because development is controlled by the correct timing of expression of particular genes and by the correct timing of cell division, disruption of underlying timing mechanisms has great potential to affect larval development.

Increasing photophase significantly retards the development of *Discoglossus pictus* larvae (Gutierrez et al. 1984). Larvae reared under longer scotophase (1L:23D, 6L:18D) developed faster (grew larger and reached further developmental stages) than larvae reared under shorter scotophase (24L:0D, 14L:10D). It is important to note, however, that larvae reared under constant light did not differ obviously in development from those reared under the more natural (14L:10D) lighting condition. Delgado et al. (1987) also found that increasing the length of photophase tends to retard larval *Xenopus laevis* development such that larvae grown in constant light (24L:0D) were smaller than larvae reared under a photoperiod with a longer scotophase (12L:12D). *Xenopus laevis* larvae developed more slowly and metamorphosed later when reared under a short scotophase (23L:1D) than under a long scotophase (1L:23D) (Edwards and Pivorun 1991). *Rana pipiens* larvae appeared to experience accelerated metamorphosis when exposed to continuous lighting rather than normal cyclical illumination, although no statistical analyses were performed (Eichler and Gray 1976). Because none of these studies included details concerning the irradiance or illumination (they all used fluorescent bulbs of various wattages) during photophase, it is unclear how much light is necessary to elicit such effects.

To predict the effects of artificial night lighting on larval development in nature, it will be necessary to determine how much of a shift in illumination is needed to act as a *Zeitgeber* or to affect development directly. It may be that artificial night lighting in natural habitats is insufficient to elicit these effects, that all species are affected at some threshold illumination, or that sensitivity to particular levels of artificial night lighting is species specific. It is clear that reducing the duration of scotophase has the potential to cause serious changes to frog physiology and development, but much more research is needed on this topic.

Effects of Dynamic Shifts in Illumination

When illumination changes over brief periods of time, frogs must adapt to the changes or suffer reduced visual capabilities. Dark adaptation and light adaptation occur when a frog moves from light to dark areas or from dark to light areas, respectively, or when the illumination where the frog resides changes in either of those ways (Fain et al. 2001, Fite 1976). Dark adaptation is accomplished largely through dilation of the pupils, exposing more photopigment to the lesser amount of light in the environment. When dark-adapted eyes are suddenly exposed to great increases in illumination, more light enters the eye than is needed to form an image because the pupil dilated before the rapid increase in illumination. The excessive amount of light entering the eye causes excess photopigment to be bleached (i.e., chemically altered by the light). Humans experience this effect when moving from a dark room into bright sunshine or when exposed to a bright light at night (Figure 9.2). Bleached photopigments cannot respond to light again until their original chemical structure is restored. In dark-adapted nocturnal frogs, returning the eyes to a dark-adapted state after photopigment bleaching caused by a brief, bright flash of light can require hours, depending on the intensity of the light relative to ambient illumination and the prestimulus adaptational state of the eye (Cornell and Hailman 1984, Donner and Reuter 1962).

Green rods, red rods, and cones in the retina all respond differently to different wavelengths of light and adapt to changes in illumination differ-

"See, Frank? Keep the light in their eyes and you can bag them without any trouble at all."

Figure 9.2. A rapid increase in illumination temporarily blinds anuran amphibians in the same manner it does humans.

ently (Besharse and Witkovsky 1992, Donner and Reuter 1962, Fain 1976), suggesting that color vision and monochromatic low-illumination vision may recover differently after photopigment bleaching. Dark adaptation also shifts light intensity preferences to lower values, potentially affecting the magnitude of phototactic movements in dark-adapted and light-adapted frogs (Hailman and Jaeger 1976, 1978, Hartman and Hailman 1981).

Rapid increases in illumination such as those created by researchers entering a breeding aggregation of frogs with a headlamp can negatively affect the frogs' visual capabilities (Buchanan 1993a). In that study, individuals of *Hyla chrysoscelis* were exposed to live prey at ambient illuminations of 0.001 lux (control), 3.8 lux (= dim headlamp), or 12 lux (= bright headlamp). Frogs required significantly longer to detect or to attempt to capture prey after rapid increases in illumination than under the control treatment when there was no increase in illumination. These results suggest that the frogs were temporarily "blinded" by rapid increases in illumination and needed a substantial poststimulus time for their eyes to adapt to the light so that they could identify and capture food. In nature, a delay in attempting to capture a moving prey item could mean a lost opportunity to forage. Although I originally performed that study as a means of demonstrating that researchers' methods of observing nocturnal frogs could bias observations of visually mediated behaviors such as visual mate choice, the results are directly applicable to the study of the effects of light pollution.

Rapid shifts in illumination are not entirely novel to frogs; some frogs may be exposed to brief flashes of light from lightning during thunderstorms. Frogs living along roadways or waterways with substantial automobile or boat traffic could be exposed to frequent rapid shifts in illumination and resultant photopigment bleaching caused by the moving lights and therefore may never achieve optimal light adaptation (Jaeger 1981). Such frogs probably would have difficulty capturing fast-moving prey or prey with low contrast relative to the background at low, nocturnal illuminations.

Frogs that breed in roadside ditches may be particularly susceptible to headlight-induced photopigment bleaching. Many night-breeding anurans tend to aggregate in choruses in the early night and therefore may be exposed to peak nocturnal traffic activity at that time. Frogs crossing roadways may be temporarily blinded by oncoming headlights through excessive photopigment bleaching in the dark-adapted state. If such frogs reduce their movements as a response to lost visual capabilities or disruption of phototaxis, they may suffer greater risk of mortality from being struck by vehicles (Hels and Buchwald 2001). Larvae that use rapid changes in light levels as an indicator of the position of shadows (and therefore floating objects) in the environment (Jamieson and Roberts 2000) may experience disruption to normal patterns of swimming or attachment, potentially exposing them to greater levels of predation.

Importance of Spectral Composition of Artificial Light Sources

Different sources of artificial light produce light of differing spectral composition depending on the type of reactive material or coatings on each lamp. For example, high pressure sodium vapor lamps produce peak outputs in the red, yellow, and green wavelengths; low pressure sodium vapor lamps produce light output in the yellow wavelengths; and mercury vapor lamps produce strong blue, yellow, and red emissions. Natural sunlight and moonlight contain energy at many wavelengths and thus appear white to humans with color vision (Dusenbery 1992). Most frogs studied have trichromatic color vision (Chapman 1966, Donner and Reuter 1962, 1976, Hailman and Jaeger 1974, Muntz 1962) and possibly tetrachromatic color vision with sensitivity in the ultraviolet wavelengths (Govardovskiĭ and Zueva 1974, Honkavaara et al. 2002), at least at illuminations that probably would seem abnormally bright for nocturnal species. At lower nocturnal illuminations, color vision probably disappears, although spectral preferences, particularly the "blue preference," may be retained (Donner and Reuter 1962, 1976).

Frogs typically exhibit strong spectral preferences that may result in phototaxis. Muntz (1962) suggested that the preference of frogs to move toward blue light (less than 500 nm) is an adaptive response that causes semiaquatic frogs to jump toward the pond rather than toward the shore when they detect a predator. Although this idea has never been tested experimentally, most frogs consistently exhibit the "blue preference," whether or not they are semiaquatic. Hailman and Jaeger (1974) suggested that the "blue-mode" spectral response might be a way to increase available illumination by avoiding heavily vegetated areas and moving into clearings. Regardless of the ultimate causation, most frogs presented with blue light at an intensity higher than ambient illumination are likely to exhibit phototaxis and move toward the light. A few species tested by Hailman and Jaeger (1974) exhibited a "U-shaped" spectral response curve (preferring light less than 475 nm and greater than 600 nm), suggesting that they are more likely to exhibit positive phototaxis to light that is violet or red (Jaeger and Hailman 1971). The spectral qualities of light and the intensity of the light source interact to determine the preference of the frog (Hartman and Hailman 1981). When dark-adapted, individuals of *Rana pipiens* seem to prefer blue, yellow, orange, and red light at lower illuminations over the same lights at higher illuminations. Dark-adapted frogs preferred violet, blue, and green wavelengths of light at intermediate illuminations over the same range of wavelengths at low illuminations. For light-adapted frogs, all preferences were reversed except for that of green light, which was unchanged by adaptational state. Dark-adapted frogs at low illuminations are more likely to be attracted to blue light and to light of longer wavelengths (yellow, orange, and red), whereas those same frogs are likely to be attracted to green or violet wavelengths at slightly higher illuminations. Thus, dark-adapted frogs may respond differently to lamps with different spectral outputs, preferring dim long-wavelength light (yellow, orange, red) over bright long-wavelength light. Light-adapted frogs already near lights may prefer brighter long-wavelength light over dimmer long-wavelength light. There appear to be strong interaction effects between intensity and wavelength preference; the difficulty of making predictions concerning the behavior of frogs moving through an artificially illuminated habitat should be apparent. Field studies of intensity-dependent and spectrally dependent phototaxis are needed.

Spectral information in the natural environment may be affected by artificial lighting that does not mimic the spectral distribution of lighting from natural sources or that enhances illumination sufficiently to enable color vision. For example, Buchanan (1994) suggested that visible colors or patterns (a sexual dimorphism in *Hyla squirella*) might be important in mate choice or male–male competition if color vision is possible at the nocturnal illuminations under which mate choice is occurring. In that species, males have labial and lateral stripes that are more yellow than those of females. Under monochromatic sources such as low pressure sodium vapor lights, the differences in coloration between males and females may not be obvious, disrupting behaviors that rely on this sexual dimorphism, should such behaviors exist. Visual cues are important in mate choice under nocturnal conditions in *Hyla squirella* (Taylor 2004), but it is not yet known whether coloration plays a role in mate choice. Alternatively, if color vision is possible only at higher than normal nocturnal illuminations, information based on coloration could become available to frogs that have not had such information available previously, potentially affecting intersexual and intrasexual interactions. Far too little research has been conducted on the nocturnal visual capabilities of anurans to predict all of the ways in which artificial night lighting may affect them.

Conclusion

Although an impressive body of literature demonstrates that changes in natural night lighting affect the biology of anuran amphibians, much more research is needed before detailed predictions can be made about the effects of artificial night lighting on particular anuran species. Specifically, research is needed on early life stages (embryos and larvae) because it is in these stages that a majority of natural mortality usually occurs. Studies of the effects of constant illumination on early larval development, metamorphosis, and larval community ecology would be particularly useful.

Many of the effects of artificial night lighting on frogs may be subtle or complex and not easily predictable. Changes in lighting may cause a cascade of effects in natural communities containing frogs if frog prey or predators change in abundance in response to changes in natural lighting. One might imagine a scenario in a constantly illuminated roadside ditch, for example, where constant light encourages greater than average algal growth, resulting in greater than average tadpole survivorship, size at metamorphosis, and juvenile recruitment. Insects, attracted to the streetlamps, may leave the forest and be killed at lights in great numbers, reducing the availability of food for the increased number of anuran offspring, resulting in extreme competition for food by frogs in the forest but perhaps less competition for food near the streetlamps, where food is abundant. Alternatively, constant illumination of the anuran breeding habitat may slow development in the larvae or may improperly regulate fat deposition and reproductive activity in adults, resulting in lowered recruitment and reproduction by the population. Long-term field studies and mesocosm experiments are needed to clarify the effects of artificial night lighting at the population and community levels.

Notwithstanding the need to conduct further research, nocturnal amphibians appear to be sensitive to changes in lighting, and it is therefore important to reduce the amount of artificial light entering their habitat whenever possible. The most obvious ways to limit anurans' exposure to artificial night lighting are to remove lights that are unnecessary, turn off lights except when necessary, reduce the amount of light that lamps produce, and direct light only where it is needed. Increases in illumination in amphibian habitats from sky glow can be reduced dramatically through the use of full cutoff luminaires that direct light output toward the ground, eliminating light broadcast into the sky. This also reduces the wattage necessary to illuminate the target area, because less light energy is wasted, and reduces the amount of light reflected into the sky. Direct glare can be reduced by removing unnecessary lights and using motion detectors that turn lights on only when necessary.

Another management option is to block light from entering amphibian habitats. Blocking light may be problematic, however, because solid, opaque surfaces that block light could also limit animal movements. But thick hedges of native plants between illuminated areas and critical anuran habitat may suffice to reduce glare into the habitat (although the light being blocked may negatively affect the plants and any animals that make use of the hedge, and the plantings would not reduce sky glow). Coupled with reduction in lamp brightness, use of full cutoff luminaires, and reduction of sky glow, hedge barriers of native plants should be investigated as a means to protect natural light regimes within amphibian habitats. Likewise, hedges or barriers could be effective at reducing the exposure of frogs to rapidly shifting illumination associated with vehicles moving along roadways at night.

Changing or narrowing the spectral output of lamps is not likely to ameliorate the effect of adding light to anurans' natural habitats because frogs and toads possess color vision and are affected differently by different wavelengths of light. It seems likely that the use of lights near amphibian habitats that best match normal, nocturnal spectra would have the least effect on anurans, although little is known about the effect of different wavelengths of light on frogs in nature.

Without further research, it is difficult to predict the specific outcome of exposure of frogs to artificial light based solely on the results of disparate laboratory and field studies, although available evidence suggests that artificial night lighting can have substantial effects. The effects of artificial night lighting on frogs can reasonably be categorized as either direct or indirect. Direct effects are likely to be easily observable and testable in laboratory or field experiments and may include reduced or enhanced foraging, developmental anomalies, and altered mate choice or hormone levels. Some direct effects may be measurable only over multiple frog generations and may include population declines or increases that are difficult to detect without long-term studies.

Indirect effects are more difficult to quantify and include both long-term and short-term consequences. Classic examples of indirect effects might be community-level effects such as increased primary production in lighted areas that drives greater tadpole survival, which in turn manifests as greater competition in the adult environment. Just as easily, artificial light could allow visually foraging predators to consume one species of frog preferentially, reducing their population size while releasing another species from competition. My point is not to weave hypothetical "just-so" stories but to emphasize the necessity of taking physiological, population, and community-level approaches to studying the effects of artificial night lighting.

Likewise, although conserving population sizes, the genetic structure of populations, and the diversity of communities are primary goals in conservation biology, it is also important to recognize and protect the evolutionary process. By altering anuran habitat through the addition of light, humans are changing the evolutionary trajectories of those affected species, causing them to adapt to new sets of conditions. Simply conserving species richness or population sizes does not conserve the evolutionary and behavioral diversity contained in those taxa.

It is clear that direct effects of artificial night lighting already exist and probably will be more commonly demonstrated as researchers develop interests in the effects of light pollution on wildlife. From a conservation standpoint, it would be prudent to adopt a precautionary approach and limit anurans' exposure to artificial night lighting whenever possible while biologists continue to investigate the effects of lighting on wetland communities.

Acknowledgments

I thank S. Wise, R. Jaeger, T. Longcore, and C. Rich for their comments on drafts of this manuscript. I particularly appreciate the efforts of C. Rich and T. Longcore for organizing the Ecological Consequences of Artificial Night Lighting conference in 2002, for arranging to have this book published, and for their exceptional editorial skills. Support was provided by The Urban Wildlands Group, the UCLA Institute of the Environment, and Utica College. Support and equipment were provided by Mountain Lake Biological Station both directly and through a Pratt Postdoctoral Fellowship.

Literature Cited

Aho, A.-C., K. Donner, C. Hydén, L. O. Larsen, and T. Reuter. 1988. Low retinal noise in animals with low body temperature allows high visual sensitivity. *Nature* 334:348–350.

Alford, R. A., and S. J. Richards. 1999. Global amphibian declines: a problem in applied ecology. *Annual Review of Ecology and Systematics* 30:133–165.

Alonso-Gómez, A. L., N. de Pedro, B. Gancedo, M. Alonso-Bedate, A. I. Valenciano, and M. J. Delgado. 1994. Ontogeny of ocular serotonin *N*-acetyltransferase activity daily rhythm in four anuran species. *General and Comparative Endocrinology* 94:357–365.

Baggs, J. E., and C. B. Green. 2003. Nocturnin, a deadenylase in *Xenopus laevis* retina: a mechanism for posttranscriptional control of circadian-related mRNA. *Current Biology* 13:189–198.

Baker, J. 1990. Toad aggregations under street lamps. *British Herpetological Society Bulletin* 31:26–27.

Basinger, S. F., and M. T. Matthes. 1980. The effect of long-term constant light on the frog pigment epithelium. *Vision Research* 20:1143–1149.

Beiswenger, R. E. 1977. Diel patterns of aggregative behavior in tadpoles of *Bufo americanus*, in relation to light and temperature. *Ecology* 58:98–108.

Besharse, J. C., and P. Witkovsky. 1992. Light-evoked contraction of red absorbing cones in the *Xenopus* retina is maximally sensitive to green light. *Visual Neuroscience* 8:243–249.

Binkley, S., K. Mosher, F. Rubin, and B. White. 1988. *Xenopus* tadpole melanophores are controlled by dark and light and melatonin without influence of time of day. *Journal of Pineal Research* 5:87–97.

Biswas, N. M., J. Chakraborty, S. Chanda, and S. Sanyal. 1978. Effect of continuous light and darkness on the testicular histology of toad (*Bufo melanostictus*). *Endocrinologia Japonica* 25:177–180.

Buchanan, B. W. 1992. Bimodal nocturnal activity pattern of *Hyla squirella*. *Journal of Herpetology* 26:521–522.

Buchanan, B. W. 1993a. Effects of enhanced lighting on the behaviour of nocturnal frogs. *Animal Behaviour* 45:893–899.

Buchanan, B. W. 1993b. *The influence of the competitive environment on the expression of satellite behavior in anuran amphibians*. Ph.D. dissertation, University of Southwestern Louisiana, Lafayette.

Buchanan, B. W. 1994. Sexual dimorphism in *Hyla squirella*: chromatic and pattern variation between the sexes. *Copeia* 1994:797–802.

Buchanan, B. W. 1998. Low-illumination prey detection by squirrel treefrogs. *Journal of Herpetology* 32:270–274.

Bush, F. M. 1963. Effects of light and temperature on the gross composition of the toad, *Bufo fowleri*. *Journal of Experimental Zoology* 153:1–13.

Camargo, C. R., M. A. Visconti, and A. M. L. Castrucci. 1999. Physiological color change in the bullfrog, *Rana catesbeiana*. *Journal of Experimental Zoology* 283:160–169.

Chapman, R. M. 1966. Light wavelength and energy preferences of the bullfrog: evidence for color vision. *Journal of Comparative and Physiological Psychology* 61:429–435.

Cinzano, P., F. Falchi, and C. D. Elvidge. 2001. The first world atlas of the artificial night sky brightness. *Monthly Notices of the Royal Astronomical Society* 328:689–707.

Conant, R., and J. T. Collins. 1998. *A field guide to reptiles and amphibians: eastern and central North America*. Third edition. Houghton Mifflin, Boston.

Cornell, E. A., and J. P. Hailman. 1984. Pupillary responses of two *Rana pipiens*-complex anuran species. *Herpetologica* 40:356–366.

da Silva Nunes, V. 1988. Vocalizations of treefrogs (*Smilisca sila*) in response to bat predation. *Herpetologica* 44:8–10.

Delgado, M. J., P. Gutiérrez, and M. Alonso-Bedate. 1983. Effects of daily melatonin injections on the photoperiodic gonadal response of the female frog *Rana ridibunda*. *Comparative Biochemistry and Physiology* 76A:389–392.

Delgado, M. J., P. Gutiérrez, and M. Alonso-Bedate. 1987. Melatonin and photoperiod alter growth and larval development in *Xenopus laevis* tadpoles. *Comparative Biochemistry and Physiology* 86A:417–421.

d'Istria, M., P. Monteleone, I. Serino, and G. Chieffi. 1994. Seasonal variations in the daily rhythm of melatonin and NAT activity in the Harderian gland, retina, pineal gland, and serum of the green frog, *Rana esculenta*. *General and Comparative Endocrinology* 96:6–11.

Donner, K. O., and T. Reuter. 1962. The spectral sensitivity and photopigment of the green rods in the frog's retina. *Vision Research* 2:357–372.

Donner, K. O., and T. Reuter. 1976. Visual pigments and photoreceptor function. Pages 251–277 in R. Llinás and W. Precht (eds.), *Frog neurobiology: a handbook*. Springer-Verlag, Berlin.

Duellman, W. E., and L. Trueb. 1994. *Biology of amphibians*. Johns Hopkins University Press, Baltimore.

Dusenbery, D. B. 1992. *Sensory ecology: how organisms acquire and respond to information*. W. H. Freeman and Company, New York.

Edwards, M. L. O., and E. B. Pivorun. 1991. The effects of photoperiod and different dosages of melatonin on metamorphic rate and weight gain in *Xenopus laevis* tadpoles. *General and Comparative Endocrinology* 81:28–38.

Eichler, V. B., and L. S. Gray Jr. 1976. The influence of environmental lighting on the growth and prometamorphic development of larval *Rana pipiens*. *Development Growth & Differentiation* 18:177–182.

Fahrig, L., J. H. Pedlar, S. E. Pope, P. D. Taylor, and J. F. Wegner. 1995. Effect of road traffic on amphibian density. *Biological Conservation* 73:177–182.

Fain, G. L. 1976. Sensitivity of toad rods: dependence on wave-length and background illumination. *Journal of Physiology* 261:71–101.

Fain, G. L., H. R. Matthews, M. C. Cornwall, and Y. Koutalos. 2001. Adaptation in vertebrate photoreceptors. *Physiological Reviews* 81:117–151.

Fite, K. V. (ed.). 1976. *The amphibian visual system: a multidisciplinary approach*. Academic Press, New York.

Frank, K. D. 1988. Impact of outdoor lighting on moths: an assessment. *Journal of the Lepidopterists' Society* 42:63–93.

Gancedo, B., A. L. Alonso-Gómez, N. de Pedro, M. J. Delgado, and M. Alonso-Bedate. 1996. Daily changes in thyroid activity in the frog *Rana perezi*: variation with season. *Comparative Biochemistry and Physiology* 114C:79–87.

Gerhardt, H. C., and F. Huber. 2002. *Acoustic communication in insects and anurans: common problems and diverse solutions*. University of Chicago Press, Chicago.

Goin, C. J., and O. B. Goin. 1957. Remarks on the behavior of the squirrel treefrog, *Hyla squirella*. *Annals of the Carnegie Museum* 35:27–36.

Goin, O. B. 1958. A comparison of the nonbreeding habits of two treefrogs, *Hyla squirella* and *Hyla cinerea*. *Quarterly Journal of the Florida Academy of Sciences* 21:49–60.

Govardovskiĭ, V. I., and L. V. Zueva. 1974. Spectral sensitivity of the frog eye in the ultraviolet and visible region. *Vision Research* 14:1317–1321.

Green, C. B., and J. C. Besharse. 1996. Identification of a novel vertebrate circadian clock–regulated gene encoding the protein nocturnin. *Proceedings of the National Academy of Sciences* 93:14884–14888.

Green, C. B., M.-Y. Liang, B. M. Steenhard, and J. C. Besharse. 1999. Ontogeny of circadian and light regulation of melatonin release in *Xenopus laevis* embryos. *Developmental Brain Research* 117:109–116.

Gutierrez, P., M. J. Delgado, and M. Alonso-Bedate. 1984. Influence of photoperiod and melatonin administration on growth and metamorphosis in *Discoglossus pictus* larvae. *Comparative Biochemistry and Physiology* 79A:255–260.

Hailman, J. P. 1982. Extremely low ambient light levels of *Ascaphus truei*. *Journal of Herpetology* 16:83–84.

Hailman, J. P. 1984. Bimodal nocturnal activity of the western toad (*Bufo boreas*) in relation to ambient illumination. *Copeia* 1984:283–290.

Hailman, J. P., and R. G. Jaeger. 1974. Phototactic responses to spectrally dominant stimuli and use of colour vision by adult anuran amphibians: a comparative survey. *Animal Behaviour* 22:757–795.

Hailman, J. P., and R. G. Jaeger. 1976. A model of phototaxis and its evaluation with anuran amphibians. *Behaviour* 56:215–249.

Hailman, J. P., and R. G. Jaeger. 1978. Phototactic responses of anuran amphibians to monochromatic stimuli of equal quantum intensity. *Animal Behaviour* 26:274–281.

Hartman, J. G., and J. P. Hailman. 1981. Interactions of light intensity, spectral dominance and adaptational state in controlling anuran phototaxis. *Zeitschrift für Tierpsychologie* 56:289–296.

Hels, T., and E. Buchwald. 2001. The effect of road kills on amphibian populations. *Biological Conservation* 99:331–340.

Henderson, R. W., and R. Powell. 2001. Responses by the West Indian herpetofauna to human-influenced resources. *Caribbean Journal of Science* 37:41–54.

Herman, C. A. 1992. Endocrinology. Pages 40–54 in M. E. Feder and W. W. Burggren (eds.), *Environmental physiology of the amphibians*. University of Chicago Press, Chicago.

Higginbotham, A. C. 1939. Studies on amphibian activity. I. Preliminary report on the rhythmic activity of *Bufo americanus* Holbrook, and *Bufo fowleri* Hinckley. *Ecology* 20:58–70.

Hödl, W., and A. Amézquita. 2001. Visual signaling in anuran amphibians. Pages 121–141 in M. J. Ryan (ed.), *Anuran communication*. Smithsonian Institution Press, Washington, D.C.

Honkavaara, J., M. Koivula, E. Korpimäki, H. Siitari, and J. Viitala. 2002. Ultraviolet vision and foraging in terrestrial vertebrates. *Oikos* 98:504–510.

Jaeger, R. G. 1981. Foraging in optimum light as a niche dimension for neotropical frogs. *National Geographic Society Research Reports* 13:297–302.

Jaeger, R. G., and J. P. Hailman. 1971. Two types of phototactic behaviour in anuran amphibians. *Nature* 230:189–190.

Jaeger, R. G., and J. P. Hailman. 1973. Effects of intensity on the phototactic responses of adult anuran amphibians: a comparative survey. *Zeitschrift für Tierpsychologie* 33:352–407.

Jaeger, R. G., and J. P. Hailman. 1976a. Ontogenetic shift of spectral phototactic preferences in anuran tadpoles. *Journal of Comparative and Physiological Psychology* 90:930–945.

Jaeger, R. G., and J. P. Hailman. 1976b. Phototaxis in anurans: relation between intensity and spectral preferences. *Copeia* 1976:92–98.

Jaeger, R. G., and J. P. Hailman. 1981. Activity of Neotropical frogs in relation to ambient light. *Biotropica* 13:59–65.

Jaeger, R. G., J. P. Hailman, and L. S. Jaeger. 1976. Bimodal diel activity of a Panamanian dendrobatid frog, *Colostethus nubicola*, in relation to light. *Herpetologica* 32:77–81.

Jamieson, D., and A. Roberts. 2000. Responses of young *Xenopus laevis* tadpoles to light dimming: possible roles for the pineal eye. *Journal of Experimental Biology* 203:1857–1867.

Joshi, B. N., and K. Udaykumar. 2000. Melatonin counteracts the stimulatory effects of blinding or exposure to red light on reproduction in the skipper frog *Rana cyanophlyctis*. *General and Comparative Endocrinology* 118:90–95.

Larsen, L. O., and J. N. Pedersen. 1982. The snapping response of the toad, *Bufo bufo*, towards prey dummies at very low light intensities. *Amphibia-Reptilia* 2:321–327.

Lee, J. H., C. F. Hung, C. C. Ho, S. H. Chang, Y. S. Lai, and J. G. Chung. 1997. Light-induced changes in frog pineal gland *N*-acetyltransferase activity. *Neurochemistry International* 31:533–540.

McDiarmid, R. W., and R. Altig (eds.). 1999. *Tadpoles: the biology of anuran larvae.* University of Chicago Press, Chicago.

Morgan, W. W., and S. Mizell. 1971a. Daily fluctuations of DNA synthesis in the corneas of *Rana pipiens*. *Comparative Biochemistry and Physiology* 40A:487–493.

Morgan, W. W., and S. Mizell. 1971b. Diurnal fluctuation in DNA content and DNA synthesis in the dorsal epidermis of *Rana pipiens*. *Comparative Biochemistry and Physiology* 38A:591–602.

Muntz, W. R. A. 1962. Effectiveness of different colors of light in releasing positive phototactic behavior of frogs, and a possible function of the retinal projection to the diencephalon. *Journal of Neurophysiology* 25:712–720.

Pearse, A. S. 1910. The reactions of amphibians to light. *Proceedings of the American Academy of Arts and Sciences* 45:159–208.

Rand, A. S., M. E. Bridarolli, L. Dries, and M. J. Ryan. 1997. Light levels influence female choice in Túngara frogs: predation risk assessment? *Copeia* 1997:447–450.

Riley, C. F. C. 1913. Responses of young toads to light and contact. *Journal of Animal Behavior* 3:179–214.

Ryan, M. J. (ed.). 2001. *Anuran communication*. Smithsonian Institution Press, Washington, D.C.

Schlaepfer, M. A., M. C. Runge, and P. W. Sherman. 2002. Ecological and evolutionary traps. *Trends in Ecology & Evolution* 17:474–480.

Stebbins, R. C. 1985. *A field guide to western reptiles and amphibians*. Second edition. Houghton Mifflin, Boston.

Stebbins, R. C., and N. W. Cohen. 1997. *A natural history of amphibians*. Princeton University Press, Princeton, New Jersey.

Steenhard, B. M., and J. C. Besharse. 2000. Phase shifting the retinal circadian clock: *xPer2* mRNA induction by light and dopamine. *Journal of Neuroscience* 20:8572–8577.

Stuart, S. N., J. S. Chanson, N. A. Cox, B. E. Young, A. S. L. Rodrigues, D. L. Fischman, and R. W. Waller. 2004. Status and trends of amphibian declines and extinctions worldwide. *Science* 306:1783–1786.

Tárano, Z. 1998. Cover and ambient light influence nesting preferences in the Túngara frog *Physalaemus pustulosus*. *Copeia* 1998:250–251.

Taylor, R. C. 2004. *Factors influencing female choice in the squirrel treefrog,* Hyla squirella. Ph.D. dissertation, University of Louisiana at Lafayette, Lafayette.

Vanecek, J. 1998. Cellular mechanisms of melatonin action. *Physiological Reviews* 78:687–721.

Verheijen, F. J. 1985. Photopollution: artificial light optic spatial control systems fail to cope with. Incidents, causations, remedies. *Experimental Biology* 44:1–18.

Warrant, E. J. 1999. Seeing better at night: life style, eye design and the optimum strategy of spatial and temporal summation. *Vision Research* 39:1611–1630.

Wells, K. D. 1977. The social behaviour of anuran amphibians. *Animal Behaviour* 25:666–693.

Wilbur, H. M. 1980. Complex life cycles. *Annual Review of Ecology and Systematics* 11:67–93.

Woolbright, L. L. 1985. Patterns of nocturnal movement and calling by the tropical frog *Eleutherodactylus coqui*. *Herpetologica* 41:1–9.

Wright, A. H., and A. A. Wright. 1949. *Handbook of frogs and toads of the United States and Canada*. Comstock Publishing Company, Ithaca, New York.

Chapter 10

Influence of Artificial Illumination on the Nocturnal Behavior and Physiology of Salamanders

Sharon E. Wise and Bryant W. Buchanan

Salamanders, like other amphibians, are particularly sensitive to environmental perturbations and are suffering global population declines (Alford and Richards 1999, Stuart et al. 2004). Salamanders often have complex life cycles including aquatic and terrestrial stages; such species are exposed to a range of habitats and potentially to multiple environmental stressors. As we document in this chapter, salamanders appear vulnerable to disruption of their environments by artificial night lighting.

Salamanders are important components of many forest and aquatic ecosystems. The greatest influence of salamanders on forest ecosystems may be as predators of invertebrates that are responsible for decomposition of leaf litter (Wyman 1998). They may have such a large effect on invertebrate populations because in North American forests the density of at least some populations of terrestrial plethodontid salamanders is extremely high (Jaeger 1979, Hairston 1987, Ovaska and Gregory 1989), with a collective biomass that is higher than that of other vertebrates

(Burton and Likens 1975b). They reach such a high biomass because they are extremely efficient at assimilating food, producing new tissue at a rate greater than or equal to that of other small vertebrates (Burton and Likens 1975a).

The life cycles of salamanders are complex and vary between species, although almost all species of salamanders lay eggs (Stebbins and Cohen 1997, Petranka 1998). Typically, the adults are terrestrial and return to the aquatic habitat when breeding, or they are semiterrestrial, living in or near the aquatic site. The adults lay eggs in or around fresh water, such as ponds, lakes, rivers, streams, or seeps. The larvae hatch from these eggs and remain in the aquatic environment from a few weeks to a few years (Petranka 1998). In most species, these larvae gradually metamorphose into the adult form. However, especially in harsh terrestrial environments, individuals of some species (such as *Notophthalmus* spp. and *Ambystoma* spp.) bypass the terrestrial phase (Duellman and Trueb 1994, Petranka 1998). The larvae instead transition into adults that retain some larval characteristics such as gills. In newts, *Notophthalmus viridescens*, for example, the life cycle can include four stages. Larvae hatch from eggs laid in the water. After approximately two to five months (Petranka 1998), the larvae metamorphose into terrestrial juveniles called red efts. The efts disperse throughout the terrestrial habitat for two to seven years, depending on the locality (Petranka 1998), then return to water as mature adults. Adults are aquatic but may leave ponds if the ponds dry or freeze. Some species of salamanders are completely aquatic, such that both larvae and adults remain in the aquatic habitat. In some of these completely aquatic species, the adults may not transform from larval to adult body forms. Other species of salamanders, such as some members of the Plethodontidae, are completely terrestrial. In these terrestrial species, there is direct development, with no free-living larval stage. Instead, eggs are laid on land, and the hatchlings are morphologically similar to adults. For a more comprehensive review of the life history of individual species, see Petranka (1998).

Nocturnal and crepuscular activity in amphibians is thought to be an adaptation to the problem of water loss associated with living in a terrestrial habitat (Shoemaker et al. 1992). Many terrestrial salamanders are active nocturnally above ground (Ralph 1957, Keen 1984), although some species are active diurnally in the leaf litter (e.g., Jaeger 1980b). Aquatic salamanders, such as *Notophthalmus viridescens* (Petranka 1998) and *Triturus vulgaris* (Griffiths 1985), often exhibit more diurnal or crepuscular activity. Such a trend in nocturnal activity is not necessarily related to habitat. For example, aquatic adults of *Cryptobranchus alleganiensis* (Petranka 1998) and aquatic larvae of *Triturus cristatus* (Dolmen 1983) are active mostly at night, illustrating that other life history characteristics and factors such as predation pressure and prey type can also influence diel activity.

Salamanders, perhaps even more than anurans, are sensitive to changes in the environment and therefore are important indicators of the health of some ecosystems (deMaynadier and Hunter 1998, Welsh and Droege 2001). Salamander populations, like anuran populations, have been declining worldwide (Blaustein and Wake 1990, Alford and Richards 1999, Stuart et al. 2004) as a result of environmental perturbations. Although no single factor seems to be the cause of these declines and some species are in decline for unknown reasons, several anthropogenic factors may be responsible for declines in at least some salamander populations, including ozone depletion and a consequential increase in UV-B radiation, global warming and climatic change, habitat loss and destruction, and acidification of terrestrial and aquatic habitats caused by acid deposition (reviewed by Petranka 1998, Alford and Richards 1999, Stuart et al. 2004).

Because many salamanders are nocturnally active or may have endogenous rhythms regulated by light, salamanders may also be sensitive to changes in the properties of ambient light resulting from artificial night lighting. This chapter summarizes research examining some aspects of the physiology and behavior of salamanders that may be influenced by artificial night lighting. Generally, the studies reviewed in this chapter have either experimentally altered the nocturnal environment of salamanders in the laboratory (e.g., by exposing salamanders to continuous light, increased ambient light, or monochromatic light) or examined the nocturnal activity under natural or artificial lighting in the field. Because of the breadth of this topic, this review of the literature is not intended to be complete but instead provides examples of potential ways in which artificially altering the nocturnal light environment may affect populations of salamanders. These alterations may include increased ambient illumination, increased or disrupted photoperiod, and changes in the spectral properties of ambient light. The potential effects of artificial night lighting on populations of salamanders have not been addressed widely in the literature; our current research is focused on the effects of artificial night lighting on individual-level activity and behavior of salamanders.

In this chapter we consider the effect of artificially increasing ambient illumination on salamander phototaxis, foraging, predator avoidance, and territorial behavior. We discuss the potential effects of modification of photoperiod on endocrine function, metabolic activity, and diel cycles. We then report on the effects of spectral composition of light on the orientation and movement of salamanders. Our conclusion considers the challenges of studying the effects of artificial night lighting on populations of salamanders.

Increases in Ambient Illumination

Modern human development has resulted in the introduction of artificial light to natural habitats during normally dark periods at night (Cinzano et al. 2001). The ambient illumination to which nocturnally active salamanders are exposed in natural habitats can be increased in two ways, through sky glow and point sources. Sky glow, resulting from anthropogenic light that is emitted from populated areas and reflected off particles in the atmosphere such as water vapor, dust, or smog, can increase ambient illumination over a large area. As a result, habitats close to highly populated areas may be affected by chronic increases in nocturnal illuminations. Direct glare from individual light sources (e.g., streetlamps, residential lighting, and sports field lighting), can create microhabitat variation such that in a smaller, more localized area there may be a large increase in illumination compared with that in nearby areas less affected by these light sources. Such variation results in brighter areas alternating with darker areas, which may negatively affect salamanders adapted to either light or dark conditions as they move between these lighted and unlighted areas. Increased illumination from sky glow and direct glare may influence populations of salamanders in various ways, resulting in benefits, costs, or both, depending on the species and situation. The following is a discussion of the potential effects of artificial night lighting on salamander behavior as a result of increased illumination from sky glow or point sources.

Phototaxis

Salamanders can be affected by artificial illumination through their fixed action patterns for orientation to light (phototropism), movement to light (positive phototaxis), or movement away from light (negative phototaxis). Controlled experiments without serious flaws in design demonstrating phototaxis by salamanders are rare, and for many nocturnally active salamanders, phototactic behavior is known only anecdotally. Based on published studies, most nocturnally active salamanders probably are negatively phototactic (Table 10.1), although this is an oversimplification. Few species tested seem to be monotonically photopositive or photonegative. Most of these studies did not test salamanders over a range of illuminations but generally compared behavior in extremely bright or extremely dark conditions (but see Test 1946, Wood 1951). Also, at least some salamanders have color vision (e.g., *Salamandra salamandra*; Przyrembel et al. 1995) such that the wavelength of ambient light may influence phototactic responses. Finally, some species of salamanders exhibit an ontogenetic switch in photic response, being positively phototactic as larvae and negatively phototactic as adults (e.g., *Ambystoma opacum* and *Eurycea bislineata* or *Eurycea cirrigera*; Table 10.1). Such an ontogenetic switch may be a response to change from an aquatic habitat for larvae to a more terrestrial habitat for adults. Marangio (1975) suggested that positive phototaxis may be beneficial to larvae living in cooler environments. Lighted areas of ponds may be slightly warmer, and warmer areas may increase the rate of growth in larvae. In *Ambystoma talpoideum*, faster growth increased fitness of individuals by reducing predation pressure from predators that can consume only small prey, decreasing larval growth period so that larvae emerged from temporary ponds sooner, and increasing body size and fecundity as adults (Semlitsch 1987a, Semlitsch et al. 1988). The switch to a photonegative response by terrestrial adults, at least to bright lights, may be a response to desiccation and predation pressures, which are signaled by increased light (Marangio 1975, Sugalski and Claussen 1997).

Negative phototaxis and nocturnal behavior in terrestrial salamanders may not necessarily be the result of an avoidance of light but may be an avoidance of harmful temperature and moisture levels that are more likely to occur in lighted areas than in unlighted areas (Heatwole 1962, Roth 1987, Schneider 1968). Indeed, using the term *nocturnal* to describe individual species of salamanders may not be accurate because many salamanders are active diurnally under the leaf litter (e.g., *Plethodon cinereus*; Fraser 1976, Jaeger 1980a, 1980b) or under aquatic conditions of low visibility (e.g., *Ambystoma tigrinum*; Rodda 1986). Many species of salamanders probably are active both during the day and at night under suitable environmental conditions unrelated to light, although some studies of photic responses controlled for temperature and humidity and still demonstrated phototactic responses to light (e.g., Sugalski and Claussen 1997).

Variation in illuminance from artificial light sources also had varying effects on the phototactic behavior of salamanders. Adults of *Necturus maculosus* exposed to white light of 1,875 or 8,000 lux moved away from

Table 10.1. Phototaxis in salamanders.

Species	Developmental Stage	Photic Response	References
Ambystomatidae			
Ambystoma macrodactylum croceum	Larval	Negative	Anderson 1972
	Adult	Negative	Anderson 1972
Ambystoma macrodactylum sigillatum	Larval	Positive	Anderson 1972
	Adult	Negative	Anderson 1972
Ambystoma maculatum	Larval	Positive or neutral	Schneider 1968, Schneider et al. 1991
	Adult	Negative	Pearse 1910
Ambystoma mexicanum	Larval	Neutral	Schneider et al. 1991
Ambystoma opacum	Larval	Positive[1]	Marangio 1975
	Adult	Negative	Marangio 1975
Ambystoma ordinarium	Larval	Positive	Anderson and Worthington 1971
	Adult	Positive	Anderson and Worthington 1971
Ambystoma tigrinum	Larval	Negative	Schneider et al. 1991
	Juvenile[2]	Negative	Ray 1967, 1970
Cryptobranchidae			
Cryptobranchus alleganiensis	Adult	Negative	Reese 1906, Pearse 1910
Plethodontidae			
Eurycea bislineata or *Eurycea cirrigera*[3]	Larval	Positive[4]	Wood 1951
	Adult	Negative	Wood 1951
Eurycea longicauda	Adult	Negative	Hutchison 1958
Eurycea lucifuga	Larval	Negative	Banta and McAtee 1906
	Adult	Neutral[5]	Hutchison 1958
Plethodon cinereus	Adult	Negative	Pearse 1910, Test 1946, Sugalski and Claussen 1997, Vernberg 1955
Plethodon glutinosus	Adult	Negative	Vernberg 1955
Proteidae			
Necturus maculosus	Adult	Negative	Reese 1906, Pearse 1910, Cole 1922
Proteus anguinus	Adult	Negative	Hawes 1945

(*continues*)

Table 10.1. Continued

Species	*Developmental Stage*	*Photic Response*	*References*
Salamandridae			
Notophthalmus viridescens	Adult	Negative[6]	Pearse 1910, Reese 1917
Taricha granulosa	Adult	Negative[7]	Kimeldorf and Fontanini 1974

[1]Large larvae are negatively phototactic (Marangio 1975).

[2]Not fully grown but terrestrial.

[3]Cannot be determined whether salamanders were *Eurycea bislineata* or *Eurycea cirrigera*.

[4]Wood (1951) found that larvae moved to areas of dim light more than bright or dark areas but were found more often in bright areas than dark areas. Additionally, larger (older) larvae moved to areas of dim light more than bright or dark areas but were found equally in light and dark areas (indicating a switch in preferences from lighter to darker areas, typical of adults of these species).

[5]Hutchison (1958) concluded that *Eurycea lucifuga* was only slightly negatively phototactic, but assuming independent events for each salamander observation recorded in the study, *Eurycea lucifuga* actually showed no statistically significant preference for lighted or dark areas ($n_{dark} = 31$, $n_{light} = 24$, $\chi^2 = 3.31$, $p > 0.05$, our analysis). Based on observational, not experimental, evidence, Banta and McAtee (1906) stated that *Eurycea lucifuga* was slightly negatively phototactic.

[6]Although negatively phototactic, adults orient to the light and thus are positively phototropic (Pearse 1910, Reese 1917).

[7]Salamanders avoided areas lighted with visible light (mercury vapor lamp or tungsten incandescent lamp with filter blocking < 420 nm light) or near-UV light (about 365 nm).

the brighter area sooner than they moved away from the dimmer area, although there was not a large difference in response (Cole 1922). Negatively phototropic salamanders (*Plethodon cinereus* and *Plethodon glutinosus*) also differed in their responses to varying intensities of light. Individuals of *Plethodon cinereus* showed no difference in reaction time to white light of 10.8 or 96.9 lux, whereas individuals of *Plethodon glutinosus* reacted almost twice as quickly to the brighter light than to the dimmer light (Vernberg 1955). Wood (1951) found that larval salamanders of *Eurycea bislineata* or *Eurycea cirrigera* preferred dim areas to bright or dark areas, whereas the adults preferred dark areas to dim or bright areas.

Studies of phototaxis generally have examined short-term responses to light, that is, salamanders were exposed to light at one end of a test chamber, and the position of the salamanders was recorded for minutes or up to several days after the initial exposure. Also, many of these studies were conducted during daylight hours (but see Test 1946, Sugalski and Claussen 1997). In the laboratory experiment of Sugalski and Claussen

(1997), phototactic behavior in individuals of *Plethodon cinereus* was observed during day and night. They found a negative phototactic response during two diurnal observations, but no light or dark preferences were exhibited in the nocturnal observation; that is, salamanders were found more often in dimmer areas (650 lux) than in lighter areas (1,400 lux) during photophase but not during scotophase, although both light levels were much brighter than what would normally be found under the forest canopy during diurnal periods (see Chapter 9, Figure 9.1, this volume). The photic response of salamanders to chronically lighted areas during nocturnal peak hours of activity is largely unknown and perhaps difficult to predict. Further studies are needed to examine preferences for dark or light areas during nocturnal periods.

Another confounding factor making it difficult to determine the general phototactic responses of salamanders is that most studies of phototaxis suffered from design flaws. In these studies, pseudoreplication was common; more than one salamander was exposed to the treatments simultaneously, or one animal was observed multiple times over the duration of the experiment and the behavior recorded as independent observations (Reese 1917, Test 1946, Wood 1951, Hutchison 1958). Also, researchers often did not include suitable controls or monitor environmental conditions that may have affected microhabitat choice, including temperature, which may vary with the addition of light sources that emit heat (Pearse 1910, Reese 1906, 1917, Schneider et al. 1991). Of the studies presented in Table 10.1, only Test (*Plethodon cinereus*; 1946), Wood (*Eurycea bislineata* or *Eurycea cirrigera*; 1951), and Sugalski and Claussen (*Plethodon cinereus*; 1997) reported temperatures in their test chambers. Schneider et al. (1991) attempted to control for potential variation in temperature by removing the eyes of some of the test salamanders so that some of the larvae tested in the chambers would be able to feel temperature differences but not visually detect light; the larvae without eyes did not avoid lighted areas even though temperatures may have been higher in these lighted conditions. Additionally, many of these researchers did not report light levels, and many of those who did used diurnal light levels when testing salamanders (e.g., 1,875 and 8,000 lux, Cole 1922; 650 and 1,400 lux, Sugalski and Claussen 1997; 10.8 and 96.9 lux, Vernberg 1955). Under brighter illuminations (those found during the day) salamanders were found to be photonegative; however, under darker illuminations (those found during the night), salamanders can be photopositive. The context-dependent responses of salamanders in the studies presented here make it difficult to predict how salamanders in their natural habitat, under nocturnal conditions, might respond to point sources of ambient light.

Foraging Behavior

Increases in nocturnal illumination resulting from artificial night lighting, including sky glow, continuous point sources (e.g., streetlamps), or intermittent sources (e.g., vehicle headlights), influence fundamentally important behaviors such as foraging. Artificial night lighting can benefit foraging success of salamanders that are not nocturnally photonegative by increasing the light available for visually guided foraging at night, when humidity and temperature are suitable, and by attracting photopositive insects, making prey more abundant in lighted areas. Conversely, artificial night lighting can hinder foraging by creating shadows in nocturnal habitats, such that prey in darker areas are less visible because the eyes of the salamander do not have time to adequately adapt between areas of light and dark, by making the foraging salamander more visible to predators, by driving photonegative insects away from lighted areas, and by pulling photopositive insects away from unlighted areas.

Rapid shifts in illumination can result from intermittent light sources such as lightning, vehicle headlights, or lights switching on or off after dark. The effects of such rapid shifts in illumination have not been widely studied, but our laboratory study (Buchanan and Wise, unpublished data) indicates that rapid shifts may influence foraging success under some conditions. Adults of *Plethodon cinereus* that were dark adapted to illuminations of 0.001 lux were exposed to either no shift in illumination (0.001 lux as a control) or immediate increases in illuminations to 3.8 or 12 lux. When chemical cues were available to the salamanders, the response to live prey was unaffected under rapidly increased illuminations compared with the control. However, when we removed chemical cues by placing live prey behind a clear plastic barrier, it took salamanders significantly less time to orient toward prey and to attempt to capture prey in the rapidly illuminated treatments. That is, rapid increases in illumination aided visual foraging in these salamanders. This result is different from what might be expected based on photopigment bleaching that should occur in the dark-adapted eyes of salamanders that are suddenly exposed to rapid increases in illumination (Fain et al. 2001, Lasansky and Marchiafava 1974). For example, using a similar experimental design, Buchanan (1993) found that foraging in gray treefrogs (*Hyla versicolor*) was significantly hindered, not improved as in the salamanders, when frogs were exposed to the same rapid increases in illumination to which the salamanders were exposed. Frogs may be more sensitive to light than most salamanders because of their relatively larger eye diameter and greater proportion of rods to cones (Duellman and Trueb 1994, Warrant 1999).

The effect of ambient illumination on foraging during nocturnal periods may depend on the intensity of the ambient light. Buchanan (unpublished data) observed the foraging behavior of *Plethodon cinereus* acclimated for at least 60 minutes to total darkness (no detectable light) or white light of varying intensities (10^{-5}–10^{-3} lux) but constant wavelength. With no olfactory cues, salamanders responded significantly sooner to visual cues of live prey with increasing light intensity. Thus, adding light facilitated foraging at normal, nocturnal illuminations. Placyk and Graves (2001) acclimated adults of *Plethodon cinereus* for one minute to light of varying intensity (illumination values not reported) and wavelength. Salamanders did not respond to visual cues of live prey differently between treatments, but the salamanders ate significantly fewer prey over a 15-minute period in the bright light treatment (100-W incandescent light bulb) than in the dim (7-W incandescent light bulb) or ultraviolet "black" light (4–400 nm) treatment. Because salamanders used by Placyk and Graves (2001) may not have had time to adapt to the bright or dim treatments, the results are difficult to interpret. Cordova, Buchanan, and Wise (unpublished data) examined the effect of nocturnal illumination on foraging and activity in *Plethodon cinereus*. Salamanders were habituated to 1 lux (photophase) and 10^{-3} lux (scotophase). After seven days, salamanders were moved to experimental chambers before the onset of scotophase, where they were exposed to ambient illuminations of approximately 10^{-3} lux (moonlight), 10^{-1} lux (twilight), or 1 lux (dim daylight under the forest canopy). Salamanders were significantly more active at 10^{-1} and 1 lux than at 10^{-3} lux, although there was no difference in number of prey eaten at all three light levels. This difference may be caused by an intrinsic increase in activity or by the absence of cover in the experimental chambers. Conversely, Placyk and Graves (2001) found more activity in bright light than in dim light, but the dim light treatment in their study may have been brighter, depending on reflectance of the background, than the 1-lux treatment that we used. Placyk and Graves (2001) suggested that in bright light salamanders might be actively searching for refugia, resulting in increased activity. It seems that under laboratory conditions, unusually high nocturnal ambient illuminations (such as those used in Placyk and Graves 2001) negatively affect foraging behavior, whereas less extreme illuminations (such as those used in our unpublished studies) enhance visually guided foraging.

Predator Avoidance

In the presence of visually guided predatory fishes, larval salamanders may switch from more diurnal to more nocturnal activity (Taylor 1983, Semlitsch 1987b, Stangel and Semlitsch 1987, Sih et al. 1992). For example, in the absence of fish, larval *Ambystoma maculatum* were equally active during day and night in the water column, where prey is abundant, but in the presence of the bluegill (*Lepomis macrochirus*), larvae became more active at night than during the day (Semlitsch 1987b). This decrease in feeding activity resulted in reduced body size for *Ambystoma maculatum*, possibly resulting from changes in the diet of larvae when confined to poorer foraging areas in the presence of fish.

Larvae that switch to more nocturnal activity in the presence of fish predators still may not avoid predation if artificial light allows fish to see at night. Additionally, increased light levels at night can increase the foraging periods of diurnal predators, forcing larvae to remain inactive for longer periods of time. Increased periods of inactivity or reduced foraging time in optimal habitats (e.g., open water) may result in decreased growth and survivorship. These predictions have yet to be tested.

Territorial Behavior

Territoriality, common in many populations of terrestrial salamanders, influences the distribution and abundance of individuals in a particular habitat (Hairston 1987). In some populations of *Plethodon cinereus*, adults use a variety of visual displays and chemical cues for marking and defending territories. Such displays convey information about body size and condition, willingness to fight, and submissiveness during agonistic encounters (Jaeger 1984, Mathis et al. 1995, Townsend and Jaeger 1998, Wise and Jaeger 1998).

Buchanan (unpublished data) examined the effect of increased nocturnal illumination on the territorial behavior in *Plethodon cinereus* under controlled laboratory conditions. Territorial residents were acclimated to nocturnal illuminations of less than 10^{-5} lux (total darkness, no detectable light), 10^{-4} lux (starlight), or 10^{-2} lux (moonlight). Then, intruding salamanders were introduced into the test chambers of the residents, and the behavior of residents was monitored under the three lighting conditions. Residents exhibited significantly more visual threat displays at higher light levels than at lower light levels. Although higher nocturnal illuminations increased the use of visual displays in the laboratory, the effect of such a change in agonistic behavior on the population dynamics of these territorial salamanders in natural habitats is still unknown.

Alteration of Photoperiod

Artificial night lighting not only may increase the illumination of the environment, as discussed in the previous section, but also may alter photoperiod by increasing perceived daylength. However, it is unknown how much artificial illumination is necessary to eliminate the *Zeitgeber* associated with normal, daily changes in illumination. To understand fully the implications of artificial night lighting for populations of salamanders, it will be necessary for researchers to examine the effect of varying light levels on fundamental physiological functions regulated by photoperiod such as hormone production and metabolic rate, as well as on behavioral responses such as locomotor activity and foraging behavior.

Several important hormones and fixed action patterns may have endogenous rhythms that are entrained by photoperiod. It has been hypothesized that the onset of scotophase is important in phase-setting circadian rhythms because of its consistency and predictability (Adler 1970, Gern et al. 1983). If this hypothesis is correct, then the addition of artificial night lighting, even if less intense than daylight, could mask the photic signal of onset of scotophase or increase the duration of photophase if the artificial light brightens the dark phase sufficiently. This section introduces literature examining the effect of photoperiod on endogenous rhythms and diel patterns associated with endocrine function, metabolic rates, activity, and behavior.

Endocrine Function

In vertebrates, the production of many hormones is modulated by photoperiod, resulting in cyclic patterns of plasma levels of hormones such as melatonin (Vanecek 1998) and prolactin (Freeman et al. 2000). Prolactin, an anterior pituitary hormone (although also produced outside the pituitary; Bole-Feysot et al. 1998), has an endogenous rhythm with light as the *Zeitgeber* (Freeman et al. 2000). Melatonin is a hormone produced by the retina (Quay 1965) and the pineal gland (Gern and Norris 1979, Gern et al. 1983). In salamanders (Gern and Norris 1979, Gern et al. 1983, Rawding and Hutchison 1992), as in other vertebrates that have been studied so far (Vanecek 1998), melatonin production is cyclical and modulated by photoperiod.

In larval amphibians such as frog tadpoles, pineal melatonin causes the aggregation of pigment granules, which contain melanin, in the skin, lightening the skin color (Vanecek 1998). Larvae of *Ambystoma tigrinum* reared under a natural photoperiod or under dark conditions in the laboratory or in a cave were lighter in color in the dark laboratory and cave treatments than in the natural photoperiod treatment (Banta 1912). Although melatonin levels were not measured, such differences in pigmentation probably resulted from differences in melatonin production. Melatonin may also inhibit mitosis, resulting in lower mitotic activity during scotophase, shown for onion root tips by Banerjee and Margulis (1973). Melatonin also seems to enhance the sensitivity of rods, as occurs during dark adaptation (Vanecek 1998). Although few studies have explored the influence of melatonin on the physiology of salamanders, especially terrestrial salamanders, melatonin functions in thermoregulation by lowering body temperature and reducing tolerance to high temperatures (Erskine and Hutchison 1982).

In amphibians, as in other vertebrates, melatonin production increases during the dark phase and decreases during daylight. Rawding and Hutchison (1992) examined the influence of photoperiod on plasma melatonin levels in aquatic adult *Necturus maculosus*. When salamanders were housed on a 12L:12D photoperiod, plasma melatonin levels were found to be higher during scotophase than photophase. When the photoperiod was reversed, plasma melatonin levels shifted with photoperiod, remaining higher during the shifted dark phase than during the shifted photophase. Aquatic adults of *Ambystoma tigrinum* also had significantly higher plasma melatonin levels during scotophase than during photophase (Gern and Norris 1979). Adults of *Ambystoma tigrinum* kept on a continuous light cycle had significantly lower melatonin levels during what would normally be scotophase than salamanders kept on a 12L:12D photoperiod, but melatonin levels during photophase did not change (Gern et al. 1983). Constant light probably influenced melatonin production, resulting in lower plasma melatonin during normal scotophase similar to levels at photophase.

Gern et al. (1983) did not state the illuminations used in their experiment under the light and dark phase of the photoperiod, although the same light source and illumination probably were used during day and night for the constant light treatment. If so, then we can predict that nocturnal light sources that are as strong as daytime sources may have a large effect on melatonin levels. When nocturnal illumination is less than that during the day, which probably would be the situation resulting from artificial light sources, the reduction in melatonin production during scotophase resulting from the addition of artificial light may be less than that seen in Gern et al. (1983). Although not tested, one prediction is that melatonin production under artificial night lighting would be lower than that under natural nocturnal ambient illuminations, and the reduction is influenced by the intensity of the light. Alternatively, if a particular shift in illumination acts as the *Zeitgeber*, then the magnitude of the absolute illumination may not affect plasma melatonin levels; that is, there would not be a graded response in plasma melatonin levels, as may be the case in mammals (e.g., rats; Dauchy et al. 1999). To determine the effect of artificial night lighting on melatonin production in salamanders, future studies must address the intensity of light necessary to inhibit melatonin production or the magnitude of the shift in illumination needed to stimulate melatonin production.

Prolactin and photoperiod have been hypothesized to act independently in stimulating limb regeneration in salamanders (Schauble 1972, Schauble and Tyler 1972, Maier and Singer 1981), although production of prolactin is regulated by photoperiod (Freeman et al. 2000). Maier and Singer (1977) removed forelimbs from aquatic adult newts (*Notophthalmus viridescens*) that were kept at constant temperature on a 15L:9D, continuous light, or continuous dark photoperiod. Limb regeneration was significantly faster in newts kept under continuous light, significantly slower in newts kept under continuous dark, and intermediate in newts kept at 15L:9D. When the pineal gland was covered with opaque paint, limb regeneration was slowed, even under constant light, indicating that the pineal gland was involved in this regulatory function (Maier and Singer 1982). Maier and Singer (1977) suggested that the increased regeneration rates in newts exposed to greater periods of light may have been the result of increased mitotic activity in the regenerating limbs; such increased mitotic activity may have been the result of decreased melatonin, which inhibits mitosis (Banerjee and Margulis 1973), or increased levels of prolactin, which increases epidermal mitotic activity (Hoffman and Dent 1977). In regenerating limbs of newts, the epithelium had an increased affinity for prolactin (Furlong et al. 1987). Additionally, amputated forearms regenerated faster in *Notophthalmus viridescens* given prolactin injections than in control newts (Schauble and Tyler 1972). However, regeneration was found to be faster only when newts were kept in continuous darkness; newts in continuous light with and without prolactin injections showed equal rates of limb regeneration (Maier and Singer 1981). Therefore, increased daylength may have interacted with melatonin and prolactin to increase mitotic activity, allowing for increased rates of regeneration with greater periods of light exposure. Although increasing daylength has a profound effect in the laboratory, it is difficult to predict the exact effects of artificial night lighting on limb or epithelial regeneration or the overall effect of such changes in prolactin levels on the growth and maintenance of individuals in a natural population.

Metabolic Rates

Metabolic rates are highly dependent on ambient temperature in salamanders (Gatten et al. 1992). Thermoregulation by behavioral modification (i.e., moving to areas with more optimal temperatures; Anderson and Graham 1967, Marangio 1975) and by physiological modification (Hutchison and Dupré 1992) are important regulators of metabolic rate in amphibians. Many salamanders found at higher latitudes exhibit seasonal variation in metabolic rates, with lower metabolic rates during winter (Pinder et al. 1992). Lower metabolic rates are associated with reduced locomotor speed and activity (Rome et al. 1992) and hibernation in some species (Pinder et al. 1992). However, other seasonal variables, such as reproductive condition (Finkler and Cullum 2002) and photoperiod (Whitford and Hutchison 1965), can also influence metabolic rates.

To test directly the influence of photoperiod on the metabolism of salamanders, Whitford and Hutchison (1965) measured metabolic rates in adults of *Ambystoma maculatum* at constant temperature (15° ± 1°C) after the salamanders were acclimated for 2–4 weeks on different photoperiods: 16L:8D, 8L:16D, or complete darkness. During photophase, salamanders were exposed to four 15-W fluorescent bulbs (although no illumination values were given), and during scotophase the salamanders were kept in complete darkness. After acclimation, pulmonary, cutaneous, and total oxygen consumption were significantly lower when salamanders were kept on the 8L:16D photoperiod than when kept on the 16L:8D photoperiod. A similar result was found for carbon dioxide production, although there was no significant difference in pulmonary carbon dioxide production in salamanders kept on 8L:16D and 16L:8D photoperiods. Salamanders kept in total darkness showed significantly higher pulmonary, cutaneous, and total rates of oxygen consumption and carbon dioxide production than salamanders kept on either an 8L:16D or 16L:8D photoperiod. In adults of *Desmognathus ochrophaeus*, temperature, but not photoperiod, affected oxygen consumption (Fitzpatrick 1973), but Fitzpatrick used a different species of salamander than Whitford and Hutchison (1965), kept salamanders under photoperiods (10L:14D or 14L:10D) that were not as disparate as those used by Whitford and Hutchison (1965), and did not feed salamanders before testing, as did Whitford and Hutchison (1965). Fasting may have lowered metabolic rates (see Gatten et al. 1992), potentially overriding any effects of photoperiod.

Several results from the experiment of Whitford and Hutchison (1965) are of particular interest in determining the effect of artificial night lighting on the metabolic rates of salamanders. First, metabolic rates increased with increased daylength. Because photoperiod and temperature, as well as respiratory rates, are predicted to vary seasonally, Whitford and Hutchison suggested that temperature and photoperiod might regulate metabolic rates synergistically. Increasing the photophase artificially through night lighting may disrupt the normal seasonal cycling of metabolic rates or increase metabolic rates during all seasons, which could change energetic demands and other physiological functions that occur before hibernation, when long periods without food may occur (Pinder et al. 1992). Whitford and Hutchison (1965) also found higher metabolic rates in dark-adapted salamanders exposed to complete darkness than in salamanders kept at 8L:16D or 16L:8D. At least three hypotheses could explain this result. First, salamanders kept artificially in complete darkness may have higher metabolic rates than salamanders kept on natural light cycles as a result of chronic stress. This hypothesis, however, seems unlikely because terrestrial salamanders often are fossorial and may be exposed naturally to periods of constant darkness with limited access to prey (Adler 1970, Fraser 1976). Second, metabolic rates may be higher naturally during dark periods, when salamanders are expected to be most active. In *Necturus maculosus*, metabolic rates were higher during scotophase than during photophase when these salamanders were subjected to photoperiods with a light and dark phase (12L:12D) at 5°C or 15°C (Miller and Hutchison 1979). Based on this hypothesis, introducing light during normal dark periods may actually reduce metabolic rates and thereby perhaps reduce activity that would be associated with higher metabolic activity. Third, higher metabolic rates of these dark-adapted salamanders may be the result of going from a dark environment immediately into a lighted room where metabolic rates were measured, causing a stress-induced increase in respiratory rate. Although this third hypothesis has not yet been tested, the implications from this hypothesis may apply to salamanders in habitats exposed to intermittent increases in illumination or sudden increases in illumination occurring after onset of scotophase (e.g., near roadsides, buildings with security lighting, or sports complexes).

Diel Activity

Many larval salamanders including *Ambystoma opacum*, *Ambystoma tigrinum*, *Ambystoma jeffersonianum* (Anderson and Graham 1967), and *Ambystoma talpoideum* (Stangel and Semlitsch 1987) show a diel pattern of vertical migration in ponds, although some species do not (e.g., *Ambystoma maculatum* and *Notophthalmus viridescens*; Anderson and Graham 1967). The timing of emergence from refugia and vertical migration seem to be influenced by ambient illumination (Anderson and Graham 1967), although other factors including temperature (Anderson and Graham 1967, Stangel and Semlitsch 1987), competition (Anderson and Graham 1967), and predation risk (Stangel and Semlitsch 1987) also may influence timing of vertical migration. Larvae exhibiting diel migrations tend to emerge from leaf litter and vegetation in the benthic layer after dark where they can forage on abundant, planktonic prey in the water column (Anderson and Graham 1967) and move to warmer areas that may increase body temperature (Hutchison and Dupré 1992). Such vertical migrations may speed growth and time of emergence from ponds (Semlitsch 1987b).

Dolmen (1983) examined activity of *Triturus vulgaris* and *Triturus cristatus* adults and larvae. In August and September, adults of *Triturus vulgaris* and *Triturus cristatus* exhibited mostly nocturnal activity, whereas the larvae of *Triturus cristatus* were diurnally active and larvae of *Triturus vulgaris* exhibited a somewhat irregular activity pattern. In October, the larvae shifted their activity, exhibiting mostly nocturnal activity. Such a shift apparently was induced when photophase became shorter than scotophase. Dolmen (1983) suggested that the phase shifts in larval *Triturus cristatus* were related to changes in length of time that the larvae were able to forage each day, such that larvae would be more successful at obtaining food if hunting occurred during the phase of longest duration (photophase in late summer and scotophase in late autumn).

Although not yet tested experimentally, disturbance of the normal photoperiod by the introduction of artificial light at night may influence vertical migrations. Anderson and Graham (1967:371) noted that light from a headlamp disturbed migrating larvae "if kept focused on the individual." They also noted a reduction in vertical migration on nights that were naturally "very bright." Such a reduction in vertical migration may reduce the size of metamorphosis or survival of larval salamanders (Semlitsch 1987b).

For many adult salamanders, it is widely accepted that forest floor or open water activity occurs mainly during scotophase (Conant and Collins

1998, Petranka 1998), although some salamanders are diurnal or crepuscular (e.g., *Notophthalmus viridescens* and *Triturus vulgaris*). Diel activity patterns have been demonstrated in several species of salamanders, including *Plethodon cinereus* (Buchanan and Wise, unpublished data, Taub 1961), *Plethodon glutinosus* or a species in the *Plethodon glutinosus* complex (Adler 1969), *Desmognathus fuscus* (Keen 1984), *Triturus vulgaris* (Dolmen 1983, Griffiths 1985), and *Triturus cristatus* (Dolmen 1983). These rhythms may be influenced by light (i.e., entrained by photoperiod) or other factors.

Adler (1969) examined the role of photoperiod on locomotor activity in terrestrial *Plethodon glutinosus* with eyes intact and without eyes (surgically removed, to test for extraoptic photoreceptors). Under constant temperature (15°C) and humidity, salamanders with and without intact eyes initiated activity around the onset of scotophase. When the photoperiod was shifted by one-hour increments, the salamanders shifted their activity to coincide with the onset of scotophase. The pattern of activity continued for salamanders kept one night in constant darkness but did not persist in individuals kept in constant darkness for several weeks, demonstrating photoperiod-regulated locomotor activity in these salamanders. Such locomotor rhythms may be related to surface activity in terrestrial salamanders, although such a relationship has not yet been demonstrated. In aquatic adults of *Notophthalmus viridescens*, locomotor activity was entrained with photoperiod when the eyes were removed but not when the eyes and the pineal gland were removed (Demian and Taylor 1977). The pineal gland therefore is responsible for modifying activity based on photoperiod in at least some species of salamanders.

The diel activity of terrestrial and aquatic adult *Triturus vulgaris* in both the laboratory and the field was influenced by photoperiod (Griffiths 1985). In the natural environment, newt activities including sexual behavior, horizontal movement, and rising to the surface to breathe were crepuscular, whereas feeding was more nocturnal. In the laboratory under constant temperature, aquatic and terrestrial adults exhibited a rhythmicity of locomotor activity that coincided with a 12L (400 lux):12D (0.02 lux) cycle; the aquatic adults were more active diurnally than the terrestrial adults, which were more crepuscular. With entrainment on a photoperiod, rhythmicity in activity continued even when newts were exposed to constant darkness, suggesting an endogenous rhythm with photoperiod as the *Zeitgeber*. This evidence suggests that photoperiod is important in entraining diel locomotor activity in salamanders, although this activity also may be regulated to some degree by temperature.

Ralph (1957) also examined diel activity in terrestrial *Plethodon cinereus* by monitoring the number of minutes salamanders moved per hour in a controlled laboratory setting. When individuals were kept on a 12L:12D cycle (light on at 0600 h, light off at 1800 h) for four days, individuals were more active at night than during the day; activity increased the most between 1700 h and 1800 h, one to two hours after dark. After this test period, salamanders were kept under continuous light (less than 10.8 lux) for 29 days. They continued to be most active from 1800 h to 0600 h under continuous light, although activity levels were lower during scotophase under continuous light than during scotophase under the 12L:12D photoperiod. Additionally, this activity pattern followed a lunar cycle, with activity suppressed at times of the lunar zenith (when the zenith occurred at night). Ralph (1957) suggested that under bright moonlight, activity in *Plethodon cinereus* was reduced as a result of this lunar influence. Although not addressed by Ralph (1957), these results indicate that photoperiod may modify surface activity, such that salamanders become more active sooner when light cues are available. Although untested, it is likely that increased nocturnal illumination from bright moonlight or artificial sources reduces activity in salamanders.

Artificial illumination during scotophase has been demonstrated to influence diel rhythms in the laboratory, but the effect of artificial illumination on diel activity of salamanders in the natural environment is not well known. Artificial night lighting may brighten the ambient environment during scotophase, modifying the endogenous rhythm and affecting nocturnal activity of salamanders. It is also possible that artificial illumination at night is not bright enough to simulate photophase, such that endogenous locomotor rhythms are not affected.

Using a vertical maze under controlled conditions, Taub (1961) determined that surface activity of adult *Plethodon cinereus* on the forest floor occurred only during night observations. Moisture and temperature only slightly affected this surface activity, supporting the hypothesis that time of day was the most important factor in determining surface activity. Besides surface activity, Taub (1961) did not find any pattern of vertical migration related to photoperiod, although she did not conduct censuses continually over day and night periods. Keen (1984) conducted all-night censuses of surface activity in *Desmognathus fuscus* in experimental enclosures that were kept dry (no rainfall directly into enclosures) or moist (rain allowed to fall in enclosures). Surface activity increased greatly around 2100 h (approximately 1–2 hours after dark), regardless of moisture conditions, but was modulated by moisture levels as well. Surface activity by salamanders dropped significantly between 2400 h and 0200 h in the dry enclosures but not until 0600–0700 h in moist enclosures. Buchanan and Wise (unpublished data) conducted all-night censuses of salamanders in the field. We found that the surface activity of *Plethodon cinereus* also was closely related to ambient illumination (more than to temperature), such that surface activity increased greatly 1–2 hours after sunset, and salamanders remained active for longer periods when humidity or moisture levels were higher. Thus, it seems that light levels determined the onset of surface activity. To test this hypothesis, other environmental conditions, especially temperature and humidity, would need to be controlled in the laboratory or factored into experimental designs in the field.

Even small increases in illumination, comparable to those caused by sky glow, may affect the surface activity of terrestrial salamanders. Buchanan and Wise (unpublished data) searched transects that were artificially illuminated (with white minilamps, 10^{-2} lux at the surface of the leaf litter) or were not artificially illuminated (as a control, 10^{-4} lux). We found that significantly more salamanders were active on the forest floor in the unlighted transects than in the lighted transects during the earliest sampling period (2200–2310 h, 1–2 hours after dark), but there was no difference in the number of salamanders active on the forest floor in the lighted and unlighted transects in a later sampling period (2320–0030 h). Increasing nocturnal light levels seemed to delay the time of emergence from leaf litter and refugia. Such a delay may be the result of a perceived shifting of photoperiod that stimulates increased locomotor activity (Adler 1969, Griffiths 1985). Delaying the time of emergence may reduce the amount of food these salamanders are able to capture on a given night. Because terrestrial plethodontids are limited in their foraging to periods when the forest floor is moist (Fraser 1976, Jaeger 1980a, 1980b, Keen 1984), artificial night lighting has the potential to shorten foraging periods, limit food intake, and depress rates of growth, reproduction, and survival during hibernation.

Spectral Variation in Ambient Light

The spectral properties of many artificial lights differ from those of the normal night sky, which has a fairly broad, evenly distributed spectrum of visible light, with somewhat more energy in the red wavelengths (Broadfoot and Kendall 1968, Massey et al. 1990). For example, primarily yellow light is emitted from both low pressure (589-nm) and high pressure

(540- to 630-nm) sodium vapor lamps (Massey et al. 1990, Cinzano et al. 2001). Mercury vapor lamps emit yellow light at 545–575 nm (Cinzano et al. 2001) and blue light at 405–436 nm (Massey et al. 1990). The spectral sensitivities of salamander eyes to particular wavelengths are based on the intensity of light striking the photoreceptors in the retina at each wavelength and the response of the visual system to these wavelengths (determined experimentally by action spectra; DeVoe et al. 1997); however, in some species the spectral sensitivity may change from larval to adult stages or from terrestrial to aquatic stages (Roth 1987). Additionally, salamanders may show spectral sensitivity that is similar to or different from that expressed in the eye, using extraocular photoreceptors probably located in the pineal organ (Deutschlander et al. 1999a, Phillips et al. 2001).

Phototaxis and Activity

Very few studies have addressed the effect of the spectral distribution of ambient light on the behavior of salamanders. Vernberg (1955) attempted to compare the phototactic response of *Plethodon cinereus* and *Plethodon glutinosus* to red (650–750 nm), green (450–550 nm), and blue (containing green and red wavelengths) light with their response to white light or darkness. Individuals preferred the dark side of the test chamber over the lighted side (10.8 or 96.9 lux), regardless of the color of the light. When salamanders were given a choice of white light (10.8 lux) and red (650–750 nm), green (450–550 nm), or blue (unknown wavelength) light of unknown illumination, individuals of *Plethodon glutinosus* preferred the side of the chamber with red, green, or blue light over the side with white light, whereas individuals of *Plethodon cinereus* showed a preference for the side with red or blue light but not green light. Generally, these salamanders were negatively phototactic and avoided white light over more monochromatic light. However, Vernberg (1955) did not provide values of light intensity for the monochromatic lamps, making it difficult to interpret the results of his experiments.

Placyk and Graves (2001) examined foraging behavior of individuals of *Plethodon cinereus* under three different light treatments including 7-W and 100-W incandescent bulbs and a black light (reported spectrum of 4–400 nm, qualifying as ultraviolet light). Salamanders exhibited more exploratory behavior when tested under the black light than under incandescent light. Salamanders tested with black light also differed in amount of activity and number of attacks on prey when tested with incandescent lights. However, the illumination and spectral properties of all three light treatments were not given, and salamanders were acclimated to light treatments for only one minute. To determine the overall effect of light on salamander behavior, researchers must be careful to measure both the illuminance (or irradiance when considering ultraviolet wavelengths) and spectral properties of lights used in experiments, to consider the spectral sensitivities of the photoreceptors in relation to the intensity of the wavelengths that are tested (i.e., action spectra), and to allow salamanders sufficient time for light or dark adaptation before testing, because all of these factors may influence salamander behavior.

Orientation and Homing

Adult red-spotted newts (*Notophthalmus viridescens*) are aquatic, living in ponds, rivers, and lakes. In some populations individuals migrate to and from ponds as a result of local biotic and environmental conditions. Adults leave ponds in winter to hibernate in terrestrial refugia, when water levels fall and ponds dry, when temperatures in ponds become too high, or when ponds are infested by ectoparasites such as leeches (Gill 1978). Newts show high fidelity to ponds, and adults will return to their home ponds even if displaced to other ponds (Gill 1979). The ability to home toward familiar ponds without gaining directional information during displacement (true navigation) has been demonstrated in adults of *Notophthalmus viridescens* (Phillips et al. 1995). Such map-based homing results from an ability to determine direction using a light-dependent magnetic compass (Phillips and Borland 1992a, 1992b, 1992c, Deutschlander et al. 1999b) and to gain spatial information at the displacement site, establishing geographic position (Phillips and Borland 1994). Polarized light detection has also been shown to be involved in spatial orientation in both *Notophthalmus viridescens* (Phillips et al. 2001) and *Ambystoma tigrinum* (Taylor and Adler 1973). The photoreceptors involved in these navigational systems and in spatial orientation and homing probably are extraocular, involving the pineal gland (Adler 1970, Deutschlander et al. 1999a, Phillips et al. 2001, Taylor 1972).

Compass orientation and homing are affected by the spectral properties of light. Phillips and Borland (1992a, 1992b, 1992c) trained newts under natural skylight to recognize shoreward orientation in a given direction (north, south, east, or west). In a test arena with an artificial magnetic field equal to that of Earth, newts were unable to orient toward shore (in the appropriate north, south, east, or west direction) in the absence of visible light (only infrared more than 700 nm available) but were able to orient correctly when tested under a full-spectrum 150-W xenon arc lamp (Phillips and Borland 1992b). The ability of newts to orient varied with wavelength of light; newts oriented correctly toward shore when the wavelength of the light was 400 or 450 nm (Phillips and Borland 1992a, 1992c) but oriented randomly when tested under 475-nm light and 90° counterclockwise to the direction of shore when tested under 500-, 550-, and 600-nm light (Phillips and Borland 1992a, 1992c). Using a design similar to that described for testing compass orientation, Phillips and Borland (1994) found that newts, transported to the lab in light-safe boxes with continuously changing magnetic fields to remove directional cues, were able orient toward their home ponds when tested under full-spectrum, 400-nm, or 450-nm light. When tested under 550-nm or 600-nm light, the newts were unable to orient toward their home ponds, and they oriented in random directions. Phillips and Borland (1992a, 1994) found similar results when newts were trained with long-wavelength light instead of natural, full-spectrum light. Phillips and Borland (1994) and Deutschlander et al. (1999b) proposed that photoreceptors with two antagonistic spectral absorption systems sensitive to short- and long-wavelength visible light modulated the magnetoreceptors, with the short-wavelength absorption system more sensitive than the long-wavelength absorption system.

Compass and home orientation in newts was disrupted by monochromatic, long-wavelength light (Phillips and Borland 1992a, 1992c, 1994). Although not yet tested in the field, artificial illumination may affect the ability of newts and other pond-breeding salamanders to navigate to home ponds. Many outdoor lights, including high and low pressure sodium vapor lamps, emit mostly long-wavelength light, and low pressure sodium vapor lamps are nearly monochromatic (Massey et al. 1990). Direct glare from point sources, especially from those using sodium vapor lamps, are more likely to have greater negative effects on orientation than sky glow.

Conclusion

Few studies have addressed the effect of artificial night lighting on salamander populations in the field or over long periods of time, but many laboratory studies have examined the short-term effects of artificial lighting, including increases in ambient illumination, increased daylength, and spectral quality of ambient illumination, on salamander behavior and physiology. Most of these studies demonstrated a change in behavior and physiological function in the presence of artificial night light compared with more natural light conditions, including changes in hormone production, metabolism, activity patterns, and homing ability. Unfortunately, in many of these studies researchers used light levels higher than would result from commonly used sources of exterior night lighting or did not report intensity or wavelength of light used in their studies. Such high levels of light and omissions in reporting light levels or wavelengths used in experiments limit the ability of others to interpret the results of experiments in relation to the effect of artificial light on salamanders in more natural habitats. Researchers are encouraged to report the intensity and spectral properties of light used in their experiments and to use light levels that would be experienced by animals in the field, from natural or artificial sources.

To understand fully the effects that light pollution may have on salamanders, more studies in both the laboratory and the field are needed, especially long-term studies of single populations. Laboratory studies should concentrate on determining the potential effect on behavior and physiological functions under controlled conditions at levels that would be found in salamander habitat. Field studies are needed to address the short- and long-term effects of artificial nocturnal illumination on salamander activity, population distribution, and population density. By combining well-designed laboratory and field studies, researchers will be able to determine more clearly the ecological consequences of artificial night lighting for salamander populations.

Even though continuing research is essential to increase knowledge of the effects of artificial night lighting on salamanders, the current literature, summarized in this chapter, documents various behaviors and physiological functions that are disrupted sufficiently by artificial lighting to conclude that artificial lighting is highly likely to have adverse consequences for salamanders. Endocrine function, foraging behavior, territoriality, and activity patterns have evolved under natural light photoperiods. Only recently, and in a very short time evolutionarily, these natural light cycles have been disrupted by artificial night lighting. Although species certainly can thrive in environments with some degree of change in nocturnal lighting (e.g., lightning, lunar phases), the permanent alteration of natural light cycles from artificial lighting or more frequent changes in illumination probably has adverse consequences for individual species and has the potential to change population and community structure. When undertaking projects in which the conservation of populations, species, or communities of salamanders is a concern, planners should include an assessment of artificial night lighting and assume that less artificial light at night is always better.

Acknowledgments

We thank V. H. Hutchison for critically reviewing this manuscript and providing suggestions for additional references. We thank C. Rich and T. Longcore for organizing the symposium on the Ecological Consequences of Artificial Night Lighting and for publishing this chapter. Additionally, we acknowledge the financial support of The Urban Wildlands Group and the UCLA Institute of the Environment to attend the symposium and Utica College to produce this chapter. We thank H. M. Wilbur, director of the Mountain Lake Biological Station, for financial support and permission to conduct some of the unpublished research at the station. Finally, we appreciate the work of J. Cordova, J. M. Frye, H. Gordon, and T. Clary in conducting some of the unpublished research presented in this chapter.

Literature Cited

Adler, K. 1969. Extraoptic phase shifting of circadian locomotor rhythm in salamanders. *Science* 164:1290–1292.

Adler, K. 1970. The role of extraoptic photoreceptors in amphibian rhythms and orientation: a review. *Journal of Herpetology* 4:99–112.

Alford, R. A., and S. J. Richards. 1999. Global amphibian declines: a problem in applied ecology. *Annual Review of Ecology and Systematics* 30:133–165.

Anderson, J. D. 1972. Phototactic behavior of larvae and adults of two subspecies of *Ambystoma macrodactylum*. *Herpetologica* 28:222–226.

Anderson, J. D., and R. E. Graham. 1967. Vertical migration and stratification of larval *Ambystoma*. *Copeia* 1967:371–374.

Anderson, J. D., and R. D. Worthington. 1971. The life history of the Mexican salamander *Ambystoma ordinarium* Taylor. *Herpetologica* 27:165–176.

Banerjee, S., and L. Margulis. 1973. Mitotic arrest by melatonin. *Experimental Cell Research* 78:314–318.

Banta, A. M. 1912. The influence of cave conditions upon pigment development in larvae of *Ambystoma tigrinum*. *American Naturalist* 46:244–248.

Banta, A. M., and W. L. McAtee. 1906. The life history of the cave salamander, *Spelerpes maculicaudus* (Cope). *Proceedings of the United States National Museum* 30:67–83.

Blaustein, A. R., and D. B. Wake. 1990. Declining amphibian populations: a global phenomenon? *Trends in Ecology and Evolution* 5:203–204.

Bole-Feysot, C., V. Goffin, M. Edery, N. Binart, and P. A. Kelly. 1998. Prolactin

(PRL) and its receptor: actions, signal transduction pathways and phenotypes observed in PRL receptor knockout mice. *Endocrine Reviews* 19:225–268.

Broadfoot, A. L., and K. R. Kendall. 1968. The airglow spectrum, 3100–10,000 A. *Journal of Geophysical Research, Space Physics* 73:426–428.

Buchanan, B. W. 1993. Effects of enhanced lighting on the behaviour of nocturnal frogs. *Animal Behaviour* 45:893–899.

Burton, T. M., and G. E. Likens. 1975a. Energy flow and nutrient cycling in salamander populations in the Hubbard Brook Experimental Forest, New Hampshire. *Ecology* 56:1068–1080.

Burton, T. M., and G. E. Likens. 1975b. Salamander populations and biomass in the Hubbard Brook Experimental Forest, New Hampshire. *Copeia* 1975:541–546.

Cinzano, P., F. Falchi, and C. D. Elvidge. 2001. The first world atlas of the artificial night sky brightness. *Monthly Notices of the Royal Astronomical Society* 328:689–707.

Cole, W. H. 1922. The effect of temperature on the phototropic response of *Necturus*. *Journal of General Physiology* 4:569–572.

Conant, R., and J. T. Collins. 1998. *A field guide to reptiles and amphibians: eastern and central North America*. Houghton Mifflin, Boston.

Dauchy, R. T., D. E. Blask, L. A. Sauer, G. C. Brainard, and J. A. Krause. 1999. Dim light during darkness stimulates tumor progression by enhancing tumor fatty acid uptake and metabolism. *Cancer Letters* 144:131–136.

deMaynadier, P. G., and M. L. Hunter Jr. 1998. Effects of silvicultural edges on the distribution and abundance of amphibians in Maine. *Conservation Biology* 12:340–352.

Demian, J. J., and D. H. Taylor. 1977. Photoreception and locomotor rhythm entrainment by the pineal body of the newt, *Notophthalmus viridescens* (Amphibia, Urodela, Salamandridae). *Journal of Herpetology* 11:131–139.

Deutschlander, M. E., S. C. Borland, and J. B. Phillips. 1999a. Extraocular magnetic compass in newts. *Nature* 400:324–325.

Deutschlander, M. E., J. B. Phillips, and S. C. Borland. 1999b. The case for light-dependent magnetic orientation in animals. *Journal of Experimental Biology* 202:891–908.

DeVoe, R. D., J. M. de Souza, and D. F. Ventura. 1997. Electrophysiological measurements of spectral sensitivities: a review. *Brazilian Journal of Medical and Biological Research* 30:169–177.

Dolmen, D. 1983. Diel rhythms and microhabitat preferences of the newts *Triturus vulgaris* and *T. cristatus* at the northern border of their distribution area. *Journal of Herpetology* 17:23–31.

Duellman, W. E., and L. Trueb. 1994. *Biology of amphibians*. Johns Hopkins University Press, Baltimore.

Erskine, D. J., and V. H. Hutchison. 1982. Reduced thermal tolerance in an amphibian treated with melatonin. *Journal of Thermal Biology* 7:121–123.

Fain, G. L., H. R. Matthews, M. C. Cornwall, and Y. Koutalos. 2001. Adaptation in vertebrate photoreceptors. *Physiological Reviews* 81:117–151.

Finkler, M. S., and K. A. Cullum. 2002. Sex-related differences in metabolic rate and energy reserves in spring-breeding small-mouthed salamanders (*Ambystoma texanum*). *Copeia* 2002:824–829.

Fitzpatrick, L. C. 1973. Energy allocation in the Allegheny Mountain salamander, *Desmognathus ochrophaeus*. *Ecological Monographs* 43:43–58.

Fraser, D. F. 1976. Empirical evaluation of the hypothesis of food competition in salamanders of the genus *Plethodon*. *Ecology* 57:459–471.

Freeman, M. E., B. Kanyicska, A. Lerant, and G. Nagy. 2000. Prolactin: structure, function, and regulation of secretion. *Physiological Reviews* 80:1523–1631.

Furlong, S. T., W. G. Chaney, M. K. Heideman, and S. C. Bromley. 1987. Increased prolactin binding and morphological changes in the wound epithelium of regenerating limbs of *Notophthalmus viridescens*. *Cell and Tissue Research* 249:411–419.

Gatten, R. E. Jr., K. Miller, and R. J. Full. 1992. Energetics at rest and during locomotion. Pages 314–377 in M. E. Feder and W. W. Burggren (eds.), *Environmental physiology of the amphibians*. University of Chicago Press, Chicago.

Gern, W. A., and D. O. Norris. 1979. Plasma melatonin in the neotenic tiger salamander (*Ambystoma tigrinum*): effects of photoperiod and pinealectomy. *General and Comparative Endocrinology* 38:393–398.

Gern, W. A., D. O. Norris, and D. Duvall. 1983. The effect of light and temperature on plasma melatonin in neotenic tiger salamanders (*Ambystoma tigrinum*). *Journal of Herpetology* 17:228–234.

Gill, D. E. 1978. The metapopulation ecology of the red-spotted newt, *Notophthalmus viridescens* (Rafinesque). *Ecological Monographs* 48:145–166.

Gill, D. E. 1979. Density dependence and homing behavior in adult red-spotted newts *Notophthalmus viridescens* (Rafinesque). *Ecology* 60:800–813.

Griffiths, R. A. 1985. Diel profile of behaviour in the smooth newt, *Triturus vulgaris* (L.): an analysis of environmental cues and endogenous timing. *Animal Behaviour* 33:573–582.

Hairston, N. G. Sr. 1987. *Community ecology and salamander guilds*. Cambridge University Press, Cambridge.

Hawes, R. S. 1945. On the eyes and reactions to light of *Proteus anguinus*. *Quarterly Journal of Microscopical Science* 86:1–54.

Heatwole, H. 1962. Environmental factors influencing local distribution and activity of the salamander, *Plethodon cinereus*. *Ecology* 43:460–472.

Hoffman, C. W., and J. N. Dent. 1977. Hormonal effects on mitotic rhythm in the epidermis of the red-spotted newt. *General and Comparative Endocrinology* 32:512–521.

Hutchison, V. H. 1958. The distribution and ecology of the cave salamander, *Eurycea lucifuga*. *Ecological Monographs* 28:1–20.

Hutchison, V. H., and R. K. Dupré. 1992. Thermoregulation. Pages 206–249 in M. E. Feder and W. W. Burggren (eds.), *Environmental physiology of the amphibians*. University of Chicago Press, Chicago.

Jaeger, R. G. 1979. Seasonal spatial distributions of the terrestrial salamander *Plethodon cinereus*. *Herpetologica* 35:90–93.

Jaeger, R. G. 1980a. Fluctuations in prey availability and food limitation for a terrestrial salamander. *Oecologia* 44:335–341.

Jaeger, R. G. 1980b. Microhabitats of a terrestrial forest salamander. *Copeia* 1980:265–268.

Jaeger, R. G. 1984. Agonistic behavior of the red-backed salamander. *Copeia* 1984:309–314.

Keen, W. H. 1984. Influence of moisture on the activity of a plethodontid salamander. *Copeia* 1984:684–688.

Kimeldorf, D. J., and D. F. Fontanini. 1974. Avoidance of near-ultraviolet radiation exposures by an amphibious vertebrate. *Environmental Physiology & Biochemistry* 4:40–44.

Lasansky, A., and P. L. Marchiafava. 1974. Light-induced resistance changes in retinal rods and cones of the tiger salamander. *Journal of Physiology* 236:171–191.

Maier, C. E., and M. Singer. 1977. The effect of light on forelimb regeneration in the newt. *Journal of Experimental Zoology* 202:241–244.

Maier, C. E., and M. Singer. 1981. The effect of prolactin on the rate of forelimb regeneration in newts exposed to photoperiod extremes. *Journal of Experimental Zoology* 216:395–397.

Maier, C. E., and M. Singer. 1982. The effect of limiting light to the pineal on the rate of forelimb regeneration in the newt. *Journal of Experimental Zoology* 219:111–114.

Marangio, M. S. 1975. Phototaxis in larvae and adults of the marbled salamander, *Ambystoma opacum*. *Journal of Herpetology* 9:293–297.

Massey, P., C. Gronwall, and C. A. Pilachowski. 1990. The spectrum of the Kitt Peak night sky. *Publications of the Astronomical Society of the Pacific* 102:1046–1051.

Mathis, A., R. G. Jaeger, W. H. Keen, P. K. Ducey, S. C. Walls, and B. W. Buchanan. 1995. Aggression and territoriality by salamanders and a comparison with the territorial behaviour of frogs. Pages 633–676 in H. Heatwole and B. K. Sullivan (eds.), *Amphibian biology*. Volume 2. *Social behaviour*. Surrey Beatty & Sons, Chipping Norton, New South Wales.

Miller, K., and V. H. Hutchison. 1979. Activity metabolism in the mudpuppy, *Necturus maculosus*. *Physiological Zoology* 52:22–37.

Ovaska, K., and P. T. Gregory. 1989. Population structure, growth, and reproduction in a Vancouver Island population of the salamander *Plethodon vehiculum*. *Herpetologica* 45:133–143.

Pearse, A. S. 1910. The reactions of amphibians to light. *Proceedings of the American Academy of Arts and Sciences* 45:159–208.

Petranka, J. W. 1998. *Salamanders of the United States and Canada*. Smithsonian Institution Press, Washington, D.C.

Phillips, J. B., K. Adler, and S. C. Borland. 1995. True navigation by an amphibian. *Animal Behaviour* 50:855–858.

Phillips, J. B., and S. C. Borland. 1992a. Behavioural evidence for use of a light-dependent magnetoreception mechanism by a vertebrate. *Nature* 359:142–144.

Phillips, J. B., and S. C. Borland. 1992b. Magnetic compass orientation is eliminated under near-infrared light in the eastern red-spotted newt *Notophthalmus viridescens*. *Animal Behaviour* 44:796–797.

Phillips, J. B., and S. C. Borland. 1992c. Wavelength specific effects of light on magnetic compass orientation of the eastern red-spotted newt *Notophthalmus viridescens*. *Ethology Ecology & Evolution* 4:33–42.

Phillips, J. B., and S. C. Borland. 1994. Use of a specialized magnetoreception system for homing by the eastern red-spotted newt *Notophthalmus viridescens*. *Journal of Experimental Biology* 188:275–291.

Phillips, J. B., M. E. Deutschlander, M. J. Freake, and S. C. Borland. 2001. The role of extraocular photoreceptors in newt magnetic compass orientation: parallels between light-dependent magnetoreception and polarized light detection in vertebrates. *Journal of Experimental Biology* 204:2543–2552.

Pinder, A. W., K. B. Storey, and G. R. Ultsch. 1992. Estivation and hibernation. Pages 250–274 in M. E. Feder and W. W. Burggren (eds.), *Environmental physiology of the amphibians*. University of Chicago Press, Chicago.

Placyk, J. S. Jr., and B. M. Graves. 2001. Foraging behavior of the red-backed salamander (*Plethodon cinereus*) under various lighting conditions. *Journal of Herpetology* 35:521–524.

Przyrembel, C., B. Keller, and C. Neumeyer. 1995. Trichromatic color vision in the salamander (*Salamandra salamandra*). *Journal of Comparative Physiology A* 176:575–586.

Quay, W. B. 1965. Retinal and pineal hydroxyindole-O-methyl transferase activity in vertebrates. *Life Sciences* 4:983–991.

Ralph, C. L. 1957. A diurnal activity rhythm in *Plethodon cinereus* and its modification by an influence having a lunar frequency. *Biological Bulletin* 113:188–197.

Rawding, R. S., and V. H. Hutchison. 1992. Influence of temperature and photoperiod on plasma melatonin in the mudpuppy, *Necturus maculosus*. *General and Comparative Endocrinology* 88:364–374.

Ray, A. J. Jr. 1967. Avoidance-learning in the tiger salamander. *American Journal of Psychology* 80:642–643.

Ray, A. J. Jr. 1970. Instrumental avoidance learning by the tiger salamander *Ambystoma tigrinum*. *Animal Behaviour* 18:73–77.

Reese, A. M. 1906. Observations on the reactions of *Cryptobranchus* and *Necturus* to light and heat. *Biological Bulletin* 11:93–99.

Reese, A. M. 1917. Light reactions of the crimson-spotted newt, *Diemyctylus viridescens*. *Journal of Animal Behavior* 7:29–48.

Rodda, G. H. 1986. A comparison of locomotor activity in neotenic and terrestrial salamanders (*Ambystoma tigrinum*). *Herpetologica* 42:483–491.

Rome, L. C., E. D. Stevens, and H. B. John-Alder. 1992. The influence of temperature and thermal acclimation on physiological function. Pages 183–205 in M. E. Feder and W. W. Burggren (eds.), *Environmental physiology of the amphibians*. University of Chicago Press, Chicago.

Roth, G. 1987. *Visual behavior in salamanders*. Springer-Verlag, Berlin.

Schauble, M. K. 1972. Seasonal variation of newt forelimb regeneration under controlled environmental conditions. *Journal of Experimental Zoology* 181:281–286.

Schauble, M. K., and D. B. Tyler. 1972. The effect of prolactin on the seasonal cyclicity of newt forelimb regeneration. *Journal of Experimental Zoology* 182:41–46.

Schneider, C. W. 1968. Avoidance learning and the response tendencies of the larval salamander *Ambystoma punctatum* to photic stimulation. *Animal Behaviour* 16:492–495.

Schneider, C. W., B. W. Marquette, and P. Pietsch. 1991. Measures of phototaxis and movement detection in the larval salamander. *Physiology & Behavior* 50:645–647.

Semlitsch, R. D. 1987a. Density-dependent growth and fecundity in the paedomorphic salamander *Ambystoma talpoideum*. *Ecology* 68:1003–1008.

Semlitsch, R. D. 1987b. Interactions between fish and salamander larvae: costs of predator avoidance or competition? *Oecologia* 72:481–486.

Semlitsch, R. D., D. E. Scott, and J. H. K. Pechmann. 1988. Time and size at metamorphosis related to adult fitness in *Ambystoma talpoideum*. *Ecology* 69:184–192.

Shoemaker, V. H., S. S. Hillman, S. D. Hillyard, D. C. Jackson, L. L. McClanahan, P. C. Withers, and M. L. Wygoda. 1992. Exchange of water, ions, and respiratory gases in terrestrial amphibians. Pages 125–150 in M. E. Feder and W. W. Burggren (eds.), *Environmental physiology of the amphibians*. University of Chicago Press, Chicago.

Sih, A., L. B. Kats, and R. D. Moore. 1992. Effects of predatory sunfish on the density, drift, and refuge use of stream salamander larvae. *Ecology* 73:1418–1430.

Stangel, P. W., and R. D. Semlitsch. 1987. Experimental analysis of predation on the diel vertical migrations of a larval salamander. *Canadian Journal of Zoology* 65:1554–1558.

Stebbins, R. C., and N. W. Cohen. 1997. *A natural history of amphibians*. Princeton University Press, Princeton, New Jersey.

Stuart, S. N., J. S. Chanson, N. A. Cox, B. E. Young, A. S. L. Rodrigues, D. L. Fischman, and R. W. Waller. 2004. Status and trends of amphibian declines and extinctions worldwide. *Science* 306:1783–1786.

Sugalski, M. T., and D. L. Claussen. 1997. Preference for soil moisture, soil pH, and light intensity by the salamander, *Plethodon cinereus*. *Journal of Herpetology* 31:245–250.

Taub, F. B. 1961. The distribution of the red-backed salamander, *Plethodon c. cinereus*, within the soil. *Ecology* 42:681–698.

Taylor, D. H. 1972. Extra-optic photoreception and compass orientation in larval and adult salamanders (*Ambystoma tigrinum*). *Animal Behaviour* 20:233–236.

Taylor, J. 1983. Orientation and flight behavior of a neotenic salamander (*Ambystoma gracile*) in Oregon. *American Midland Naturalist* 109:40–49.

Taylor, D. H., and K. Adler. 1973. Spatial orientation by salamanders using plane-polarized light. *Science* 181:285–287.

Test, F. H. 1946. Relations of the red-backed salamander (*Plethodon cinereus*) to light and contact. *Ecology* 27:246–254.

Townsend, V. R. Jr., and R. G. Jaeger. 1998. Territorial conflicts over prey: domination by large male salamanders. *Copeia* 1998:725–729.

Vanecek, J. 1998. Cellular mechanisms of melatonin action. *Physiological Reviews* 78:687–721.

Vernberg, F. J. 1955. Correlation of physiological and behavior indexes of activity in the study of *Plethodon cinereus* (Green) and *Plethodon glutinosus* (Green). *American Midland Naturalist* 54:382–393.

Warrant, E. J. 1999. Seeing better at night: life style, eye design and the optimum strategy of spatial and temporal summation. *Vision Research* 39:1611–1630.

Welsh, H. H. Jr., and S. Droege. 2001. A case for using plethodontid salamanders for monitoring biodiversity and ecosystem integrity of North American forests. *Conservation Biology* 15:558–569.

Whitford, W. G., and V. H. Hutchison. 1965. Effect of photoperiod on pulmonary and cutaneous respiration in the spotted salamander, *Ambystoma maculatum*. *Copeia* 1965:53–58.

Wise, S. E., and R. G. Jaeger. 1998. The influence of tail autotomy on agonistic behaviour in a territorial salamander. *Animal Behaviour* 55:1707–1716.

Wood, J. T. 1951. Protective behavior and photic orientation in aquatic adult and larval two-lined salamanders, *Eurycea b. bislineata* x *cirrigera*. *Virginia Journal of Science* 2:113–121.

Wyman, R. L. 1998. Experimental assessment of salamanders as predators of detrital food webs: effects on invertebrates, decomposition and the carbon cycle. *Biodiversity and Conservation* 7:641–650.

Part IV

Fishes

Night, Atlantic

I love it when you can actually see the colors. My tackle box still looks reddish. Who'd have thought moonlight could be strong enough to show the colors? But it's so dark out here, I mean, the moonlight is so pure, that the moon illuminates magnificently.

We're usually only a mile or two from shore on these full-moon tides. You go on the full because the moon gets the tide really cranking hard, and the biggest striped bass like those war-horse tides. New moons rev the tides too, but the new moon so deprives the night of light that nearly any disturbance triggers the glow of phosphorescent plankton and jellies. Your line lights up. That scares the fish. Older fishermen call this "fire in the water," and I think it sometimes makes the fish stop hunting because their slipstream glows, alerting their prey. So you go on the full, when the wash-away moonlight is bright enough to quench the fire.

At night you're looking for big fish, over 20 pounds, sometimes more than twice that. I've never caught a 40-pounder during daylight. And though nowadays I have mixed feelings about killing such survivors, I still like to take home a 20-pounder and call it food.

Up to three nights on either side of full counts as good, but there's nothing like witnessing a sundown ocean give birth to a perfect full moon, watching that massive illuminated magnet climb from horizon to sky, and feeling it start to pull the ocean until the tides really heap up in the rips and the fish filter in.

Moon fishing is an excuse to float the night. During nighttime, the normal firewall between imagination and reality dissolves. Imagination and perception merge. They have to. You can't see as well, as far, as deeply into the water, so you have to work to imagine what's going on. With visual details lost in the broad strokes of moonlight, you've tied your knots by muscle-memory, and you're mostly operating your reel by feel.

Your boat drifts, your mind drifts, you're imagining what's happening under that cloaked tide, things start seeming otherworldly. You are reminded how wide is the world, how flexible and expansive your mind can be when it's working right. And you slip your leash to explore the great spaces and the vast vault of sky.

The sea off Montauk, Long Island, is a place of big fish and big water. The tide turns the nearshore ocean into a miles-wide river of streaming current. The fish position themselves along the ridges of long-drowned hills, hunkered down out of the full blast of the tidal wind, poised to ambush prey that gets swept up and over those ridge-tops.

Some of that prey has your hook in it.

The underwater ridges deflect and roil the currents, creating lines of standing waves on the surface that shine white in the cool autumn moonlight. You run the boat uptide of those waves, drop your weighted line to the bottom 40 feet below, and release yourself to the moon-tide as it sweeps you and your bait across the slope, through the rip-line and down the other side.

The recovery of striped bass from depletion to abundance—thanks to unusually strong management—has been so dramatic that if you have two or three people fishing it's not unusual on most drifts for at least one line to get arrested by a heavy fish. People who've never done this before ask how they'll know if they get a bite. I tell them don't worry—you'll know.

And I run the boat back up for another drift and again we cast our lines to the blackness.

You sense the sweep of tide and water. You drift along imagining the bottom; visualizing the fish stemming the dark current, gathering moonlight with their special eyes and scanning upwards for vulnerable silhouettes. The whole while you are imagining the bottom, you too are looking up, making eye contact with the moon, orienting on Orion, dreaming. The boat gets tossed in the surface turbulence just past the undersea ridge-line. You feel your trailing bait entering that turbulent zone at the ridge-top because your weighted line actually lightens as the accelerating current sweeps it forward. You feel two thumps, very strong, your line stretches tight, your rod bows heavily, and as the fish takes off you can only feel the line flying from the reel, just a little out of control.

Carl Safina

Chapter 11

Artificial Night Lighting and Fishes

Barbara Nightingale, Travis Longcore,
and Charles A. Simenstad

Human activities subject many aquatic habitats to significant alterations in natural cycles of illumination. More than half of the world's human population lives within 100 km of an ocean, and most other major human developments are near rivers or lakes (Marsh and Grossa 2002). Coastal zones, lakes, rivers, ponds, and streams are all subject to artificial night lighting from cities, recreation, commerce, and industry. Permanent lights are also found at fish farms in nearshore zones, and cruise ships and recreational boats dot the seas. Fishing boats operate all night, some using high-intensity lights to attract their catch. Humans have altered natural lighting conditions in the most productive and biodiverse portions of aquatic environments.

The alteration of natural lighting regimes could be expected to have a substantial effect on aquatic organisms because light, along with temperature, structures aquatic habitats. Despite the well-known and profound influence of light on the behavior of aquatic organisms, especially invertebrate and vertebrate animals, little research has addressed the consequences of human disruption of diel, lunar, and seasonal cycles of illumination.

This chapter presents a review of the documented and predictable effects of artificial lighting on a portion of aquatic communities, the teleosts (bony fishes).

The ambient nighttime light regime, unpolluted by artificial lights, is one controlled only by the effects of moonlight, starlight, and cloud cover. Whereas aquatic biota have adapted to natural nighttime light regimes over evolutionary history, the pollution of aquatic habitats with artificial night lighting is a relatively recent phenomenon. Indeed, moonlight, starlight, and bioluminescence are the only night lighting sources to which fishes are evolutionarily adapted (Hobson et al. 1981). Light pollution has modified the intensity, spectra, frequency, and duration of nighttime light reaching and penetrating water surfaces. This is not just a localized effect—urban and other intense concentrations of nighttime lighting change the light regime for tens to hundreds of miles (Cinzano et al. 2001).

This chapter covers the observed and potential effects of artificial night lighting on teleost fishes in freshwater streams and lakes and in shallow marine and estuarine nearshore areas, with some reference to pelagic zones. These inland and nearshore areas are closest to permanent artificial night lighting associated with overwater structures, shoreline buildings, boardwalks, bridges, and other human developments. Docks and other overwater structures have been studied often because of the unique illumination situations they create. They are associated with, and partially block, artificial lights, causing abrupt changes in the light environment. Studies in the Pacific Northwest have documented fish behavior around overwater structures, mostly in "gray literature" reports (e.g., Fields 1966, Johnson et al. 2000, Prinslow et al. 1980, Ratté and Salo 1985, Taylor and Willey 2000, Weitkamp 1982, Weitkamp and Schadt 1982). The effects of artificial lighting on circadian rhythms and physiology are not reviewed here, although these topics are now the subject of considerable research in the context of aquaculture (Boeuf and Le Bail 1999). Nor is the breadth of research on light and fish behavior included. Many reviews are available that address various aspects of the broader topic (e.g., Blaxter 1975b, Reebs 2002, Woodhead 1966).

Salmonids provide many examples in this chapter. Salmon species are widespread and of significant economic value, and their anadromous life history brings them in contact with environments that may be altered by lighting from human development. Considerable research effort has concentrated on salmon because of their economic importance and the rare or endangered status of many populations. Examples from other groups within the teleosts complement this emphasis on the salmonids.

We begin with a brief description of the visual system of teleosts and the influence of species, age, ambient illumination, and lighting type on their responses to light. The potential effects of artificial lighting on fishes are divided into influences on foraging behavior, predation risk, migration, reproduction, and harvest.

Teleosts and Visual Perception

Teleosts represent 96% of all living fishes and constitute the bulk of the world's major fisheries, including herring, smelt, salmon, tuna, cod, trout, halibut, lingcod, flounder, catfish, and other commercially important species (Helfman et al. 1997). For teleosts, important behaviors such as feeding, schooling, and migration depend on specific light intensities. The visual cell layer of the teleost eye consists of two types of photoreceptors, rods (scotopic) and cones (photopic), each with different light thresholds. These photoreceptors respond to increases and decreases in light intensity by changing position within the eye relative to the light source. When light intensity is above the cone threshold, the eye assumes the light-adapted state, with the cone cells contracted and the rod cells elongated. When light intensity falls below the cone threshold level, the cones are elongated while the rods are contracted. These photoreceptor changes are in direct proportion to the logarithm of light intensity (Ali 1959). The importance of these rod and cone thresholds corresponds to the ability of fishes to school and feed. Fish schools disband and cease feeding by visual means when the light drops below the rod threshold (Ali 1959, Azuma and Iwata 1994).

Underwater light perception depends on the light-transmitting qualities of the water coupled with the spectral properties of the retinal visual pigments that function as light receptors. As these visual pigments absorb light, energy is released that electrically activates nerve cells. Differences in the capacity of visual pigments to absorb light are determined by genetics and habitat, and this capacity changes with the light spectrum available in the species' habitats (Wald et al. 1957). For example, as juvenile salmonids move from fresh to salt water, their pigments change from porphyropsin- to rhodopsin-dominated (Beatty 1966, Folmar and Dickhoff 1981). These changes alter the visual sensitivity from the red-yellow hues of freshwater streams to the blue color of estuarine and ocean waters. The positions of the smolts' visual cells are responsive to ambient light and not to internal rhythmic diurnal patterns.

Teleost Responses to Light

Responses of fish to light can be divided into two categories: the reaction to luminance (brightness of direct glare from a light in the environment) and the reaction to illumination (the amount of light per unit area incident on objects in a fish's surroundings). One is the response to the light itself, the other to the effects of the light. Studies of fish under experimental conditions have shown that this basic response varies, even within species, depending on many factors. These include characteristics of the fish (e.g., age; Byrne 1971, Fields and Finger 1954, Hoar 1953), ambient conditions (Godin 1982), and characteristics of the light (e.g., duration, intensity, and spectrum; Fields and Finger 1954, Patrick 1978, 1983, Patten 1971, Pinhorn and Andrews 1963).

Influence of Fish Age and Species

Research on salmon behavioral responses to both natural and artificial light reveals consistent differences between species and ontogenetic stages. Behavioral responses are correlated with the foraging strategy of the species. Species such as coho (*Oncorhynchus kisutch*), Atlantic salmon (*Salmo salar*), and steelhead (*Oncorhynchus mykiss*) that occupy and defend stream territories tend to be quiescent at night (Godin 1982, Hoar 1951, Northcote 1978). Although coho fry occasionally aggregate, they demonstrate no true schooling or milling behavior. Species that disperse to nursery lakes (e.g., sockeye [*Oncorhynchus nerka*]) and estuaries (e.g., chinook [*Oncorhynchus tshawytscha*], pink [*Oncorhynchus gorbuscha*], and chum [*Oncorhynchus keta*]) typically school, show nocturnal activity, and show negative phototaxis (Godin 1982, Hoar 1951). Laboratory studies illustrate the variation in responses to light. The nocturnally active pink and chum fry dart wildly about when lights are turned on after darkness, whereas diurnal coho fry move briefly or remain quiescent (Hoar et al. 1957).

Response to light changes during the life cycle of salmonids as well. Hoar (1951) observed changes in the response to visual stimuli after the transformation from fry to smolt. After visual disturbance, smolts took cover for a longer period than fry and tended to scatter wildly when light was flashed on them at night. Smolts also showed greater aggregating tendencies, lower stimulation thresholds, higher levels of general excitability, more activity at night, stronger cover reaction, and less activity during the day (Folmar and Dickhoff 1981, Hoar 1976, McInerney 1964). Other environmental factors influence responses to light as well. Some juvenile salmonids adopt a nocturnal strategy when stream temperatures are low and so exhibit different responses to light in winter and summer (Fraser and Metcalfe 1997).

Influence of Ambient Illumination

Response to light depends on the ambient conditions before introduction of a new light source (Godin 1982, Richardson and McCleave 1974). The speed of progression of physiological changes in the eye from dark adapted to light adapted and from light adapted to dark adapted is influenced by the intensity of the new light and the intensity of light to which the fish were previously exposed (Ali 1962, Protasov 1970, Puckett and Anderson 1987). For example, fish previously exposed to higher light intensities attain the dark-adapted state more slowly than those previously exposed to lower light intensities (Ali 1962). Wavelength is also believed to influence the speed of these reactions (Protasov 1970).

On the basis of studies of salmonid attraction to light sources in dark conditions, Puckett and Anderson (1987) describe the attraction (for these positive phototactic fish) to solid light in the following formula:

$$\text{Percentage attraction} = A - k\text{Abs}(\log I_s/I_a)$$

where A is a constant expressing the percentage of fish attracted under the best conditions, k is a constant, Abs is the absolute value operator, log is the logarithm to the base e, I_s is the intensity of light that fish encounter, and I_a is the intensity of light to which the fish were adapted. The maximum attraction occurs when fish are adapted to the light. Because adaptation to a new light level takes some time, the strength of positive phototaxis can be observed only after an adjustment period.

The time needed for light-adapted chum and pink fry to adapt to dark ranges from 30 to 40 minutes. The time needed for dark-adapted fry to adapt to increased light ranges from 20 to 25 minutes (Brett and Ali 1958, Protasov 1970). During these periods of transition, their visual acuity ranges from periods of blindness to slightly diminished acuity, depending on the magnitude of light intensity contrasts. As the animals mature, time needed for light adaptation generally shortens, whereas time taken for dark adaptation tends to increase. The time delay for adaptation to new lighting levels and the reduced vision during adaptation present a potential adverse effect from artificial lighting. If fish are exposed to unnatural rapid changes in the light environment, normal behavior may be interrupted (see Buchanan 1993 and Chapter 9, this volume for reference to a similar effect on amphibians).

Influence of Light Duration and Spectrum

Duration of lighting influences fish responses. Strobe lights produce extremely short flashes of light. Unlike the normal flickering light caused by wave, cloud, and sun conditions in underwater environments, the discharge of strobe light is apparently disturbing to fishes (Dera and Gordon 1968, Loew and McFarland 1990). The abrupt flashes of a strobe produce large contrasts in light intensity over a duration too short for any retinal adaptation to take place. The findings of Sager et al. (1987) with dark-adapted estuarine fishes and of Nemeth and Anderson (1992) with dark-adapted juvenile coho and chinook in freshwater conditions consistently demonstrated an avoidance reaction to strobe light in dark conditions. Fish may be exposed to strobe lights at dams, power-generating stations, and locks; strobe lights can be used to reduce entrainment of salmonids into passages where they could be harmed (Johnson et al. 2005). Other short-duration lights (e.g., vehicle headlights illuminating a shoreline) would likely have similar disruptive effects. In contrast, constant illumination allows adaptation to the lighting level and a behavioral response to the new conditions.

The response of fishes to light depends also on spectrum. Fishes in streams and lakes are sensitive to the red and yellow wavelengths more common in these environments, whereas oceanic and open-water fishes are more sensitive to blue lights (Beatty 1966, Folmar and Dickhoff 1981). Sensitivity of marine fishes in the blue spectrum corresponds to the light emitted by marine bioluminescent plankton (Hobson et al. 1981). Because bioluminescent plankton create light in response to movement, prey species can detect the bioluminescent glow caused by predators as they swim by, creating a strong selective pressure for prey to detect light in this spectrum (Hobson et al. 1981). Increased illumination in the right spectrum triggers light-dependent behaviors in some species and suppresses behaviors in light-sensitive species.

Researchers have used mercury vapor lights, which have more energy in the blue and ultraviolet spectrum, to attract fish. Wickham (1973) and Puckett and Anderson (1987) found fishes to be attracted to mercury vapor lights under certain conditions. Similarly, in a comparison of juvenile coho and chinook, Nemeth and Anderson (1992) found greater fish activity and fewer avoidance reactions to mercury vapor lights than to strobe lights, with strong dependence on the initial ambient conditions. For steelhead, Puckett and Anderson (1987) found the fish to avoid the mercury vapor light initially and then swim toward the light, probably after adaptation. The attraction of many fish species to mercury vapor lights has resulted in their installation at dams and power plants to mark the entrance to fish bypasses.

Fish rarely are exposed to fluorescent or incandescent lighting in the wild. Puckett and Anderson (1987), however, found juvenile salmon to be attracted to incandescent light when encountering a decrease in ambient light intensity. Roach (*Rutilus rutilus*), a strongly photonegative species, exhibited photonegative behavior to fluorescent light in an experimental setting (Van Anholt et al. 1998).

In sum, alterations in the light regime do not affect all fish species uniformly. Each species, and even developmental stages within a species, has a unique response to lighting conditions, the result of a long evolutionary history with predictable light regimes. The response is also plastic (Reebs 2002) and may change depending on other environmental factors and the condition of the individual.

Effects of Artificial Night Lighting on Fishes

Although responses vary greatly between species and between age classes of fishes, artificial night lighting influences fish foraging and schooling behavior, spatial distribution, predation risk, migration, and reproduction. Effects in these areas aggregate to influence community ecology of fishes and both their prey and predators across the affected aquatic landscape (e.g., Prinslow et al. 1980, Ratté and Salo 1985, Tabor et al. 2001, Yurk and Trites 2000).

Foraging and Schooling

Artificial lighting may affect foraging and schooling behavior of diurnal, crepuscular, and nocturnal fishes. Fish display large plasticity in these behaviors; some normally diurnal fish forage at night, and nocturnal fish occasionally may be active during the day (Hobson 1965, Reebs 2002). Nocturnal species respond to extremely low illumination levels; some species even exhibit negative phototaxis at illumination less than 10^{-2} lux and forage at illumination as low as 10^{-5} lux (see Chapter 15, Table 15.1, this volume; Blaxter 1975a). Other species begin to forage only above a given illumination, on average around 0.1 lux (Blaxter 1975a). Species often exhibit a preferred range of illumination for schooling and foraging behavior, with variation between individuals, between ontogenetic stages, and with other environmental conditions. For example, European minnows (*Phoxinus phoxinus*) are less active above a threshold illumination around 0.2 lux but forage at higher illumination when hungry (Harden Jones 1956, Woodhead 1956).

Schooling is generally interpreted to be an antipredator behavior by which members of a school visually detect predators and evade as a group. Schools are almost universally observed to break up as illumination decreases below a threshold (see Woodhead 1966 and references therein), which, for example, in minnows is between 2.4×10^{-2} and 3.2×10^{-3} lux (Harden Jones 1956). Individuals continue feeding alone at lower light levels but often with less efficiency (Harden Jones 1956). Artificial night lighting may induce schooling behavior as predation risk increases.

The strong lunar cycles in fish foraging behaviors are well known (Gibson 1978, Patten 1971), and chronic increases in illumination from sky glow are in the range of or exceed moonlight (Tabor et al. 2001; see Chapters 9 and 15, this volume). Single artificial lights along streams are also sufficient to increase illumination above the usable range for nocturnal fish. Contor and Griffith (1995) observed a negative relationship between illumination and foraging in juvenile rainbow trout (*Oncorhynchus mykiss*), with fewer fish foraging during the full moon or under artificial light from a large billboard by the river.

Planktivorous fishes generally forage by sight, and plankton exhibit diel vertical migration in the water column to reduce predation from fishes (Gliwicz 1986). Increased illumination during a dark night can provide foraging opportunities for some fish, which locate zooplankton that move to the surface to feed under cover of darkness. This condition occasionally is produced naturally, when a full moon rises late in the night after complete darkness. Fish can then forage under the full moon on zooplankton that came to the surface to forage in what had started as a dark night (Gliwicz 1986). Gliwicz named this phenomenon the "lunar light trap," but similar light traps could be produced by artificial lights that illuminate a dark water body in the middle of the night. Such light traps would be exploited by fish that normally forage at higher illuminations and gain an advantage from the sudden illumination (e.g., the freshwater sardine *Limnothrissa miodon*; Gliwicz 1986). But not all species may benefit—some nocturnal foragers that show little decrease in efficiency with decreased illumination could be harmed by increased competition under artificially illuminated conditions. Furthermore, such truly nocturnal species are less efficient than diurnal foragers under all conditions, by two orders of magnitude in at least one instance (Holzman and Genin 2003). The implication of these differences is that artificial lights illuminating a foraging area, either chronically or in unexpected periods, may provide a competitive advantage to fishes that normally forage at those higher illuminations, to the detriment of other species.

At least one study has documented a change in local foraging communities under artificial lights. Prinslow et al. (1980) reported changes to fish assemblages and predation rates in a study of the effects of high-intensity security lights on a naval base in Puget Sound's Hood Canal. At that site, the level of intensity of artificial night lighting on the base wharves appeared to influence the behavior of fishes along the adjacent shoreline, with light intensities of 200–400 lux attracting to forage aggregations of juvenile chum and other small fishes that ordinarily would be quiescent.

Predation Risk and Foraging Success

Elevated illumination usually increases the predation risk on fishes at night. Vertical migration of juvenile salmonids allows them to maintain a constant light environment, where foraging on zooplankton is possible while detection by predators is minimized (Scheuerell and Schindler 2003). This "antipredation window" may be eliminated or reduced in environments subject to increased artificial illumination. Minimal increases in lighting may thereby disrupt interactions between predator and prey species. Greater increases in illumination may allow normally diurnal predators to continue to forage at night, perhaps even on normally diurnal prey species (Hobson 1965). Increased light also aids predatory fish or mammals attacking from below by allowing them to distinguish the dark form of their prey against an illuminated background (Hobson 1966).

Field studies at artificial lights document altered predator–prey dynamics. Spiny dogfish (*Squalus acanthias*), a Puget Sound shark, appeared to be attracted to security lighting, probably because it illuminated aggregating prey fishes (Prinslow et al. 1980). Pacific herring (*Clupea harengus pallasi*) and sand lance (*Ammodytes hexapterus*) apparently were subjected to significant predation at security lighting (Prinslow et al.

1980). Also, 39 individual predators were observed when lights were on, whereas only two were observed when lights were off. In another instance, salmon fry were exposed to high predation from sculpins (*Cottus* spp.) at lighted areas along their migratory route (Tabor et al. 1998, 2001). The greatest numbers of salmon fry were found in stomachs of sculpins in the areas of brightest nighttime lighting (Tabor et al. 1998). First Nation representatives in British Columbia, Canada, have requested the minimization of lights at salmon farms based on the observation that herring and herring fry are attracted to lights in the cages and are eaten there by the farmed fish (British Columbia Environmental Assessment Office 1997).

Lights along a migratory watercourse may allow increased predation by other vertebrate predators. Both inmigrating adult and outmigrating juvenile salmon are captured by mammalian and avian predators, which can exert a significant pressure on depressed fish populations (Yurk and Trites 2000). On the Puntledge River in British Columbia, Canada, lights from a bridge, halogen lights from a recreational field, and halogen lights from a sawmill facilitated foraging on outmigrating smolts by harbor seals (*Phoca vitulina*; Yurk and Trites 2000). A "lights-out" experimental treatment at the bridge reduced the number of seals feeding, but on subsequent nights the seals repositioned to exploit illumination from residual urban light.

The observations at artificial lights are consistent with the history of research on natural illumination levels and predation risk (Woodhead 1966). Increased illumination is almost universally correlated with increased risk of predation for a species (Cerri 1983), largely because illumination determines the distance at which predator and prey detect each other (Howick and O'Brien 1983, Munz and McFarland 1977). The diel vertical migration of zooplankton to stay in dark conditions reduces predation risk from small predatory fish (Gliwicz 1986). Invertebrate activity is highest on the darkest nights, so planktivorous fish tend to be nocturnal (Hobson 1965). Such fishes may congregate in dense schools during the day to avoid their own predators, then disperse to forage at night (Hobson 1965). But during brightly moonlit nights, diurnal piscivorous predators can remain active (Hobson 1965), increasing predation risk for smaller planktivorous fish. Many predatory fish have their own, often diurnal terrestrial vertebrate predators (Alexander 1979) from which nocturnality is one possible evolutionary or behavioral escape. The use of a particular range of illumination by many fishes therefore is limited by predator avoidance at the upper bound and efficient foraging at the lower bound. When predation risk is very high (e.g., to torpid juvenile salmonids unable to move quickly in cold water), inefficient foraging in extremely dark conditions is tolerable to avoid predation (Fraser and Metcalfe 1997). Darkness, whether at night or caused by turbidity, may be an essential component of habitat for young fishes that must escape predation (Grecay and Targett 1996). Low prey concentrations and potentially lower foraging efficiency are offset by the predation refuge provided by darkness (Fraser and Metcalfe 1997). Furthermore, because at low illumination visual detection depends on the size of the prey item, altered illumination regimes change the size distribution of prey consumed—brighter conditions lead to consumption of smaller prey and vice versa (Elston and Bachen 1976, Holzman and Genin 2003, Mills et al. 1986). For some other species, capture efficiency is unchanged in complete darkness. Two examples out of what probably are many are roach (*Rutilis rutilis*; Diehl 1988) and the zooplanktivorous coral-reef fish *Apogon annularis* (Holzman and Genin 2003). Such nocturnal specialist species probably will suffer disproportionately in environments subjected to chronically elevated nighttime illumination.

Disruption of Migration

Salmonids migrate from spawning areas through streams, rivers, and estuaries to the ocean, often moving at night. Returning adult fish also migrate at night. Altered light environments along these routes may interrupt movement, increase predation on migrating fish, and ultimately reduce the number of successful migrants.

Juvenile Pacific salmon begin downstream migration while their eyes are adapting to the dark (Brett and Ali 1958). This process takes between 30 and 60 minutes depending on species and age, during which time individuals are unable to orient to fixed objects and either swim with the current or are displaced downstream (Ali 1959). The synchronized response of individuals to lighting conditions results in a peak in migration, with many individuals moving during a short period, which provides the advantage of minimizing contact with predators (Hoar 1958). Under natural lighting conditions, this mass migration is well synchronized, with its "obvious advantages" (Ali 1959). Recent research has documented adverse effects of artificial night lighting on this migratory behavior. In a study of lighted and nonlighted areas along the Cedar River in Renton, Washington, Tabor et al. (2001) found increased nighttime lighting intensities to have a profound effect on sockeye (*Oncorhynchus nerka*) fry. Increased nighttime light intensity, measured at lighted building and bridge sites, caused the fry to delay migration and move to the low-velocity and lighted shoreline habitats. Downstream migration of sockeye fry each night was initiated after light intensity was less than 1 lux, a finding consistent with Ali's (1959) earlier research. When the light level was artificially increased to 32 lux, however, migration almost completely stopped. Similarly, Prinslow et al. (1980) previously showed that security lighting caused a delay in chum outmigrating through a canal. Tabor et al. (2001) also documented increased predation on migrating fry by sculpin under artificial lights.

Artificial lights have been used to attract fish to ladders that bypass dams (Larinier and Boyer-Bernard 1991a, 1991b) and to prevent them from being sucked into pipes at power plants (Haymes et al. 1984). The response of fish depends on the light spectrum and varies by species. Alewife (*Alosa pseudoharengus*) and yellow perch (*Perca flavescens*) responded negatively to red lights (Patrick 1978), ruffe (*Gymnocephalus cernua*) and Eurasian perch (*Perca fluviatilis*) were repelled by mercury vapor lights (Hadderingh and Kema 1982), alewife were attracted to mercury vapor lights filtered to allow only blue light (Haymes et al. 1984), silver eels (*Anguilla anguilla*) avoided sodium vapor lights (Cullen and McCarthy 2000), juvenile coho and chinook avoided bright mercury vapor and strobe lights, and juvenile chinook were attracted to dim mercury vapor light (Nemeth and Anderson 1992). In one instance, intermittent lights increased capture of fish in the first portion of the lights-off phase (Croze et al. 1999). So although lights certainly are useful on fish bypass structures, no single combination of spectrum, intensity, and duration attracts or repels all species.

Lights can affect the dispersal and larval recruitment of other fish groups. Munday et al. (1998) investigated the possibility of increased recruitment of reef fish by using a light-attractor device. This exploits the attraction of larval and juvenile stages of many pelagic fishes to bright lights (Choat et al. 1993, Doherty 1987, Munday et al. 1998). Reefs with a fluorescent light attractor device had three times more settlement than control reefs (Munday et al. 1998). Researchers did not conclude whether increased attraction resulted in long-term increases in population size and diversity.

Reproductive Behavior

Illumination levels influence reproductive behavior, such as courtship and spawning. Courtship displays are affected by ambient illumination, although such relationships have received relatively little attention in the literature (Endler 1987, Long and Rosenqvist 1998). Display distances between guppies (*Poecilia reticulata*) vary with light intensity, and changing light spectrum at dawn and dusk alters perception of markings on males (Long and Houde 1989). The effects, if any, of artificial illumination on fish courtship behavior have not yet been documented.

Spawning, without visual courtship displays, can occur at any time of day and may also depend on visual cues. Salmonids tend to migrate upstream and spawn at night (Evans 1994), but effects of artificial light on salmonid spawning behavior have not been recorded. There is some evidence from pelagic fish that artificial lighting may disrupt spawning behaviors. For example, cod (*Gadus* spp.) usually spawn at night, but when a bright light was shone on captive fish they stopped spawning and began to exhibit aggressive behavior (Woodhead 1966). In the wild, spawning cod moved into deeper water when a light was shone on them (Woodhead 1966). Many other pelagic fishes spawn at dusk or at night, including flying fishes (Exocoetidae), mullet (*Mugil* spp.), anchovies (*Anchoa* spp.), herring (Clupeidae), and grunion (*Leuresthes tenuis*) (Woodhead 1966). Of these, the littoral spawning habits of grunion expose them to the greatest possibility of disruption by artificial lighting. Anecdotal reports indicate that grunion do not spawn, or will cease spawning, in areas with flashing lights associated with activity and movement (e.g., flashlights), but constant illumination may not be as problematic if it is not too bright (K. Martin, personal communication, 2005). On beaches with both dark and light areas, however, grunion may prefer the darker areas.

Harvest

Lights have long been used as a method of attracting fishes for harvest (Ben-Yami 1976, Woodhead 1966). Many species of fish are attracted to and "held" by bright night lights in a manner similar to that seen in birds (Verheijen 1958). Fishes may be disoriented under bright lights, or the lights may allow normal orientation as if it were day (Woodhead 1966). This technique is most effective when fishing lights present a stark contrast to background illumination; that is, light fishing works best on dark nights away from other light pollution (Woodhead 1966).

Smaller fishes may be attracted to lights on boats for nontarget species (e.g., extremely bright lights from squid boats), subjecting them to increased predation (Ben-Yami 1976). Some planktivorous fishes have been observed to school and forage within the lighted zone of a boat

(Woodhead 1966). Larger fishes and other vertebrate predators become secondarily attracted to lights from boats and ships. Population-level effects have not been documented, but this phenomenon should be investigated in heavily fished areas. Lights to attract fishes or other target species also can have spillover effects on seabirds and adjacent terrestrial habitats (see Chapter 5, this volume).

Lights may reduce recreational fishing opportunities. Lights from an adjacent tennis court eliminated the seatrout (*Cynoscion* sp.) fishing opportunity in the River Cowie in Stonehaven, Scotland. This nocturnal forager becomes active only below 0.5 lux, which was exceeded by adjacent artificial lighting. Angling for the species is never attempted during a full moon, and local anglers successfully sued to have the tennis court lights shut off (Stonehaven & District Angling Association, http://www.sana.org.uk/light.htm).

Conclusion

Complex sets of light preferences add an important dimension to the interpretation of habitat for fishes. Too often a species' niche is interpreted only in space and not in time (Aschoff 1964). Because lighting conditions change constantly, they present a dynamic and complex niche dimension that allows coexistence of many species sympatrically. Disruption of that natural lighting regime may have significant consequences for species richness and community composition. Chronic artificial lighting eliminates part of the range of variation of illumination conditions within an area and changes the spectral characteristics of the light. Long-established patterns of niche partitioning may then break down. The importance of natural patterns of illumination to the life history, management, and conservation of fishes is clear, but far too little effort has been directed to assessing and mitigating the growing influence of artificial lights on fishes.

Further research is needed to understand the extent and significance of observed fish responses to artificial light cast into the underwater environment. Risks of increased mortality or decreased fitness posed by artificial night lighting include delays and changes in migratory behavior caused by changes in direction and disorientation induced by artificial night lighting, temporary blindness induced by artificial night lighting that could increase the risk of predation, attraction of predators and disruption of predator–prey interactions at artificially lighted areas, and loss of opportunity for dark-adapted behaviors, including foraging and migration. These behavioral changes are consistent with the documented studies in both marine and freshwater habitats (Fields 1966, Prinslow et al. 1980, Ratté and Salo 1985, Weitkamp 1982). Given the extensive knowledge of the role of light in structuring aquatic communities, marine and freshwater ecologists should consider the effects of artificial night lighting on these sensitive ecosystems.

Literature Cited

Alexander, G. R. 1979. Predators of fish in coldwater streams. Pages 153–170 in H. Clepper (ed.), *Predator–prey systems in fisheries management*. Sport Fishing Institute, Washington, D.C.

Ali, M. A. 1959. The ocular structure, retinomotor and photo-behavioral responses of juvenile Pacific salmon. *Canadian Journal of Zoology* 37:965–996.

Ali, M. A. 1962. Influence of light intensity on retinal adaptation in Atlantic salmon (*Salmo salar*) yearlings. *Canadian Journal of Zoology* 40:561–570.

Aschoff, J. 1964. Survival value of diurnal rhythms. *Symposia of the Zoological Society of London* 13:79–98.

Azuma, T., and M. Iwata. 1994. Influences of illumination intensity on the nearest neighbour distance in coho salmon *Oncorhynchus kisutch*. *Journal of Fish Biology* 45:1113–1118.

Beatty, D. D. 1966. A study of the succession of visual pigments in Pacific salmon (*Oncorhynchus*). *Canadian Journal of Zoology* 44:429–455.

Ben-Yami, M. 1976. *Fishing with light*. FAO Fishing Manuals. Fishing News Books, Surrey, England.

Blaxter, J. H. S. 1975a. Fish vision and applied research. Pages 757–773 in M. A. Ali (ed.), *Vision in fishes: new approaches in research*. Plenum Press, New York.

Blaxter, J. H. S. 1975b. The role of light in the vertical migration of fish: a review. Pages 189–210 in G. C. Evans, R. Bainbridge, and O. Rackham (eds.), *Light as an ecological factor: II*. Blackwell Scientific Publications, Oxford.

Boeuf, G., and P.-Y. Le Bail. 1999. Does light have an influence on fish growth? *Aquaculture* 177:129–152.

Brett, J. R., and M. A. Ali. 1958. Some observations on the structure and photomechanical responses of the Pacific salmon retina. *Journal of the Fisheries Research Board of Canada* 15:815–829.

British Columbia Environmental Assessment Office. 1997. *The salmon aquaculture review final report*. British Columbia Environmental Assessment Office, Vancouver, British Columbia.

Buchanan, B. W. 1993. Effects of enhanced lighting on the behaviour of nocturnal frogs. *Animal Behaviour* 45:893–899.

Byrne, J. E. 1971. Photoperiodic activity changes in juvenile sockeye salmon (*Oncorhynchus nerka*). *Canadian Journal of Zoology* 49:1155–1158.

Cerri, R. D. 1983. The effect of light intensity on predator and prey behaviour in cyprinid fish: factors that influence prey risk. *Animal Behaviour* 31:736–742.

Choat, J. H., P. J. Doherty, B. A. Kerrigan, and J. M. Leis. 1993. A comparison of towed nets, purse seine, and light-aggregation devices for sampling larvae and pelagic juveniles of coral reef fishes. *Fishery Bulletin* 91:195–209.

Cinzano, P., F. Falchi, and C. D. Elvidge. 2001. The first world atlas of the artificial night sky brightness. *Monthly Notices of the Royal Astronomical Society* 328:689–707.

Contor, C. R., and J. S. Griffith. 1995. Nocturnal emergence of juvenile rainbow trout from winter concealment relative to light intensity. *Hydrobiologia* 299:179–183.

Croze, O., M. Chanseau, and M. Larinier. 1999. Efficiency of a downstream bypass for Atlantic salmon (*Salmo salar* L.) smolts and fish behaviour at the Camon hydroelectric powerhouse water intake on the Garonne River. *Bulletin Français de la Pêche et de la Pisciculture* 353/354:121–140.

Cullen, P., and T. K. McCarthy. 2000. The effects of artificial light on the distribution of catches of silver eel, *Anguilla anguilla* (L.), across the Killaloe eel weir in the lower River Shannon. *Biology and Environment* 100B:165–169.

Dera, J., and H. R. Gordon. 1968. Light field fluctuations in the photic zone. *Limnology and Oceanography* 13:697–699.

Diehl, S. 1988. Foraging efficiency of three freshwater fishes: effects of structural complexity and light. *Oikos* 53:207–214.

Doherty, P. J. 1987. Light-traps: selective but useful devices for quantifying the distributions and abundances of larval fishes. *Bulletin of Marine Science* 41:423–431.

Elston, R., and B. Bachen. 1976. Diel feeding cycle and some effects of light on feeding intensity of the Mississippi silverside, *Menidia audens*, in Clear Lake, California. *Transactions of the American Fisheries Society* 105:84–88.

Endler, J. A. 1987. Predation, light intensity and courtship behaviour in *Poecilia reticulata* (Pisces: Poeciliidae). *Animal Behaviour* 35:1376–1385.

Evans, D. M. 1994. Observations on the spawning behaviour of male and female adult sea trout, *Salmo trutta* L., using radio-telemetry. *Fisheries Management and Ecology* 1:91–105.

Fields, P. E. 1966. *Final report on migrant salmon light guiding studies (Contract No. D.A.-45-108 CIVENG-63-29) at Columbia River dams*. Fisheries Engineering Research Program, U.S. Army Engineer Division, North Pacific, Corps of Engineers, Portland, Oregon.

Fields, P. E., and G. L. Finger. 1954. *The reaction of five species of young Pacific salmon and steelhead trout to light*. Technical Report No. 7, School of Fisheries, University of Washington, Seattle.

Folmar, L. C., and W. W. Dickhoff. 1981. Evaluation of some physiological parameters as predictive indices of smoltification. *Aquaculture* 23:309–324.

Fraser, N. H. C., and N. G. Metcalfe. 1997. The costs of becoming nocturnal: feeding efficiency in relation to light intensity in juvenile Atlantic salmon. *Functional Ecology* 11:385–391.

Gibson, R. N. 1978. Lunar and tidal rhythms in fish. Pages 201–213 in J. E. Thorpe (ed.), *Rhythmic activity of fishes*. Academic Press, London.

Gliwicz, Z. M. 1986. A lunar cycle in zooplankton. *Ecology* 67:883–897.

Godin, J.-G. J. 1982. Migrations of salmonid fishes during early life history phases: daily and annual timing. Pages 22–50 in E. L. Brannon and E. O. Salo (eds.), *First International Salmon and Trout Migratory Behavior Symposium*. Seattle, Washington.

Grecay, P. A., and T. E. Targett. 1996. Effects of turbidity, light level and prey concentration on feeding of juvenile weakfish *Cynoscion regalis*. *Marine Ecology Progress Series* 131:11–16.

Hadderingh, R. H., and N. V. Kema. 1982. Experimental reduction of fish impingement by artificial illumination at Bergum Power Station. *Internationale Revue der Gesamten Hydrobiologie* 67:887–900.

Harden Jones, F. R. 1956. The behaviour of minnows in relation to light intensity. *Journal of Experimental Biology* 33:271–281.

Haymes, G. T., P. H. Patrick, and L. J. Onisto. 1984. Attraction of fish to mercury vapour light and its application in a generating station forebay. *Internationale Revue der Gesamten Hydrobiologie* 69:867–876.

Helfman, G. S., B. B. Collette, and D. E. Facey. 1997. *The diversity of fishes*. Blackwell Science, Oxford.

Hoar, W. S. 1951. The behaviour of chum, pink and coho salmon in relation to their seaward migration. *Journal of the Fisheries Research Board of Canada* 8:241–263.

Hoar, W. S. 1953. Control and timing of fish migration. *Biological Reviews of the Cambridge Philosophical Society* 28:437–452.

Hoar, W. S. 1958. The evolution of migratory behaviour among juvenile salmon of the genus *Oncorhynchus*. *Journal of the Fisheries Research Board of Canada* 15:391–428.

Hoar, W. S. 1976. Smolt transformation: evolution, behavior, and physiology. *Journal of the Fisheries Research Board of Canada* 33:1234–1252.

Hoar, W. S., M. H. A. Keenleyside, and R. G. Goodall. 1957. Reactions of juvenile Pacific salmon to light. *Journal of the Fisheries Research Board of Canada* 14:815–830.

Hobson, E. S. 1965. Diurnal–nocturnal activity of some inshore fishes in the Gulf of California. *Copeia* 1965:291–302.

Hobson, E. S. 1966. Visual orientation and feeding in seals and sea lions. *Nature* 210:326–327.

Hobson, E. S., W. N. McFarland, and J. R. Chess. 1981. Crepuscular and nocturnal activities of Californian nearshore fishes, with consideration of their scotopic visual pigments and the photic environment. *Fishery Bulletin* 79:1–30.

Holzman, R., and A. Genin. 2003. Zooplanktivory by a nocturnal coral-reef fish: effects of light, flow, and prey density. *Limnology and Oceanography* 48:1367–1375.

Howick, G. L., and W. J. O'Brien. 1983. Piscivorous feeding behavior of largemouth bass: an experimental analysis. *Transactions of the American Fisheries Society* 112:508–516.

Johnson, P. N., K. Bouchard, and F. A. Goetz. 2005. Effectiveness of strobe lights for reducing juvenile salmonid entrainment into a navigation lock. *North American Journal of Fisheries Management* 25:491–501.

Johnson, P. N., F. A. Goetz, and G. R. Ploskey. 2000. *Evaluation of strobe lights for vertically displacing juvenile salmon near a filling culvert intake at the Hiram M. Chittenden locks, Seattle, WA*. U.S. Army Engineer District, Seattle.

Larinier, M., and S. Boyer-Bernard. 1991a. Downstream migration of smolts and effectiveness of a fish bypass structure at Halsou hydroelectric powerhouse on the Nive River. *Bulletin Français de la Pêche et de la Pisciculture* 321:72–92.

Larinier, M., and S. Boyer-Bernard. 1991b. Smolts downstream migration at Poutès Dam on the Allier River: use of mercury lights to increase the efficiency of a fish bypass structure. *Bulletin Français de la Pêche et de la Pisciculture* 323:129–148.

Loew, E. R., and W. N. McFarland. 1990. The underwater visual environment. Pages 1–43 in R. H. Douglas and M. B. A. Djamgoz (eds.), *The visual system of fish*. Chapman and Hall, London.

Long, K. D., and A. E. Houde. 1989. Orange spots as a visual cue for female mate choice in the guppy (*Poecilia reticulata*). *Ethology* 82:316–324.

Long, K. D., and G. Rosenqvist. 1998. Changes in male guppy courting distance in response to a fluctuating light environment. *Behavioral Ecology and Sociobiology* 44:77–83.

Marsh, W. M., and J. M. Grossa Jr. 2002. *Environmental geography: science, land use, and Earth systems*. Second edition. John Wiley & Sons, New York.

McInerney, J. E. 1964. Salinity preference: an orientation mechanism in salmon migration. *Journal of the Fisheries Research Board of Canada* 21:995–1018.

Mills, E. L., J. L. Confer, and D. W. Kretchmer. 1986. Zooplankton selection by young yellow perch: the influence of light, prey density, and predator size. *Transactions of the American Fisheries Society* 115:716–725.

Munday, P. L., G. P. Jones, M. C. Öhman, and U. L. Kaly. 1998. Enhancement of recruitment to coral reefs using light-attractors. *Bulletin of Marine Science* 63:581–588.

Munz, F. W., and W. N. McFarland. 1977. Evolutionary adaptations of fishes to the photic environment. Pages 193–274 in F. Crescitelli (ed.), *The visual system in vertebrates*. Springer-Verlag, Berlin.

Nemeth, R. S., and J. J. Anderson. 1992. Response of juvenile coho and chinook salmon to strobe and mercury vapor lights. *North American Journal of Fisheries Management* 12:684–692.

Northcote, T. G. 1978. Migratory strategies and production in freshwater fishes. Pages 326–359 in S. D. Gerking (ed.), *Ecology of freshwater fish production*. Blackwell Scientific Publications, Oxford.

Patrick, P. H. 1978. *Responses of fish to light*. Report No. 78-516-K. Ontario Hydro Research Division, Toronto.

Patrick, P. H. 1983. *Responses of alewife and gizzard shad to flashing light*. Report No. 82-442-K. Ontario Hydro Research Division, Toronto.

Patten, B. G. 1971. Increased predation by the torrent sculpin, *Cottus rhotheus*, on coho salmon fry, *Oncorhynchus kisutch*, during moonlight nights. *Journal of the Fisheries Research Board of Canada* 28:1352–1354.

Pinhorn, A. T., and C. W. Andrews. 1963. Effect of photoperiods on the reactions of juvenile Atlantic salmon (*Salmo salar* L.) to light stimuli. *Journal of the Fisheries Research Board of Canada* 20:1245–1266.

Prinslow, T. E., C. J. Whitmus, J. J. Dawson, N. J. Bax, B. P. Snyder, and E. O. Salo. 1980. *Effects of wharf lighting on outmigrating salmon, 1979*. FRI-UW-8007. Fisheries Research Institute, University of Washington, Seattle.

Protasov, V. R. 1970. *Vision and near orientation of fish*. Israel Program for Scientific Translations, Jerusalem.

Puckett, K. J., and J. J. Anderson. 1987. *Behavioral responses of juvenile salmonids to strobe and mercury lights*. FRI-UW-8717. Fisheries Research Institute, University of Washington, Seattle.

Ratté, L. D., and E. O. Salo. 1985. *Under-pier ecology of juvenile Pacific salmon (*Oncorhynchus *spp.) in Commencement Bay, Washington*. FRI-UW-8508. Fisheries Research Institute, University of Washington, Seattle.

Reebs, S. G. 2002. Plasticity of diel and circadian activity rhythms in fishes. *Reviews in Fish Biology and Fisheries* 12:349–371.

Richardson, N. E., and J. D. McCleave. 1974. Locomotor activity rhythms of juvenile Atlantic salmon (*Salmo salar*) in various light conditions. *Biological Bulletin* 147:422–432.

Sager, D. R., C. H. Hocutt, and J. R. Stauffer Jr. 1987. Estuarine fish responses to strobe light, bubble curtains and strobe light/bubble-curtain combinations as influenced by water flow rate and flash frequencies. *Fisheries Research* 5:383–399.

Scheuerell, M. D., and D. E. Schindler. 2003. Diel vertical migration by juvenile sockeye salmon: empirical evidence for the antipredation window. *Ecology* 84:1713–1720.

Tabor, R., G. Brown, A. Hird, and S. Hager. 2001. *The effect of light intensity on predation of sockeye salmon fry by cottids in the Cedar River*. U.S. Fish and Wildlife Service, Western Washington Office, Fisheries and Watershed Assessment Division, Lacey, Washington.

Tabor, R. A., G. Brown, and V. T. Luiting. 1998. *The effect of light intensity on predation of sockeye salmon fry by prickly sculpin and torrent sculpin*. U.S. Fish and Wildlife Service, Western Washington Office, Aquatic Resources Division, Lacey, Washington.

Taylor, W. J., and W. S. Willey. 2000. *Pier 64/65 short-stay moorage facility: qualitative fish and avian predator observations*. Port of Seattle Fish Migration Studies. Taylor Associates, Seattle, Washington.

Van Anholt, R. D., G. Van der Velde, and R. H. Hadderingh. 1998. Can roach (*Rutilus rutilus* (L.)) be deflected by means of a fluorescent light? *Regulated Rivers: Research & Management* 14:443–450.

Verheijen, F. J. 1958. The mechanisms of the trapping effect of artificial light sources upon animals. *Archives Néerlandaises de Zoologie* 13:1–107.

Wald, G., P. K. Brown, and P. S. Brown. 1957. Visual pigments and depths of habitat of marine fishes. *Nature* 180:969–971.

Weitkamp, D. E. 1982. *Juvenile chum and chinook salmon behavior at Terminal 91, Seattle, Washington*. Document No. 82-0415-013F. Parametrix, Seattle.

Weitkamp, D. E., and T. H. Schadt. 1982. *1980 juvenile salmonid study: Port of Seattle, Washington*. Document No. 82-0415-012F. Parametrix, Seattle.

Wickham, D. A. 1973. Attracting and controlling coastal pelagic fish with night-lights. *Transactions of the American Fisheries Society* 102:816–825.

Woodhead, P. M. J. 1956. The behaviour of minnows (*Phoxinus phoxinus* L.) in a light gradient. *Journal of Experimental Biology* 33:257–270.

Woodhead, P. M. J. 1966. The behaviour of fish in relation to light in the sea. *Oceanography and Marine Biology: An Annual Review* 4:337–403.

Yurk, H., and A. W. Trites. 2000. Experimental attempts to reduce predation by harbor seals on out-migrating juvenile salmonids. *Transactions of the American Fisheries Society* 129:1360–1366.

Part V

Invertebrates

Night, Tropics

Long ago I discovered my inability to use a library or a dictionary efficiently. I am too easily attracted by words and images. They deflect me toward paths and directions already taken or suggested by others, particularly when the words and images concern tropical forests. I recall the first books on the topic I ever held in my hands. They filled my mind with strong imagery of trees, animals, insects, people, and landscapes bathed in a special light. I kept seeking more books until a time when there was a change; a desire burned within to discover tropical forest in person.

As a tropical biologist I have been privileged with a wealth of opportunities to study insects, particularly butterflies, in tropical forests. I've shared field experiences with exceptional people from all walks of life who have left me with their own distinct perceptions about nature and our place in it. I've covered a lot of ground: North, Central, and South America, Africa, Southeast Asia, Europe, and various islands. I've aged in the field, and the body doesn't spring back as quickly as it once did. Nevertheless, through time my desire to continue this wandering lifestyle has not diminished. Rather, the quest to experience more burns the brighter.

There is an utterly magical time that occurs at tropical dusk. It is when the calls of birds wheeling overhead recede into the distance, and the constant pulse of insect and frog calls fills the air. The inflection point where both sounds are equal in volume coincides with a time when the failing light is ethereal. This heralds the other half of biodiversity, the nocturnal. Within the forest the phosphorescent light of *Pyrhophorus* beetles leaves green trails in their wake to tempt would-be mates to follow. The eerie glow of bioluminescent fungi astonishes but vanishes instantly in the light of a headlamp. The pale moonlight gives a reflected glimpse of bats trolling the surface of oxbow lakes. Overhead one can gaze into black velvet sky to see stars and comets and the cosmos beyond. This is the stuff dreams are made of.

Other forms of light are less benign to the magic of the forest. The first lights that send their electrical call in wild places draw myriad insects. A riot of color, form, and diversity that is impossible to imagine in advance. But the insects attracted to the electrical beacons will dwindle over time. Every week there will be fewer and fewer. This is because a great many die at dawn. Birds, toads, and mammals quickly learn that there is a ready meal at the lights every morning and that there is nowhere for the transfixed nocturnal denizens to hide. Ants too are regulars at the lights. With organized effectiveness they incessantly carry away the disoriented, the wounded, and the dead. There are further consequences of artificial light as well.

Even the most urbanized person cannot fail to pause at the sight of butterflies. Butterflies are insects that require the light of the sun to fly, to reproduce, and to flourish. Daylight is their realm. Nonetheless, a major part of their life cycle, the caterpillar, is a creature often active only at night. To find many caterpillars one must be armed with a flashlight and use the cover of night. The introduction of artificial lights in natural areas has a substantial impact on the diversity, distribution, and abundance of butterflies. With electric lights come the roads. With roads come vehicles, people, habitat destruction, and more lights. This is quickly attended by a reduction in the species of both adult butterflies and the food plants their caterpillars depend on for survival. The area becomes the realm of common weeds, and this reduces butterfly diversity even more. Fewer plant species equates with fewer butterfly species. This is not illusion or fancy but common sense that even a child can grasp and measure its truth.

I never thought that so many places dreamed of in my youth could be marked so deeply by the human hand. Crucial details embodied in the concept of forest held by our predecessors are lost by each passing human generation. I understand that during my grandparents' lifetime large carnivores, herds of elephants, and vast expanses of tropical wilderness were common. My experience has been less rich. Many times I've tried to imagine the tropical forests experienced by naturalists a century ago and concluded that they would be shocked at the current scale of decimation and the intruding pervasiveness of electrical light. In their eyes, our concept of forest would lack depth and vitality. Where is the tropical wilderness? When its absence is finally recognized, will we try to reconstruct it like historians who earnestly, but vainly, attempt to recreate the vital spark of a culture that has passed from living memory? How will we account for and connect all the parts? That is to say, once the concept of forest with all its components is lost, it can never be fully regained, merely reconstructed from partial memories that are not our own.

My travels have convinced me that we are among the last generation who will be able to experience tropical forests, think about them, and be illuminated by them. Humanity has developed with wilderness and the cover of night. A nocturnal world without escape from the glare of electrical lights is disturbing. The concept of future generations not being able to discover the elegant beauty of a dark, starry night accompanied by the sound of nature is profound tragedy.

Philip J. DeVries

Chapter 12

Artificial Night Lighting and Insects: Attraction of Insects to Streetlamps in a Rural Setting in Germany

Gerhard Eisenbeis

Until about three decades ago, high pressure mercury vapor lamps generated light for most street lighting in Germany. Since that time, jurisdictions have been replacing them in increasing numbers with high pressure sodium vapor lamps. During this same period, the total amount of artificial lighting has increased as well, corresponding to an increase in the amount of land developed, estimated at about 1 km^2 or more each day (Haas et al. 1997). Several investigators have examined possible adverse effects of artificial lighting on people and nature (Schmiedel 2001, De Molenaar et al. 1997, Health Council of the Netherlands 2000) and especially on nocturnal insects (Frank 1988, Schanowski and Späth 1994, Steck 1997). Because insects are critically important as pollinators and members of food webs in terrestrial ecosystems, adverse effects of street lighting on insects theoretically could have serious ecological consequences. Conversion to high pressure sodium vapor streetlamps has saved energy, but how has it affected nocturnal insects?

Light trap sampling provides a semiquantitative method for studying responses of insects to artificial light (Williams 1936, 1939, 1940; see Frank 1988 and Chapter 13, this volume for review). Light traps have functioned as instruments for faunal surveys (Malicky 1974, Taylor et al. 1978) and agricultural pest control (Hamilton and Steiner 1939, Graham et al. 1964, Hollingsworth et al. 1968, Hartstack et al. 1971, Schütte 1972, Mitchell and Agee 1981). Their efficiency has been studied in relation to moonlight intensity and phase (Bowden 1973, 1981, 1982, Bowden and Church 1973, Bowden and Morris 1975), moonlight polarization (Nowinszky et al. 1979), and other environmental factors (Mikkola 1972, Kurtze 1974, Blomberg et al. 1978, Bowden 1982, Kolligs 2000). Ordinarily, light traps used to capture insects contain their own light sources, but in Germany light traps have been designed to exploit streetlamps themselves as light sources.

In this chapter I review the behavior of insects around artificial lights and the potential effect of lighting on insect diversity. I then present for the first time in English the results of a study that used light traps at streetlamps in Germany to measure attraction of insects to different kinds of high pressure vapor discharge streetlamps. The study investigated high pressure mercury vapor and sodium vapor lamps as well as other lamps. The investigation quantified the numbers of insects trapped and analyzed the numbers according to insect order, lamp location, nocturnal temperature, and lunar phase. The remainder of the chapter compares these results with other research and discusses the implications for choice of streetlamp type.

Insect Behavior Around Streetlamps

Flight-to-light behavior of insects around streetlamps disturbs the ecology of insects in many ways and often leads to high mortality. Bowden (1982) distinguished "near" from "far" effects for the approaching behavior of insects to lamps. Most studies have focused on "near" effects, within the zone of attraction. "Far" effects are derived from a changing background illumination by the moon or other light sources. In this section I draw on all published observations of insect behavior near lamps to classify three different situations in which flight-to-light behavior manifests itself. The physiological mechanism for this behavior is reviewed in depth by Frank (1988; Chapter 13, this volume).

In the first situation, insects are disturbed from their normal activity by contact with one or more artificial illumination sources. For example, a moth crosses a meadow searching for flowers (Figure 12.1a). When it comes into the zone of attraction of a streetlamp it approaches the lamp, where different interactions are possible. The insect may fly directly onto the hot glass cover of the lamp and die immediately. Far more frequently the insect orbits the light endlessly until it is caught by predators or falls exhausted to the ground, where it dies or is caught by predators. But other insects are able to leave the nearest light zone and fly back, seeking the shelter of the darker zone. There they rest on the ground or in the vegetation. It is assumed that the trigger for this behavior is a strong dazzling effect of the lamp. Some insects are able to recover and fly back to the lamp once more, and others remain inactive, exposing them to a higher risk of predation. Many insects may fail to reach the light because they become dazzled and immobilized during their approach and rest on the ground or in the vegetation. Hartstack et al. (1968) showed that more than 50% of moths approaching a light stopped their flight on the ground. I call all these variants of behavior the "fixation" or "captivity" effect, which means that insects are not able to escape from the near zone of lighting.

The second situation is the disturbance of long-distance movement of insects by lights encountered in their flight path. The scenario begins with three insects flying through a valley along a small stream (Figure 12.1b). They use natural landmarks such as trees, stars, the moon, or the profile of the horizon to orient. The dotted arrows mark their intended route of flight. The course of the flight is then intersected by a street and a row of streetlamps. The lights prevent the insects from following their original flyway. They fly directly to a lamp and are unable to leave the illuminated zone, suffering the same fate as described above (Figure 12.1a). I call this the "crash barrier" effect because of the interruption of movement across the landscape.

The third situation shown is what I call the "vacuum cleaner" effect (Figure 12.1c). Insects that otherwise are not moving (either foraging or migrating) are drawn to their deaths by lights. Insects are sucked out of habitat areas as if by a vacuum, which may deplete local populations. In a study of insects attracted to an illuminated greenhouse, Kolligs (2000) found that the insects captured on each side of the greenhouse reflected habitat conditions on that side, providing an example of the "vacuum cleaner" effect.

The magnitude of each of the effects on insect behavior depends on background illumination. During the full moon, moonlight always competes with artificial light sources. Insects therefore perceive artificial lights only from a shorter distance, and consequently fewer insects are attracted to any given light. Lower flight activity is indicated by light

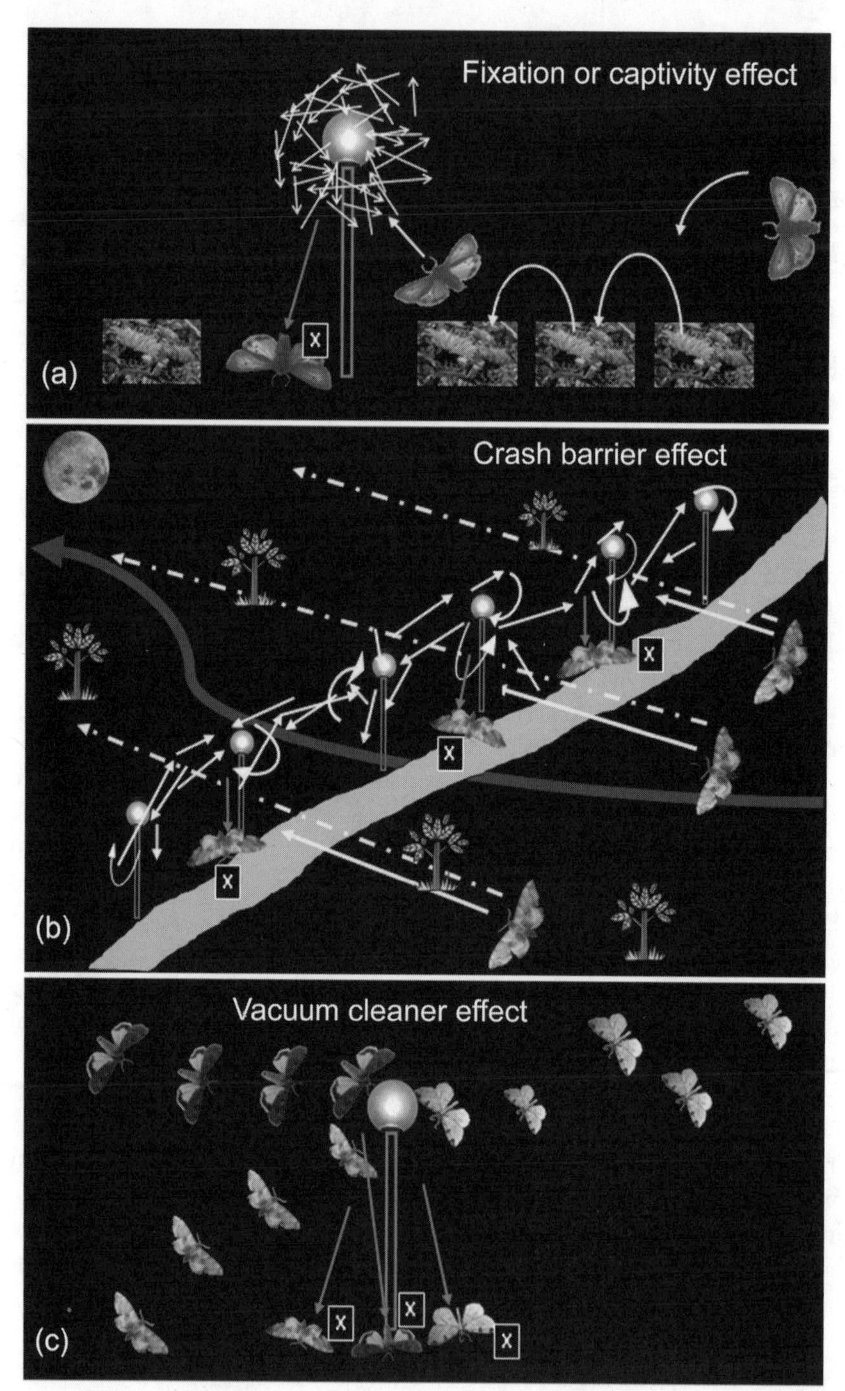

Figure 12.1. Three main effects on insect behavior in the "near" zone of artificial lights: *(a)* "fixation" or "captivity" effect, *(b)* "crash barrier" effect, and *(c)* "vacuum cleaner" effect.

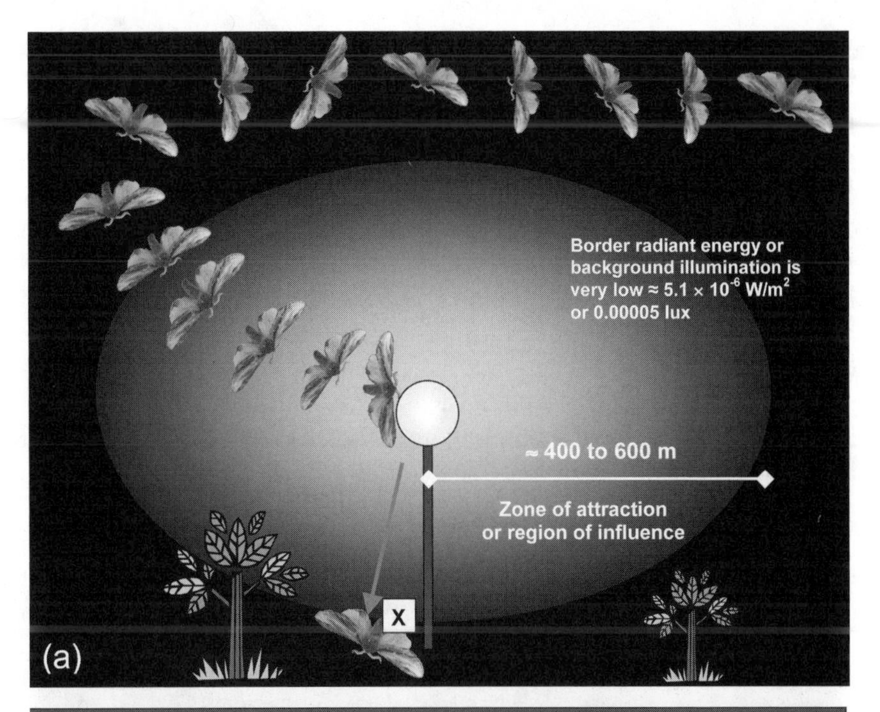

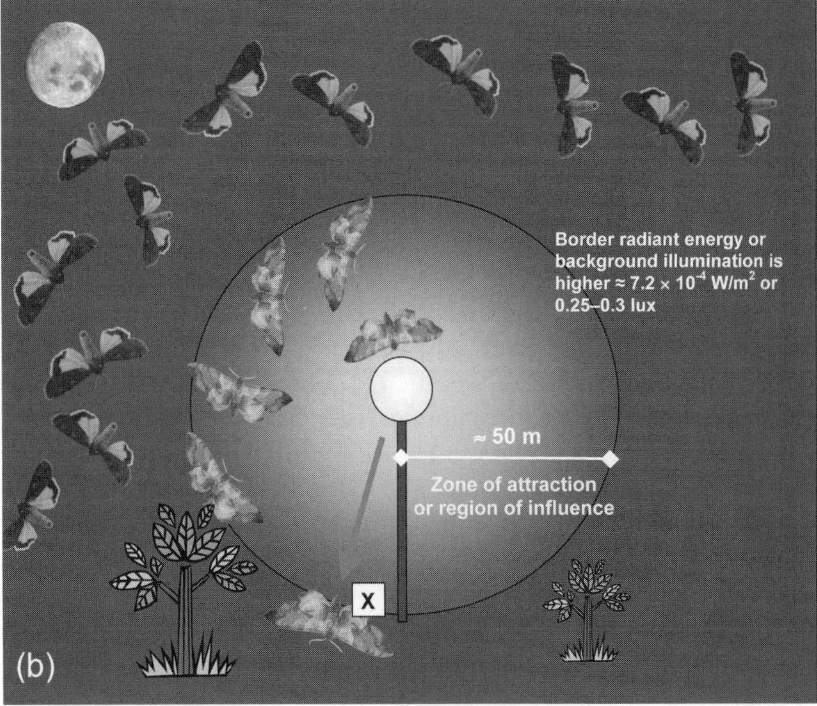

Figure 12.2. Zone of attraction of a streetlight with *(a)* a dark sky and *(b)* a moonlit sky.

trapping results, but other findings obtained with suction traps indicate that the general flight activity can be even greater at full moon (Bowden 1981, Danthanarayana 1986), thus indicating that moonlight can stimulate insects to be active for mating and reproduction. For example, the zone of attraction of a light with a dark sky can extend to several hundred meters (Figure 12.2). The maximum radius of attraction is then defined as the place where the radiant energy (illumination coming from the lamp) is equal to the background radiant energy (background illumination coming from the sky, about 5.1×10^{-6} W/m^2 or 5×10^{-5} lux). If an insect enters the zone of attraction, it will be exposed to increasing light intensity and, if physiologically susceptible to attraction, will approach nearer and nearer to the lamp.

The length of the radius of attraction is regularly discussed in the literature and is controversial. Baker and Sadovy (1978) favored very short distances of 3 m (9.8 ft) for light trap response of moths, but the test lamp was only 60 cm (2 ft) above ground. If lamps were exposed 9 m (30 ft) above ground level the radius increased to 10–17 m (33–56 ft). Based on moon phase and changing background illumination, researchers have proposed different radii of the attraction zone for a 125-W mercury vapor light source: 50–700 m (164–2,297 ft; Dufay 1964), 35–519 m (115–1,703 ft; Bowden and Morris 1975), and 57–736 m (187–2,415 ft; Bowden 1982). Kolligs (2000) found a maximum radius of 130 m (427 ft) testing different moth species using the mark–recapture method around a greenhouse. Each of three species exhibited a unique distance of attraction. Consequently the radius of the zone of attraction with a dark sky is reduced to 400–600 m (1,312–1,969 ft) in this schematic example (Figure 12.2a).

Less flight-to-light behavior occurs with a full moon (Figure 12.2b). The radius of attraction is reduced approximately one order of magnitude because the background illumination is about 7.2×10^{-4} W/m^2 or 0.25–0.3 lux. Under natural conditions, therefore, the zone of attraction changes during a lunar cycle. Additionally, changes may occur during a single night depending on weather, for example, by changing from clear to cloudy sky (see Chapter 15, Figure 15.3, this volume). Consequently, the effectiveness of catches around lamps also depends on background illumination. Sky glow from artificial lighting can create nearly the same illumination as a full moon. Illumination at the city center of Kiel, Germany was measured at 0.5 lux (Kurtze 1974), and the urban sky glow of Vienna with a cloudy sky was measured at 0.178 lux (T. Posch, unpublished observation). As yet no data are available about insect activity in the open space of settled areas that are constantly illuminated.

Potential Reduction in Insect Diversity Around Lights

In older publications, entomologists frequently reported extremely large light trap catches of many thousand insects in a night, but more recent catches have been much smaller. Robinson and Robinson (1950) caught more than 50,000 moths in a single trap on the night of August 20–21, 1949. Worth and Muller (1979) caught 50,000 moths with a single light trap from May 2 to September 12, 1978 on an isolated farm site not close to competing lights, whereas Eisenbeis and Hassel (2000) counted only 6,205 specimens of moths using 19 light traps from May 29 to September 29, 1997. Of course such simple enumeration does not allow for statistical evaluation, but there may be derived a strong indication of a progressive decline in insect populations.

In the report "A Century of Change in the Lepidoptera," Heath (1974) described some profound changes in macrolepidoptera in Great Britain, which can be attributed mainly to changes in land use. Most changes were extinctions, declines, or restrictions of species to a few local spots but also included some examples of colonization of new species and extension of existing ranges. He notes the main causes for the change of insect habitats as clearcutting of many acres of deciduous forests and their replacement with coniferous plantations, conversion of heath lands and forests to agricultural use, the agricultural revolution and changes in woodland management, use of chemicals (e.g., herbicides) in the environment, urban expansion, construction of motorways, human recreational pressure on the countryside, and periods of climatic change. There was no discussion at that time of light pollution as a serious new hazard for insects, but astronomers were pointing out deleterious effects of lighting (Riegel 1973). But Malicky (1965) reported from observations that at newly built and strongly illuminated fuel stations there was a high initial flight activity of insects during the first two years, which diminished rapidly in subsequent years. In my opinion such observations must be considered first indicators of a significant change of a local insect population caused by the "vacuum cleaner" effect.

Taylor et al. (1978) reported on the Rothamsted insect survey, which was based on a light-trapping network, in relation to the urbanization of land in Great Britain. The industrial region of middle England and the London area were clearly identified on faunal maps as islands of low diversity and density. The authors used light trapping as their basic method, but they offered no comments about the possible role of increasing artificial lighting in the decline in diversity.

Bauer (1993) investigated the insect activity around three housing areas normally illuminated by streetlamps and a seminatural habitat that was not regularly illuminated before the study. He used light traps in the light field of streetlamps in the suburban area of Konstanz, a midsized town in southern Germany. In the illuminated areas, the catch rates (5, 29, and 47 insects per trap per night in city central and two housing areas) were about two to five times lower than in the seminatural nonilluminated habitat (143 insects per trap per night), but together the results from the illuminated areas were heterogeneous. The dominance pattern of moths showed an average proportion of 14.9% at the illuminated sites and 34% at the nonilluminated site, but the differences between illuminated sites were high (2.7%, 11.6%, and 30.5%). For this reason, such data should be regarded only as a first indicator of changes in the insect population. Furthermore, it must be noted that light in the three illuminated areas was from a mix of three lamp types (high pressure mercury vapor, high pressure sodium vapor, and fluorescent), whereas the catches in the nonilluminated site were taken with high pressure mercury vapor alone. For this reason, the differences between sites actually should be somewhat smaller than was shown.

Scheibe (1999) investigated insect diversity using suction trapping along a wooded stream bank far from any artificial lighting in a low mountain range of the Taunus area in Germany. On eight nights he caught on average 2,600 insects per trap per night, and the maximum catches were 11,229 and 5,020 insects per trap per night. These data of flight activity outnumber all other data recently reported from illuminated areas in Germany. The results must be regarded as further evidence that the dark zones in the landscape have a much richer insect fauna than do lighted zones. In his Ph.D. thesis, Scheibe (2000) tried to determine the capacity of a light trap to catch insects flying within the zone of attraction of a single streetlamp. He counted all aquatic insects (e.g., mayflies, caddisflies, dipterans) emerging from a small stream in the same mountain range as his previous study, standardized as the number of emerging insects per 72 hours per 1 m (3 ft) of stream bank. During the night after such a measurement of insect emergence, he determined the number of aquatic insects flying to a streetlamp positioned near the bank. He found that different taxa of aquatic insects reacted differently, but in many instances light catches significantly outnumbered the number of emerging insects. For example, the number of caddisflies caught on an August night by the lamp was approximately the same as the number of caddisflies emerging along 200 m (656 ft) of the stream bank. Therefore, it can be concluded that the lamp has a long-distance effect for light-susceptible insect species and that many more insects are attracted than potentially would be found in the immediate surroundings of a lamp. By extrapolation, if there were a row of streetlamps along a stream, a species could become locally extinct in a short time, which can be explained by the "vacuum cleaner" effect of streetlamps.

Another example of attraction of large numbers of insects around lamps is reported from mayflies along riversides and bridges. The swarming of the species *Ephoron virgo* (or other species) is described as "summer snow" drifting (Kureck 1992, Tobias 1996) because the insects are attracted in such masses that the ground near lights is covered a centimeter thick with them. An estimated 1.5 million individuals have been recorded in one night on an illuminated road surface of a bridge. Each female loses her egg cluster upon first contact with an object. Eggs that are not released into water must be regarded as a loss for the population, with potentially significant local effects.

As discussed by Frank (1988; Chapter 13, this volume), rare species are vulnerable to population effects from artificial lighting. Kolligs (2000) reported capturing endangered Red List species as single individuals in a large study of assimilation lighting at a greenhouse. Such species can be regarded as endangered by artificial lighting. K-selected species with specialized habitat requirements and stable population sizes are most likely to be disrupted by artificial lighting (see also Eisenbeis 2001a, 2001b).

Given this understanding of insect behavior around artificial lights and the potential consequences for insect diversity and abundance, our study investigated insect attraction across different streetlamp types, comparing insect group abundance and diversity and the presumed benefit of converting to illumination sources attracting fewer insects.

Attraction of Insects to Streetlamps in Rural Germany

My research team monitored insect activity around streetlamps in the nearly treeless rural landscape of Rheinhessen in southwest Germany, which is characterized by viticulture and cultivation of cereals and sugar beet (see Eisenbeis and Hassel 2000 for details of the field study). Monitoring was performed near a small village of approximately 1,000 residents. Three sites were studied: a housing area of Sulzheim village (with some garden ponds), a farmhouse site (far from any water bodies), and a road site near Sulzheim village. Light traps were mounted just below streetlamp luminaires to capture insects.

The traps consisted of two crossed baffles (netlike or platelike), one or two catch funnels, and two small or one large catch container. Insects attracted from the light source fell or flew from the baffles via catch funnel into the collecting container. At any particular site, only one kind of trap was used. The types of luminaires and lamps in the study are commonly used for outdoor lighting in Germany. The lamps were high pressure mercury vapor (80 W) or high pressure sodium vapor (70 or 50 W). Additionally, we tested high pressure sodium–xenon vapor lamps (80 W), and for special purposes some of the high pressure mercury vapor lamps were fitted with ultraviolet absorbing filters over the glass cover of luminaires.

Normally the team prepared 19 light traps each day before dusk, which remained exposed during the night until morning. Insects were trapped in receptacles containing tissues and small vials filled with chloroform. Trapping continued from June until the end of September 1997.

We collected more than 600 samples, from which 536 were counted and analyzed in detail. Collection data were averaged for the whole flying period (arithmetic means or medians). Statistical analysis was conducted with Statistica from Statsoft.

The numbers of insects caught per trap per night peaked at nearly 1,700 in July and August, but on most nights the numbers were less than 400. Collections were smaller on nights with lower temperatures and during the full moon. Therefore, catch data do not fit a normal distribution unless plotted in the $\log(x + 1)$ form, which was used for certain calculations.

Influence of Lamp Type on Number of Insects Captured

Raw counts of insects were normalized to the number of insects caught per trap per night (Figure 12.3). This number was greatest for high pressure mercury vapor, followed by high pressure sodium–xenon vapor, and then by high pressure sodium vapor lamps. The number was lowest for high pressure mercury vapor lamps with ultraviolet-absorbing filters. The catch ratio of high pressure sodium vapor to high pressure mercury vapor (unfiltered) lamps is 0.45 for all insects and 0.25 for moths. These catch ratios suggest that conversion from high pressure mercury vapor to high pressure sodium vapor lamps would reduce catches of all insects by 55% and moths by 75%.

The ratios of insects caught with different lamps in our study are consistent with those of other studies, despite differences in methods (Figure 12.4). Bauer (1993) and Eisenbeis and Hassel (2000) used similar traps, Scheibe (1999, 2000) used suction traps, and Kolligs (2000) used three types of light traps, including box traps near the ground and air traps near streetlamps. On average 57% fewer insects were caught around high

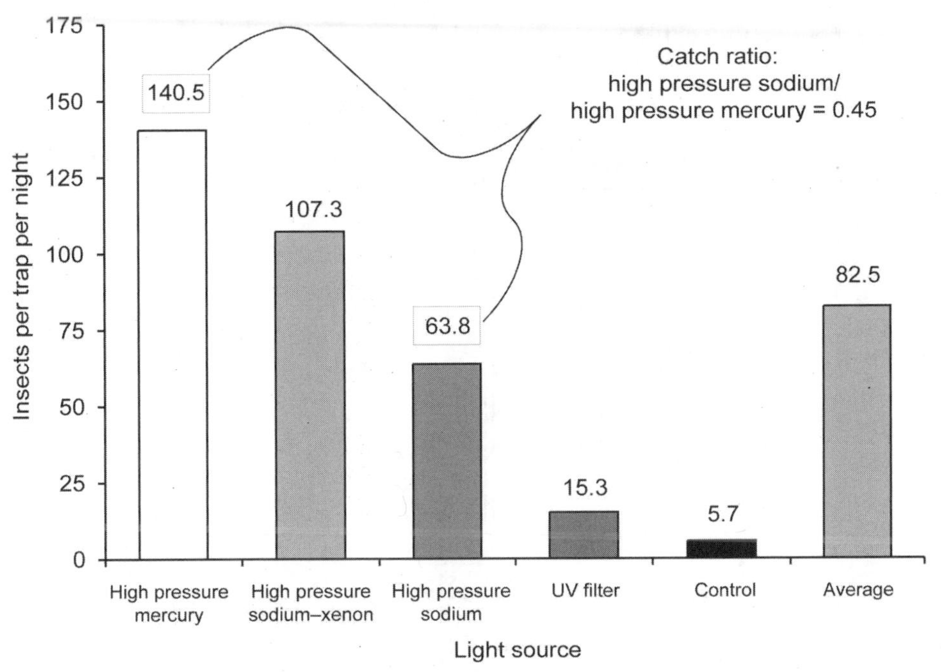

Figure 12.3. Average catch rates (arithmetic means) for the tested light sources: high pressure mercury (n = 192), high pressure sodium–xenon (n = 33), high pressure sodium (n = 201), UV filter (n = 24), control (light traps used without light source, n = 86), and average (n = 536).

pressure sodium vapor lamps than around high pressure mercury vapor lamps (Figure 12.4).

We found that the catch ratio of moths caught at high pressure sodium vapor compared to mercury vapor lamps was 0.25. Bauer (1993) reported catch ratios for moths ranging from 0.095 to 0.2. Klyuchko (1957) measured catches of noctuid moths at a 300-W incandescent lamp and a mercury vapor lamp (wattage unspecified). The ratio of individuals caught at the incandescent lamp to the mercury vapor lamp was 0.2.

High pressure mercury vapor lamps fitted with UV-blocking filters reduced catches of insects even more than high pressure sodium vapor lamps. The filters initially were intended as a possible modification of mercury vapor luminaires to protect insects. The filters reduced visible light emission to such a degree that the filtered mercury vapor streetlamps failed to satisfy German requirements (DIN No. 5044) for roadway illumination. This result may prompt light manufacturers to produce lamp cases using UV-absorbing glass.

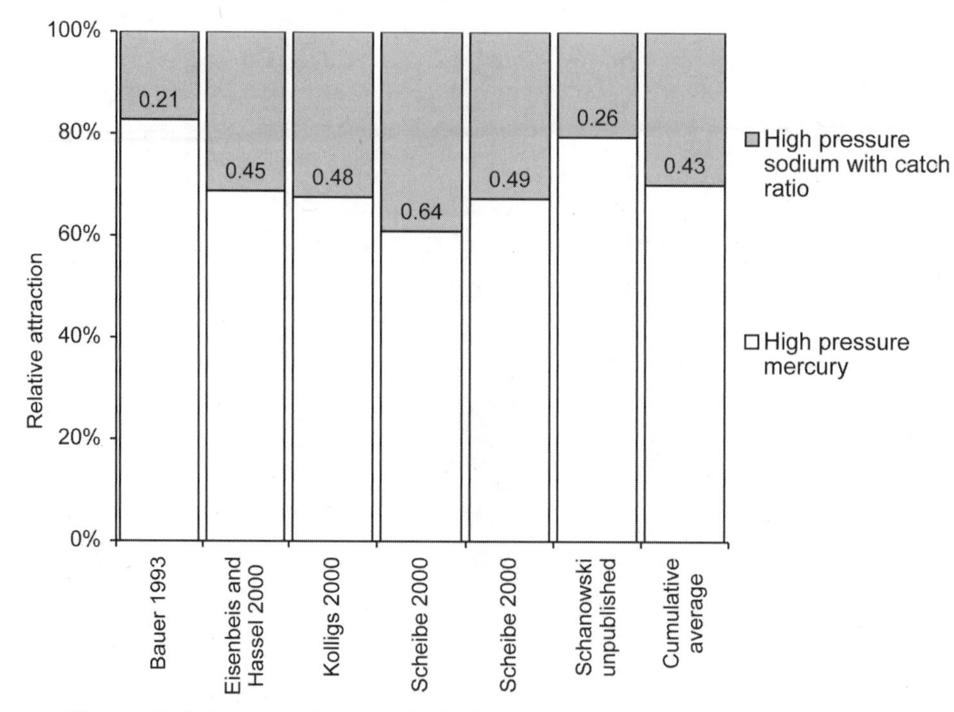

Figure 12.4. Insect catch ratios for high pressure sodium vapor and high pressure mercury vapor lamps according to different authors.

Scheibe (1999) suggested that flight of insects to high pressure mercury vapor lamps would increase only under conditions of light competition when both high pressure mercury vapor lamps and those with comparatively low ultraviolet light radiation (e.g., high pressure sodium–xenon or sodium vapor lamps) were operating simultaneously near each other. Under noncompeting conditions insects would be attracted at the same order of magnitude to lamps with low ultraviolet radiation. To test this hypothesis, lamps were operated singly and changed from night to night over a period of nearly four weeks. This experiment was conducted at a farmhouse site in a dark area without any other light sources. The catch ratio of high pressure sodium vapor lamps to high pressure mercury vapor lamps was 0.48—nearly the same value as that calculated for insects in our complete study. The findings also are consistent with the work of Bauer (1993). They demonstrate conclusively that quality of light emitted by streetlamps influences flight-to-light behavior of insects.

Influence of Lamp Location on Orders of Insects Captured

Twelve orders of insects were collected in the traps (see Eisenbeis and Hassel 2000 for complete details). The community at the road site near an open landscape with fields and vineyards was dominated by flies (Diptera, 67.6%), with the percentage of each of the other orders lower than 10%. Insects caught at the housing area of Sulzheim village were dominated by beetles (Coleoptera, 30.7%), followed by moths (Lepidoptera, 15.9%), aphids (Aphidina, 14.3%), flies (Diptera, 9.8%), caddisflies (Trichoptera, 8.1%), bugs (Heteroptera, 8.0%), and hymenopterans (Hymenoptera, 5.9%). The proportion of each of the remaining orders remained less than 5%. At the farm site three orders dominated the spectrum of insect orders: beetles (Coleoptera, 38.9%), moths (Lepidoptera, 19.4%), and bugs (Heteroptera, 12.8%). Each of the others contributed less than 10%. The aquatic caddisflies (Trichoptera) were found in high proportions (5.0%, 8.1%) at only two sites, which were near small bodies of water such as ponds in gardens. The proportion of this order was small (0.7%) at the farmhouse site, where there were no aquatic habitats nearby.

Influence of Temperature and Moonlight on Insects Captured

To investigate the importance of temperature on flight of insects to streetlamps, we compared catches of insects on warm and cool nights. We defined nights as warm if the temperature at 10:00 P.M. was higher than 19°C and cool if the temperature was lower than 17°C. On many nights temperatures were 12–14°C or higher than 21°C. The findings demonstrated that temperature strongly influenced the catches. Numbers of insects caught at streetlamps on cool nights decreased nearly to zero, and on warm nights they rose strongly, contributing to sharp peaks (see also Eisenbeis 2001a, 2001b).

To investigate the importance of moonlight on the flight of insects to streetlamps, we compared catch rates for nights around the full moon and new moon (Figure 12.5). Numbers of insects caught per night around the new moon were up to seven times higher than those around the full moon. Streetlamp type (high pressure mercury vapor versus high pressure sodium vapor) did not significantly affect these results. Nowinszky et al. (1979) calculated a ratio of 2.59, and Williams (1940) reported a ratio of 2.67 for light trap collections at new moon and full moon. Nabli et al. (1999) found no significant effect of moon phase on light trap catch rates of beneficial insects, but the sample size was too small for statistical analysis.

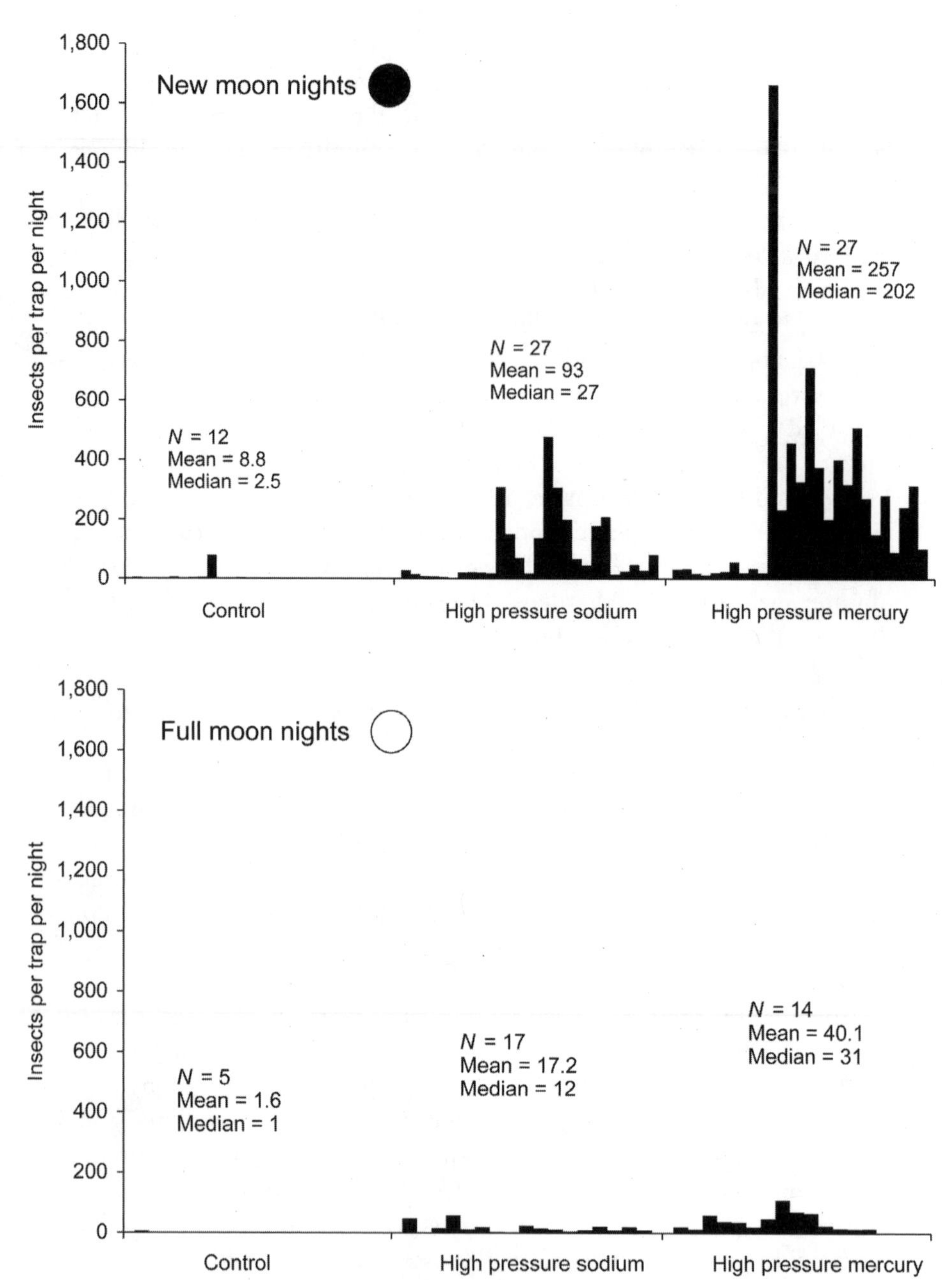

Figure 12.5. Catches of insects at streetlights in rural Germany on new moon and full moon nights.

According to Danthanarayana (1986), some insects show a trimodal flight periodicity during the lunar month, with peaks before and after the new moon and at the full moon. Other insects show a bimodal periodicity, with peaks only before and after the new moon. The drop in numbers of insects that our traps caught at streetlamps around the full moon could have resulted from decreased flight activity during this phase of the lunar cycle or from decreased attraction to artificial lighting. To differentiate between these two possibilities, insects must be collected in traps that do not depend on flight-to-light behavior, such as bait or suction traps. Using suction traps, Williams et al. (1956) were unable to confirm the existence of a regular lunar cycle in the number of insects caught.

Estimate of Total Insect Mortality Near Streetlamps

Vast numbers of insects are killed on windshields of vehicles, but measuring this destruction is problematic. Gepp (1973) attempted to calculate the number of insects killed in traffic each year in Austria and came up with an estimate of billions (116 insects/km [187/mile] for the front of a car). Similar difficulties confront attempts to quantify destruction of insects at artificial light sources. Such destruction is commonly observed when insects collide with hot lamps or are killed at the lamps by predators such as birds, spiders, and bats. Insects that are not killed may become inactive, failing to perform basic life functions such as reproduction.

Bauer (1993) compared the number of insects approaching streetlamps with the number caught in traps. The ratios found varied according to kind of insect: 1.4:1 for mayflies, 2.3:1 for caddisflies, 1.6:1 for macrolepidoptera (excluding geometrids), 11:1 for geometrid moths, 7:1 for flies (Nematocera), and 41:1 for microlepidoptera. He calculated a ratio of 3:1 for all insects. This means that for every three insects approaching a streetlamp, one will be caught in the trap. By comparing hand and trap catches, Kolligs (2000) found that traps captured about 30–40% of approaching moths.

Unfortunately, these ratios do not indicate the proportion of insects killed or incapacitated by the lamps. Bauer (1993) checked the number of dead insects found in the morning on the housing of a lamp facing vertically into the sky. Although the glass cover of the housing was not very hot, a high proportion of the insects (10–76% of the total) were dead. These mortality rates do not include insects that were eaten by predators or that flew away, were fatally injured, or survived unable to reproduce. An estimate of the death rate might reasonably be placed at 33%.

The following calculations represent an attempt to estimate the order of magnitude of insect mortality around streetlamps for a midsized city in Germany and for the whole country. These calculations are based on the following assumptions:

- The lamp types are mercury vapor or high pressure sodium vapor.
- The average catch ratio of high pressure sodium vapor to mercury vapor is about 0.4 (Figure 12.4).
- The mean number of insects approaching a lamp per night is 450 for mercury vapor and 180 for high pressure sodium vapor (as derived from our catch data in Rheinhessen).
- The mean number of insects killed at a streetlamp per night is estimated to be 150 for high pressure mercury vapor and 60 for high pressure sodium vapor.

For a hypothetical city the size of Kiel, northern Germany, which had 20,000 streetlamps and a population of 240,000 in 1998 (Kolligs 2000), the average number of insects killed at these streetlamps (assuming that all were high pressure mercury vapor) would be about 3 million per night, or 360 million per season, consisting of 120 days from June to September (including cold and warm nights).

The ratio of streetlamps to people (1:12) in Kiel is specific for that city. The ratio drops to 1:10 if the area is expanded to include a rural county (Rheinhessen, near Mainz), a mixed rural–urban county near Frankfurt (Rheingau–Taunus region), and another city, Augsburg, in southern Germany. Assuming a ratio of streetlamps to people of 1:10 for all of Germany, this country with a population of nearly 82 million would have 8.2 million streetlamps. Extrapolating the insect mortality to all of Germany, the number insects killed at streetlamps over a summer would be on the order of 10^{11}.

Selection of Streetlamp Type for Insect Conservation

Compared with other light sources, low pressure sodium vapor lighting undoubtedly attracts the fewest numbers of insects overall. Furthermore, it is most efficient in terms of visible light (lumens) in relation to energy consumption (W). Few published studies have investigated flight of insects to low pressure sodium vapor lighting (Schanowski and Späth 1994), but assessed roughly, the attraction of insects is at least an order of magnitude lower than for high pressure sodium vapor lighting and sometimes approaches zero.

Many suburban communities have introduced low pressure sodium vapor street lighting along some roads. Low pressure sodium vapor lighting often is used in industrial areas and along waterways, such as floodgates. In the United States, the roadway of the Golden Gate Bridge originally was illuminated with low pressure sodium vapor lamps, but this lighting has been converted to high pressure sodium vapor lamps with amber caps (to preserve the appearance), and additional decorative high pressure sodium vapor lighting was added in 1987.

Nostalgia for the yellow glow of the original Golden Gate Bridge lighting notwithstanding, low pressure sodium vapor lamps are not likely to gain widespread acceptance as primary light sources for illuminating streets, for several reasons. First, at higher wattage, these lamps have large dimensions. For example, a 135-W lamp is 77 cm (2.5 ft) long, and a 180-W lamp is 112 cm (3.6 ft) long. This large size makes it difficult to install these lamps in common lamp fixtures, especially those that include optical systems. Second, according to some experts, low pressure sodium vapor lamps require a larger electrical load at startup, and their lifespan is shorter than that of high pressure sodium vapor lamps. Third, their light is monochromatic, which is unacceptable when color vision is important (but color vision is possible under low pressure sodium vapor light mixed with other illumination).

High pressure sodium vapor lamps now are becoming the main source of street lighting in towns and villages for several reasons: good lighting, good color perception, moderate energy consumption, long life, and fewer problems with waste disposal of mercury (although some types of high pressure sodium vapor lamps still contain mercury).

It is often argued that lighting accounts for only a small proportion of the total energy budget of a country (e.g., 1.9% for lighting and 0.1% for street lighting). In 1996, energy consumption for street lighting was about 3.5 million MWh in Germany. The City of Osnabrück, with a population of 160,000, needs 8,000 MWh per annum (Hänel 2001), which is equivalent to 5,000 t of CO_2 emissions into the atmosphere.

In the last few years technological improvements in lighting systems and lamps have improved the energy efficiency of street lighting in Germany, but expansion of street lighting has increased total energy consumed by streetlamps. Energy savings up to 70% can be achieved by making many changes: converting lamps from high pressure mercury vapor to high pressure sodium vapor, improving mirror systems and lamp housings, optimizing distances between streetlamps, and upgrading switch gear and control units. Such improvements would make possible a reduction in average consumption of energy of about 1 kWh per day per lamp and a reduction in CO_2 production of 0.6 kg (1.3 lb) per day per lamp, based on the current mix of coal, nuclear, hydro, and wind power stations in Germany. This would result in a reduction of 5 million kg of CO_2 produced per day by the estimated 8.2 million streetlamps in Germany (Table

Table 12.1. Energy consumption and equivalent values for CO_2 output after simple and advanced retrofitting of luminaires with lamps from Philips–AEG Lighting Technique, Germany.

			Reduction	
	Old System	*New System*	*Absolute*	*%*
Simple retrofitting	2x 125-W hp mercury vapor lamps, elliptical (type HPL)	2x 70-W hp sodium lamps, elliptical (type SON)		
Energy consumption for full illumination	280 W × 11 h/d = 3.08 kWh/d	160 W × 11 h/d = 1.76 kWh/d	–1.32 kWh/d	–43
Energy consumption for reduced illumination	280 W × 5 h + 140 W × 6 h = 2.24 kWh/d	160 W × 5 h + 80 W × 6 h = 1.280 kWh/d	–0.96 kWh/d	–43
CO_2 output for full illumination	1.848 kg/d	1.056 kg/d	–0.792 kg/d	–43
CO_2 output for reduced illumination	1.344 kg/d	0.768 kg/d	–0.576 kg/d	–43
Advanced retrofitting	2x 125-W hp mercury vapor lamps, elliptical (type HPL)	1x 100-W hp sodium lamp, tubular (type SON-T Plus)		
Energy consumption for full illumination	280 W × 11 h/d = 3.08 kWh/d	115 W × 11 h/d = 1.265 kWh/d	–1.815 kWh/d	–59
Energy consumption for reduced illumination	280 W × 5 h + 140 W × 6 h = 2.24 kWh/d	115 W × 5 h + 90 W × 6 h = 1.115 kWh/d	–1.125 kWh/d	–50
CO_2 output for full illumination	1.848 kg/d	0.759 kg/d	–1.089 kg/d	–59
CO_2 output for reduced illumination	1.344 kg/d	0.669 kg/d	–0.675 kg/d	–50

Source: Robert Class, Philips–AEG Lighting Technique, Germany.

12.1). These savings have yet to be fully realized in Germany because only a few cities, such as Augsburg (Isépy 2001), have nearly completed the conversion to high pressure sodium vapor lighting.

Conclusion

High pressure mercury vapor and high pressure sodium vapor lamps are the most common light sources for public lighting in Germany. Compared with high pressure mercury vapor lamps, high pressure sodium vapor lamps reduced attraction of insects to streetlamps by 55% and attraction of moths by 75% in our investigation in rural Germany. Lamp location, lunar phase, and nocturnal temperature all affected the number of insects that flew to streetlamps. These findings are consistent with those of other studies in Germany and elsewhere. Based on six studies in Germany, the reduction for insects is 57%. Light traps have thus far been revealed to be effective tools to estimate rates of insect attraction to lamps. But light trapping must be complemented by other sampling methods to allow more exact determination of insect population dynamics.

At a global scale, outdoor lighting is increasing exponentially (Cinzano et al. 2001), which has the potential to disrupt ecosystems significantly. Insect species are especially sensitive to artificial night lighting because they often have no ability to resist the stimulus of light. Consequently, they will be disturbed in essential activities, including migration, dispersal, foraging, mating, and reproduction.

I have described three main effects of lights on insect behavior: the "fixation" or "captivity" effect, the "crash barrier" effect, and the "vacuum cleaner" effect. All these effects probably will reduce populations in the long term. It must be emphasized, however, that a robust monitoring program at different scales is necessary to detect such changes. Comparing old and new entomological data provides some indications that steep gradients in insect abundance exist between the few remaining natural habitats and urban areas, but these data remain suggestive.

Species that are K-strategists are the most endangered by artificial lighting. Also, species that exhibit mass breeding, such as many aquatic insect species, may be endangered if shores and banks are totally illuminated. Notwithstanding such spectacular cases, a constant kill of insects takes place at streetlamps and other lights every night during the summer flight period in Germany. One simple estimate of insect mortality for the area of a German town and for the country of Germany shows that lights remove vast numbers of insects from ecosystems.

Replacing older high pressure mercury vapor lamps with high pressure sodium vapor lamps and combining them with full cutoff luminaires equipped with light-guiding mirror systems reduces energy waste (and associated CO_2 emissions) and insect mortality. This information should be communicated to professionals such as landscape planners, lighting designers, and policymakers at all levels, for their support is necessary to achieve these environmental benefits.

Acknowledgments

I am indebted to the nature conservation organization Bund für Umwelt und Naturschutz Deutschland; Kreisgruppe Alzey-Worms for the initiation of the project; the Ministry for the Environment Rheinland-Pfalz for financial support; the Department of Crop Production and Crop Protection Mainz-Bretzenheim for meteorological data; the owners of the farmhouse "Eichenhof," family Kussel, for permission to install streetlamps in a special zone near the farmhouse; the mayor of Sulzheim/Rheinhessen for supporting the project; the Electric Power Station EWR of Rheinhessen (Worms) for technical support; Dr. Steck from the German Society of Lighting Technique for useful tips; and the Feldbausch Foundation from the Department of Biology at Mainz University for support.

Literature Cited

Baker, R. R., and Y. Sadovy. 1978. The distance and nature of the light-trap response of moths. *Nature* 276:818–821.

Bauer, R. 1993. *Untersuchung zur Anlockung von nachtaktiven Insekten durch Beleuchtungseinrichtungen* [Investigation of the attraction of nocturnal insects by artificial lights]. Diploma thesis, Department of Biology, University of Konstanz, Germany.

Blomberg, O., J. Itämies, and K. Kuusela. 1978. The influence of weather factors on insect catches in traps equipped with different lamps in northern Finland. *Annales Entomologici Fennici* 44:56–62.

Bowden, J. 1973. The influence of moonlight on catches of insects in light-traps in Africa. Part I. The moon and moonlight. *Bulletin of Entomological Research* 63:113–128.

Bowden, J. 1981. The relationship between light- and suction-trap catches of *Chrysoperla carnea* (Stephens) (Neuroptera: Chrysopidae), and the adjustment of light-trap catches to allow for variation in moonlight. *Bulletin of Entomological Research* 71:621–629.

Bowden, J. 1982. An analysis of factors affecting catches of insects in light-traps. *Bulletin of Entomological Research* 72:535–556.

Bowden, J., and B. M. Church. 1973. The influence of moonlight on catches of insects in light-traps in Africa. Part II. The effect of moon phase on light-trap catches. *Bulletin of Entomological Research* 63:129–142.

Bowden, J., and M. G. Morris. 1975. The influence of moonlight on catches of insects in light-traps in Africa. III. The effective radius of a mercury-vapour light-trap and the analysis of catches using effective radius. *Bulletin of Entomological Research* 65:303–348.

Cinzano, P., F. Falchi, and C. D. Elvidge. 2001. The first world atlas of the artificial night sky brightness. *Monthly Notices of the Royal Astronomical Society* 328:689–707.

Danthanarayana, W. 1986. Lunar periodicity of insect flight and migration. Pages 88–119 in W. Danthanarayana (ed.), *Insect flight: dispersal and migration*. Springer-Verlag, Berlin.

Dufay, C. 1964. Contribution a l'étude du phototropisme des Lépidoptères noctuides [Contribution to the study of phototropism of noctuid moths]. *Annales des Sciences Naturelles, Zoologie, Paris* 12e Serie 6:281–406.

Eisenbeis, G. 2001a. Künstliches Licht und Insekten: eine vergleichende Studie in Rheinhessen [Artificial light and insects: a comparative study in Rheinhessen (Germany)]. *Schriftenreihe für Landschaftspflege und Naturschutz* 67:75–100.

Eisenbeis, G. 2001b. Künstliches Licht und Lichtverschmutzung: eine Gefahr für die Diversität der Insekten? [Artificial light and light pollution: a danger to the diversity of insects?]. Pages 31–50 in S. Löser (ed.), *Verhandlungen Westdeutscher Entomologen Tag 2000*. Landeshauptstadt Düsseldorf, Löbbecke-Museum + Aquazoo, Entomologische Gesellschaft, and Düsseldorf e.V., Düsseldorf.

Eisenbeis, G., and F. Hassel. 2000. Zur Anziehung nachtaktiver Insekten durch Straßenlaternen: eine Studie kommunaler Beleuchtungseinrichtungen in der Agrarlandschaft Rheinhessens [Attraction of nocturnal insects to street lights: a study of municipal lighting systems in a rural area of Rheinhessen (Germany)]. *Natur und Landschaft* 75(4):145–156.

Frank, K. D. 1988. Impact of outdoor lighting on moths: an assessment. *Journal of the Lepidopterists' Society* 42:63–93.

Gepp, J. 1973. Kraftfahrzeugverkehr und fliegende Insekten [Roadway traffic and flying insects]. *Natur und Land (Österreich)* 59:127–129.

Graham, H. M., P. A. Glick, and D. F. Martin. 1964. Nocturnal activity of adults of six lepidopterous pests of cotton as indicated by light-trap collections. *Annals of the Entomological Society of America* 57:328–332.

Haas, R., W. Kräling, and R. Boulois 1997. Lichtkontamination [Light contamination]. *UWSF [Umweltwissenschaften und Schadstoff-Forschung]: Zeitschrift für Umweltchemie und Ökotoxikologie* 9:24.

Hamilton, D. W., and L. F. Steiner. 1939. Light traps and codling moth control. *Journal of Economic Entomology* 32:867–872.

Hänel, A. 2001. The situation of light pollution in Germany. Pages 142–146 in R. J. Cohen and W. T. Sullivan III (eds.), *Preserving the astronomical sky*. IAU [International Astronomical Union] Symposium, Volume 196. The Astronomical Society of the Pacific, San Francisco.

Hartstack, A. W. Jr., J. P. Hollingsworth, and D. A. Lindquist. 1968. A technique for measuring trapping efficiency of electric insect traps. *Journal of Economic Entomology* 61:546–552.
Hartstack, A. W. Jr., J. P. Hollingsworth, R. L. Ridgway, and H. H. Hunt. 1971. Determination of trap spacings required to control an insect population. *Journal of Economic Entomology* 64:1090–1100.
Health Council of the Netherlands. 2000. *Impact of outdoor lighting on man and nature*. Publication No. 2000/25E. Health Council of the Netherlands, The Hague.
Heath, J. 1974. A century of change in the Lepidoptera. Pages 275–292 in D. L. Hawksworth, *The changing flora and fauna of Britain*. Systematics Association Special Volume No. 6. Academic Press, London and New York.
Hollingsworth, J. P., A. W. Hartstack Jr., and D. A. Lindquist. 1968. Influence of near-ultraviolet output of attractant lamps on catches of insects by light traps. *Journal of Economic Entomology* 61:515–521.
Isépy, S. 2001. Möglichkeiten zur Verminderung von Lichtimmissionen am Beispiel der Stadt Augsburg [Possibilities for the reduction of light emissions: an example of the city of Augsburg]. *Schriftenreihe für Landschaftspflege und Naturschutz* 67:163–170.
Klyuchko, Z. F. 1957. On the flight of noctuids (Lepidoptera, Noctuidae) towards different light sources. *Doklady: Biological Sciences Sections* [Biological Sciences Sections of the Proceedings of the Academy of Sciences of the USSR] 117:951–954.
Kolligs, D. 2000. Ökologische Auswirkungen künstlicher Lichtquellen auf nachtaktive Insekten, insbesondere Schmetterlinge (Lepidoptera) [Ecological effects of artificial light sources on nocturnally active insects, in particular on moths (Lepidoptera)]. *Faunistisch–Ökologische Mitteilungen* Supplement 28:1–136.
Kureck, A. 1992. Das Massenschwärmen der Eintagsfliegen am Rhein: Zur Rückkehr von *Ephoron virgo* (Olivier 1791) [The mass swarms of mayflies on the Rhine: to the return of *Ephoron virgo* (Olivier 1791)]. *Natur und Landschaft* 67:407–409.
Kurtze, W. 1974. Synökologische und experimentelle Untersuchungen zur Nachtaktivität von Insekten [Synecological and experimental investigations on the night-activity of insects]. *Zoologische Jahrbücher: Abteilung für Systematik Ökologie und Geographie der Tiere* 101:297–344.
Malicky, H. 1965. Freilandversuche an Lepidopterenpopulationen mit Hilfe der JERMYschen Lichtfalle, mit Diskussion biozönologischer Gesichtspunkte [Outdoor exposure tests of Lepidoptera populations with the help of JERMY's light-trap, with discussion of biocenotic aspects]. *Zeitschrift für Angewandte Entomologie* 56:358–377.
Malicky, H. 1974. Der Einfluss des Standortes einer Lichtfalle auf das Anflugergebnis der Noctuidae (Lepidoptera) [The influence of location of light-traps on the resulting approach of noctuids (Lepidoptera)]. *Folia Entomologica Hungarica* Supplement 27:113–127.
Mikkola, K. 1972. Behavioural and electrophysiological responses of night-flying

insects, especially Lepidoptera, to near-ultraviolet and visible light. *Annales Zoologici Fennici* 9:225–254.

Mitchell, E. R., and H. R. Agee. 1981. Response of beet and fall armyworm moths to different colored lamps in the laboratory and field. *Journal of Environmental Science and Health, Part A: Environmental Science and Engineering* A16:387–396.

Molenaar, J. G. de, D. A. Jonkers, and R. J. H. G. Henkens. 1997. Wegverlichting en natuur. I. Een literatuurstudie naar de werking en effecten van licht en verlichting op de natuur [Road illumination and nature. I. A literature review on the function and effects of light and lighting on nature]. DWW Ontsnipperingsreeks deel 34, Delft.

Nabli, H., W. C. Bailey, and S. Necibi. 1999. Beneficial insect attraction to light traps with different wavelengths. *Biological Control* 16:185–188.

Nowinszky, L., S. Szabó, G. Tóth, I. Ekk, and M. Kiss. 1979. The effect of the moon phases and of the intensity of polarized moonlight on the light-trap catches. *Zeitschrift für Angewandte Entomologie* 88:337–353.

Riegel, K. W. 1973. Light pollution: outdoor lighting is a growing threat to astronomy. *Science* 179:1285–1291.

Robinson, H. S., and P. J. M. Robinson. 1950. Some notes on the observed behaviour of Lepidoptera in flight in the vicinity of light-sources together with a description of a light-trap designed to take entomological samples. *Entomologist's Gazette* 1:3–20.

Schanowski, A., and V. Späth. 1994. Überbelichtet: Vorschläge für eine umweltfreundliche Außenbeleuchtung [Overexposed: suggestions for environmentally friendly outside lights]. Naturschutzbund Deutschland (NABU), Kornwestheim.

Scheibe, M. A. 1999. *Über die Attraktivität von Straßenbeleuchtungen auf Insekten aus nahegelegenen Gewässern unter Berücksichtigung unterschiedlicher UV-Emission der Lampen* [On the attractiveness of roadway lighting to insects from nearby waters with consideration of the different UV-emission of the lamps]. *Natur und Landschaft* 74:144–146.

Scheibe, M. A. 2000. *Quantitative Aspekte der Anziehungskraft von Straßenbeleuchtungen auf die Emergenz aus nahegelegenen Gewässern (Ephemeroptera, Plecoptera, Trichoptera, Diptera: Simuliidae, Chironomidae, Empididae) unter Berücksichtigung der spektralen Emission verschiedener Lichtquellen* [Quantitative aspects of the attraction of roadway lighting to insects emerging from nearby waters (Ephemeroptera, Plecoptera, Trichoptera, Diptera: Simuliidae, Chironomidae, Empididae) with consideration of the spectral emission of different sources of light]. Ph.D. thesis, Fachbereich Biologie, Johannes Gutenberg-Universität, Mainz.

Schmiedel, J. 2001. Auswirkungen künstlicher Beleuchtung auf die Tierwelt – ein Überblick [Effects of artificial lighting on the animal world: an overview]. *Schriftenreihe für Landschaftspflege und Naturschutz* 67:19–51.

Schütte, F. 1972. Zum Einfluß von Licht-(Duft-)Fallen auf die Populationsdichte von *Heliothis zea* (Boddie) [On the influence of light-(smell-)traps on a closed population of *Heliothis zea* (Boddie)]. *Zeitschrift für Angewandte Entomologie* 70:302–309.

Steck, B. 1997. *Zur Einwirkung von Außenbeleuchtungsanlagen auf nachtaktive Insekten* [On the effect of nearby exterior lighting on nocturnal insects]. LiTG-Publikation 15. Deutsche Lichttechnische Gesellschaft, Berlin.

Taylor, L. R., R. A. French, and I. P. Woiwod. 1978. The Rothamsted insect survey and the urbanization of land in Great Britain. Pages 31–65 in G. W. Frankie and C. S. Koehler (eds.), *Perspectives in urban entomology*. Academic Press, London.

Tobias, W. 1996. Sommernächtliches "Schneetreiben" am Main: zum Phänomen des Massenfluges von Eintagsfliegen [Summer nights "snow flurry" at the Main: on the phenomenon of the mass flight of mayflies]. *Natur und Museum* 126(2):37–54.

Williams, C. B. 1936. The influence of moonlight on the activity of certain nocturnal insects, particularly of the family Noctuidae, as indicated by a light trap. *Philosophical Transactions of the Royal Society of London. Series B, Biological Sciences* 226:357–389.

Williams, C. B. 1939. An analysis of four years captures of insects in a light trap. Part I. General survey; sex proportion; phenology; and time of flight. *Transactions of the Royal Entomological Society of London* 89:79–131.

Williams, C. B. 1940. An analysis of four years captures of insects in a light trap. Part II. The effect of weather conditions on insect activity; and the estimation and forecasting of changes in the insect population. *Transactions of the Royal Entomological Society of London* 90:227–306.

Williams, C. B., B. P. Singh, and S. el Ziady. 1956. An investigation into the possible effects of moonlight on the activity of insects in the field. *Proceedings of the Royal Entomological Society of London, Series A. General Entomology* 31:135–144.

Worth, C. B., and J. Muller. 1979. Captures of large moths by an ultraviolet light trap. *Journal of the Lepidopterists' Society* 33:261–264.

Chapter 13

Effects of Artificial Night Lighting on Moths

Kenneth D. Frank

Flight of moths to artificial sources of light is one of the most conspicuous ecological consequences of nocturnal lighting. This behavior carries risks not only for individual moths but also for moth populations. Inventories based on light trapping in natural habitats typically document hundreds of species (Table 13.1). What is the evidence that nocturnal lighting affects moth populations?

Artificial lighting typically accompanies a host of environmental disturbances. Isolating the effects of outdoor lighting on moth populations would be achieved best with studies that systematically vary exposure of habitats to artificial lighting. Controlling lighting and other ecological variables, however, is difficult in the urban and suburban settings where outdoor lighting is concentrated. Abundance and distribution of species of moths fluctuate from year to year, particularly in urban settings (Taylor et al. 1978). Because some noctuids migrate more than a thousand kilometers (Johnson 1969), effects of lighting on dispersal could be diffused over a broad area and escape detection in short-term or geographically limited studies.

Table 13.1. Number of species of macrolepidoptera moths collected at light traps in moth surveys. Methods differ, including collection times, number of traps, type of traps, lamps, and wattage.

Location	*Habitat*	*Light Source*	*Species Trapped*	*References*
Rothamsted, England	Field, woodland	Incandescent	256	Williams 1939
Rothamsted, England	Field, woodland	Mercury vapor	579	Hosny 1959
Kiel, Germany	Suburb	High pressure sodium vapor	126	Kolligs 2000
New Brunswick, Canada	Softwood forest	Black light	311	Thomas 1996
West Virginia, United States	Deciduous forest	Black light	343	Butler et al. 1999
Maine, United States	Fields, woodland, campus	Incandescent and mercury vapor	349	Dirks 1937
Oregon, United States	Mixed coniferous forest	Black light	383	Grimble et al. 1992
Ohio, United States	Deciduous forest	Black light	521	Teraguchi and Lublin 1999
N. Queensland, Australia	Tropical forest	Black light	835	Kitching et al. 2000
Sabah, Malaysia	Secondary tropical forest	Mercury vapor	1,048	Chey et al. 1997

Few studies have systematically examined effects of artificial lighting on moths, and none has measured effects on moth populations. Many studies, however, have used light traps and laboratory light sources to study moths. These provide a basis for understanding effects of outdoor lighting on moths. In addition, recent studies from Germany have systematically investigated flight of moths to streetlights and other kinds of outdoor lighting (Eisenbeis and Hassel 2000, Hausmann 1992, Kolligs 2000; see Chapter 12, this volume). In Britain, large-scale surveys have documented the distribution of moth populations in urban environments (Plant 1993, 1999, Taylor et al. 1978), where density of artificial light sources is highest. Population studies performed in a wide variety of geographically isolated habitats in the last decade suggest how artificial lighting may affect the ability of moth species to survive habitat fragmentation, which typically accompanies outdoor illumination (Daily and Ehrlich 1996, Kozlov 1996, Nieminen 1996, Nieminen and Hanski 1998, Usher and Keiller 1998).

This chapter addresses several questions. How does flight-to-light behavior disturb vital activities of moths? How many species of moths exhibit this behavior, and how many live in illuminated environments? What factors may affect susceptibility of populations of moths to outdoor lighting? Within ecosystems, moths function as herbivores, pollinators, and prey. Based on available evidence, does artificial lighting disturb moths enough to warrant remedial action, and what might this action be?

The evidence suggests that although individuals of most species of larger moths fly to artificial light, populations of most of these species can persist near lights. The populations that artificial light is most likely to threaten are those already endangered by habitat loss and fragmentation. Restrictions on lighting may help to protect these populations and spare the unnecessary deaths of individuals of other populations.

Effects of Artificial Lighting on Individual Moths

Artificial lighting can affect almost every aspect of the life cycle of individual moths. Flight to light is a common behavior for moths, but other secondary effects such as interference with crypsis and hearing also result from lighting.

Flight to Light

When approaching lamps, moths may ignore them, circle around them, crash into them, zigzag in front of them, or loop toward them and continue past them. They may come to rest or flutter on the ground around lamps. In a study using video imaging, flight paths near lamps frequently changed angular velocity and direction (Muirhead-Thomson 1991). The maximum density of moths flying to lamps was found to be 40 cm (1.3 ft) from the midpoint of the lamps. Flight speeds usually were at their minimum when moths were flying toward the light source and often at their maximum when they were flying away from it (Muirhead-Thomson 1991). *Manduca sexta* L. (Sphingidae), on the other hand, flew faster toward lamps than away from them and approached lamps in a straight line (Spencer et al. 1997). All *Manduca sexta* landed or struck the ground while approaching lamps in a field (Hartstack et al. 1968).

Many studies have investigated the distance over which lamps elicit flight-to-light behavior, and the findings have ranged from 3 to 130 m

(10–427 ft), depending on methods and species (Baker and Sadovy 1978, Hamilton and Steiner 1939, Hartstack et al. 1971, Kolligs 2000, Plaut 1971, Robinson and Robinson 1950, Robinson 1960, Stanley 1932). Longer distances, up to 500 m (1,640 ft), have been hypothesized based on retinal sensitivity but not demonstrated (Agee 1972, Bowden and Morris 1975, Graham et al. 1961, Hsiao 1972).

Several theories try to explain the behavior of moths around artificial sources of light, but none accounts for the diversity of observed behaviors. The light-compass theory postulates that moths navigate by flying at a constant angle to a distant light source such as the moon. When near a lamp, they mistake the lamp for the distant light source and fly at a constant angle to it. This directs them along flight paths that spiral in toward the lamp or circle around it (Baker and Sadovy 1978; see Chapter 14, Figure 14.7, this volume). Experimental evidence supports the hypothesis that moths navigate by establishing a visual fix on the moon (Sotthibandhu and Baker 1979). The mach band theory postulates that artificial light produces visual artifacts of apparent darkness next to the light source. The moth flies toward the apparent darkness, which directs it into a path around the light (Hsiao 1972). Other theories postulate that moths fly to artificial light sources because the light dazzles them (Robinson and Robinson 1950), temporarily blinds them (Hamdorf and Höglund 1981), or signifies open space (McGeachie 1988).

The basis for flight-to-light behavior is difficult to understand because the circuitry and computations that moths normally use for flight control are poorly understood (Wehner 1998). For navigational guidance, light competes with other sensory cues: aerodynamic, gravitational, inertial, chemical, geomagnetic, acoustic, and imaging (Janzen 1984). Experience and memory may influence how moths process this information (Janzen 1984). The degree to which artificial light disrupts navigation of a particular moth at a particular time may depend in part on the degree to which the moth is relying on alternative cues (Janzen 1984). In the absence of sensors and systems that evolved in response to artificial light, flight-to-light behavior may best be viewed as an artifact of unknown cause.

Under some conditions, moths may avoid illuminated areas (Hsiao 1972, Robinson and Robinson 1950). A few published studies on control of agricultural pests support the possibility that light from lamps repels moths (Herms 1929, 1932, Nemec 1969, Nomura 1969). Investigating flight from light is more difficult than studying flight to light, and available information is inconclusive.

Flight Activity

Moths that fly to light may land nearby and remain quiescent, sometimes for the rest of the night. Even a few hours of lost flight time may pose a high cost for moths, most of which live for a week or less as adults (Young 1997) and fly for only part of the night (Williams 1939). Costa Rican saturniids and sphingids that come to rest at lamps are likely to stay all night (Janzen 1984). In Germany, 73% of sphingids that were marked and released after flying to a mercury vapor light were recaptured the next evening at the lamp. For noctuids, the recapture rates after one day varied from 1.9% to 43.2%, depending on the species (Kolligs 2000). Cessation of flight near light sources has interfered with light trapping of agricultural pests such as *Heliothis zea* (Boddie) (Noctuidae) and *Trichoplusia ni* (Hübner) (Noctuidae) (Hartstack et al. 1968).

Moths tend to stop flying when confined inside illuminated structures such as phone booths or garages with open doors. Moths may land in illuminated alcoves or stairwells that impede mobility or on illuminated surfaces that merely provide accessible landing sites. Collectors have exploited this behavior by capturing moths that have come to rest on illuminated sheets hung vertically near the ground (Winter 2000).

Vision

A moth flying from artificial light into darkness may be functionally blind until eye pigments have returned to their dark-adapted positions. The compound eyes of moths adapt to increases in light intensity by movement of pigment and, in some cases, by movement of cell bodies as well (Walcott 1975). In *Deilephila elpenor* L. (Sphingidae), movement of screening pigment reduces retinal sensitivity measured with electroretinograms by two to three orders of magnitude. A one-second exposure to light is sufficient to trigger these movements. The movements begin 30–60 seconds after exposure and are complete after another 60–90 seconds (Hamdorf and Höglund 1981). Return to a fully dark-adapted state takes 30 minutes of darkness in the case of *Cerapteryx graminis* L. (Noctuidae) (Bernhard and Ottoson 1960).

Some moths have simple eyes called dorsal ocelli in addition to compound eyes. Dorsal ocelli have fixed, wide-aperture optics suited for measurement of changes in light intensity rather than for resolution of images (Mizunami 1995). Signal transmission to thoracic motor systems is more rapid from these eyes than from compound eyes (Mizunami

1995). Illumination of dark-adapted *Creatonotos transiens* Walker (Arctiidae) and *Arctia caja* L. (Arctiidae) causes ultrastructural changes in the ocellar retina (Grünewald and Wunderer 1996). In bees, locusts, and dragonflies, ocelli contribute to flight control, although in moths little information exists on ocellar function (Mizunami 1995).

Artificial lighting actually may improve nocturnal vision of moths that maintain a safe distance from light sources. The response of electroretinograms of most moths studied peaks in the green and extends into the orange and ultraviolet spectral regions (Eguchi et al. 1982, Mikkola 1972, Mitchell and Agee 1981). Artificial lighting, however, may distort visual images perceived by moths. For example, mercury vapor light, which is rich in ultraviolet energy, would accentuate ultraviolet markers ("nectar guides") on flowers (Barth 1985); low pressure sodium vapor light, which contains no ultraviolet energy, would conceal them.

Hearing and Bats

Moths are the predominant or exclusive prey of common species of bats that forage around lights (Acharya and Fenton 1999, Hickey et al. 1996, Rydell 1992, Sierro and Arlettaz 1997; see Chapter 3, this volume). Bats hunt around particular streetlights in proportion to the abundance of insects around the lights (Rydell 1992; see Chapter 3, this volume). Most nocturnal moth species examined in areas with foraging bats have tympanic organs ("ears") that are used in defense against bats (Fullard 1990). These organs are located on the abdomen (Pyraloidea, Geometroidea), head (Sphingoidea), or metathorax (Noctuoidea) (Acharya and Fenton 1999, Spangler 1988). When flying moths hear signals characteristic of a bat, they loop, roll, or make unpredictable turns, or they dive or simply stop flying. When the intensity of the signal is low—indicating that the bat is distant—they fly in the opposite direction of the bat. Arctiids may also produce clicks that deter attacks (Dunning and Krüger 1996).

Moths flying around light sources have been observed to remain near the lamps despite foraging bats (Acharya and Fenton 1999). When free-flying winter moths, *Operophtera brumata* (L.) and *Operophtera fagata* (Scharfenberg) (Geometridae), were exposed to mercury vapor light, they failed about half the time to exhibit their normal evasive responses to electronically simulated ultrasonic bat signals (Svensson and Rydell 1998). Bats capture moths efficiently when light aggregates them and disables their acoustic defenses. The caloric intake of *Eptesicus nilssonii* Keyserling and Blasius (Vespertilionidae) was twice as high (0.5 kJ/minute) while hunting moths around streetlamps than while foraging dipterans in woodlands (0.2 kJ/minute; Rydell 1992).

Crypsis and Birds

Most species of moths are cryptically colored. Camouflage, however, protects moths only on suitable backgrounds. The behavior that matches moths to their backgrounds is disrupted by artificial lighting. Black moths resting in the morning on a white wall near a porch lamp illustrate the problem. By concentrating moths in a small area, artificial lighting gives foraging birds greater experience recognizing particular wing patterns, and it reinforces the association of these patterns with food. In short, areas around artificial lights function like bird feeders.

Artificial lighting may undermine protection based on other kinds of coloration. For example, underwing moths (*Catocala*; Noctuidae) have cryptic forewings that resemble tree bark and hind wings with bright red or orange bands. When an underwing moth rests on a tree trunk, the forewings overlap and completely conceal the brightly colored hind wings. If one touches an underwing moth resting during the day on a tree trunk, the moth raises its forewings, exposes the brilliant bands of its hind wings, and flies away. Several lines of evidence support the hypothesis that this sudden display of bright colors startles birds, which then pause long enough for their prey to escape (Sargent 1990). Under experimental conditions, the bird's startle response rapidly wanes with repetition (Sargent 1990). The common and visible aggregations of moths at lights probably attenuate the startle response and undermine the moth's defense mechanism.

A systematic study of birds attacking moths at light sources was performed at the Estación Biologica de Rancho Grande in Venezuela, a site famous for the vast numbers of insects that fly to light. At dawn, birds hunted moths resting on the station walls and other surfaces that the station's lights had illuminated during the night. During seven months of observation, the investigators documented 30 species of birds feeding on these moths (908 attacked and 764 eaten). The moths included ten families and an unknown number of species (Collins and Watson 1983).

Defense Against Other Predators

Dispersal of moths within a habitat ordinarily protects them from attack by predators that sit and wait for prey to approach. These predators include certain kinds of spiders. Detachable scales on the bodies and wings of moths also protect them from spiders; these scales reduce the moths' adherence to sticky threads in the web (Opell 1994). The remains of moths commonly found in webs constructed on or beside lamps are evidence of exploitation of artificial lighting by spiders. Experimentally, spiders will feed on most kinds of moths (Bristowe 1941). Spiders' selection of web locations near lamps may have nothing to do with the lamps, or it may be based on experience or genetic preference. The orb-web spider *Larinioides sclopetarius* (Clerck) (Araneidae) is genetically inclined to select illuminated sites (Heiling 1999), whereas the common house spider *Achaearanea tepidariorum* (C. L. Koch) (Theridiidae) picks illuminated sites by trial and error according to prey availability (Turnbull 1964). I have observed the jumping spider *Platycryptus undatus* (De Geer) (Salticidae) seizing moths and other insects attracted at night to an incandescent lamp mounted on a wall of a porch in Quisset, Massachusetts. Normally a diurnal predator, it hunts visually, without a web. Walls of buildings are recognized as typical habitat for this species (Kaston 1981). Regardless of the reasons that spiders position themselves near lamps, flight-to-light behavior directs moths to them.

On Caribbean islands, many tropical reptiles and amphibians sit and wait for insects at artificial light. The identities of the insects have not been reported but undoubtedly include moths. The predators include eight species of anole (Squamata; Polychrotidae), one species of gecko (Squamata; Gekkonidae), and three species of frogs (Anura; Hylidae, Leptodactylidae, and Bufonidae) (see Chapter 8, Table 8.1, this volume). Some of the anoles traditionally have been classified as diurnal hunters (Henderson and Powell 2001). In the United States, the eastern American toad (*Bufo americanus americanus* Holbrook [Bufonidae]), hunts insects at night on the ground beneath artificial light sources such as streetlamps (Hulse et al. 2001). Reptiles and amphibians appear to exploit several effects of artificial lighting: concentration of prey in the vicinity of the lamp, diversion of prey onto walls and other surfaces that support these predators, illumination of prey at night, and suppression of flight.

Courtship and Mating

Flight to light may undermine activities specific to mating. Artificial light competes with pheromones as a navigational cue for males in search of females. In a Canadian fir forest, pheromone traps and light traps captured male *Lambdina fiscellaria fiscellaria* (Guenée) (Geometridae) during the same period of night (Delisle et al. 1998). In Costa Rica, male *Roth-*

schildia lebeau (Guérin-Méneville) and *Rothschildia erycina* Shaw (Saturniidae) fly to light 2–3 hours before dawn when females are calling (Delisle et al. 1998). In an agricultural setting, male *Manduca sexta* L. (Sphingidae) caught in light traps baited with virgin females did not seek out the females (Hoffman et al. 1966). In cropland, light trap collections of *Heliothis zea* (Boddie) peaked during peak times of copulation (Graham et al. 1964, Stewart et al. 1967); only one-third to one-half of the females caught in light traps had mated (Gentry et al. 1971, Vail et al. 1968).

In the laboratory, *Heliothis zea* (Boddie) will not mate unless its eyes are dark adapted, as indicated by eye glow. Light intensity must be below 0.015 $\mu W/cm^2$ (about 0.05 lux), the illumination of a quarter moon (Agee 1969). Artificial lighting suppresses female pheromone release and male response in *Trichoplusia ni* (Hübner) (Noctuidae; Shorey and Gaston 1964, 1965, Sower et al. 1970) and *Dioryctria abietivorella* (Grote) (Pyralidae; Fatzinger 1973).

Flight to light has no effect on the mating behavior of females of some species. For example, freshly emerged female saturniids do not fly until after they have emitted pheromone and mated (Blest 1963, Nässig and Peigler 1984, Waldbauer and Sternburg 1979). Male saturniids and sphingids have been observed to ignore nearby lamps and fly to virgin females, although they later may fly to the lamps (Allen and Hodge 1955, Janzen 1984, Worth and Muller 1979). *Hyalophora cecropia* males have completed long-distance flights across illuminated urban territory to calling females (Waldbauer and Sternburg 1982).

Oviposition

Flight of gravid females to light can affect oviposition. In a light trap inventory in Maine, the proportion of females captured varied from 2% for *Nephelodes minians* (Guenée) (Noctuidae) to 82% for *Apamea finitima* [= *A. sordens*] (Guenée) (Noctuidae). Four out of five females captured at the light were gravid; the proportion ranges from half of them to all, depending on species (Dirks 1937). In Rothamsted, England (Williams 1939) and in two sites in northern Germany (Kolligs 2000), the percentage of females found at artificial lighting was similar to that found in Maine.

The proportion of females that fly to light sources increases when host plants are nearby. For example, *Hemileuca tricolor* (Packard) (Saturniidae) was observed at floodlights illuminating a school far (several hundred yards) from the nearest larval host plant, little-leaf paloverde (*Cercidium microphyllum* [Fabaceae]), in Sells, Pima County, Arizona. More than 100 individuals were found, and all were male. A like number of *Hemileuca tricolor* were found nearby on the same evening on a storefront illuminated by a 60-W incandescent bulb. The store was located in the middle of a large patch of the host plant. On the storefront the numbers of females and males were almost equal. Many of the females were gravid, and egg rings could be found on adjacent trees (Tuskes et al. 1996).

The tendency of light trap surveys to report a predominance of males probably results from the preference of investigators to position light traps in open ground where moths have unobstructed exposure to the light. This type of habitat distances the traps from larval host plants. For example, in Alton and Lymington in England, only males of *Trichiura crataegi* (L.) (Lasiocampidae) were found in light trap surveys conducted on open ground, but when the traps were moved near trees, only females were captured (Robinson and Robinson 1950).

Flight to light may shift oviposition to sites near lamps. This has been documented best in agricultural settings where egg densities may be many times higher on plants near the lamps (Beaty et al. 1951, Ficht et al. 1940, Martin and Houser 1941, Pfrimmer et al. 1955). Eggs of *Antheraea polyphemus* (Cramer) (Saturniidae; D. Schweitzer, personal communication, 2002) and *Coloradia pandora* (Blake) (Saturniidae; Brown 1984) have been found on buildings near light sources.

Alternatively, artificial lighting may suppress oviposition. The number of eggs deposited by *Heliothis* spp. on artificially illuminated cotton was 85% lower than on unilluminated cotton. Although this finding was consistent with observations in the laboratory, it contradicted earlier studies that reported higher rates of oviposition for these species on illuminated plots (Nemec 1969). Illuminating apple orchards has been reported to reduce by 30% infestation by larvae of *Cydia pomonella* (L.) (Tortricidae; Herms 1932).

Feeding

Artificial light has been noted to reduce numbers of *Catocala* moths (Noctuidae) coming to bait placed less than about 30 m (98 ft) from lamps (D. Schweitzer, personal communication, 2002). Orchard illumination has been reported to reduce feeding by *Cydia pomonella* at bait (Herms 1932) and damage by fruit-piercing noctuids (Nomura 1969). Geometrid moths carrying *Crassula fascicularis* (Lam.) (Crassulaceae) pollen on their proboscides were caught in a black light trap during the nocturnal period of peak nectar and scent production of this plant (Johnson et al. 1993).

Despite these observations, sphingids and noctuids have been observed to visit food sources in full view of electric lamps sometimes located just a few meters away. *Plusiinae* spp. (Noctuidae) especially have been observed visiting flowers exposed to artificial light at night (D. Schweitzer, personal communication, 2002). In Costa Rica, sphingids belonging to six genera were observed to drink nectar at flowers located 20–50 m (66–164 ft) from a light source. The effect of electric lighting on feeding is moot for the large number of species of moths that never feed as adults (Norris 1936).

Dispersal and Migration

Migrating or dispersing moths often fly to electric lights. Some examples are moths that fly to lights located far from habitats suitable for breeding. These locations include rocky skerries on the shores of coastal islands in Finland (Nieminen 1996, Nieminen and Hanski 1998), oil platforms in the Gulf of Mexico tens of kilometers from land (Wolf et al. 1986), urban Brno, Czech Republic (Wolda et al. 1994), Abu Dhabi's sandy deserts (Tigar and Osborne 1999), and high mountain passes in Mexico (Powell and Brown 1990). In mountain passes, geographic barriers and wind may concentrate migrants around electric lights (Beebe and Fleming 1951, Powell and Brown 1990). One study compared flight to light in pairs of light traps, one of which was located on higher ground than the other; migrant species of moths were more likely than other moths to fly to light sources on higher ground (Herczig and Mészáros 1994).

Circadian Rhythms

Flight-to-light behavior exposes moths to intense light during the moths' subjective night. Exposure to light can suppress activity that ordinarily exhibits a circadian rhythm, such as flight. Some moths fly to light sources and then rest for hours in shadows nearby, which suggests that flight to light may not only suppress flight activity but also reset the circadian clock that regulates it.

If flight to light disturbs the circadian clock that controls flight activity, it probably disturbs other clocks. Like other insects, moths have many independent photoreceptive circadian pacemakers located in different types of tissue. Light from the sun is postulated to synchronize them (Giebultowicz 2000). Brief exposure to artificial light could resynchronize or desynchronize these pacemakers, depending on their responses. For example, in the laboratory only one of two photosensitive clocks of *Pectinophora gossypiella* (Saunders) (Gelechiidae) responds to the 589-nm wavelength emitted by low pressure sodium lamps (Pittendrigh and Minis 1971). Studies have yet to investigate the possible effect of flight to light on circadian rhythms. For example, it is unknown to what extent flight-to-light behavior disturbs the timing of pheromone responsiveness of males after they have flown away from artificial light.

In the laboratory, photoperiodic manipulation has been used to control vital functions of Lepidoptera. For example, it may induce sterility in adult moths (Bebas et al. 2001) and prevent diapause in moth larvae (Saunders 1982). Although such intervention prevented diapause in larvae of *Adoxophyes orana* (Fischer von Röslerstamm) (Tortricidae) in the laboratory, however, a trial in the field failed to achieve a comparable result. Outdoors, vegetation shielded larvae from the artificial light, and low temperatures exposed them to opposing environmental stimuli (Berlinger and Ankersmit 1976).

Synchronization of life cycles with lunar cycles may help moths navigate, mate, and avoid predators. Collections of *Epiphyas postvittana* (Walker) (Tortricidae) and *Plutella xylostella* (L.) (Plutellidae) in suction traps peaked at three phases of the lunar cycle (Danthanarayana 1986). In cotton fields numbers of eggs deposited by *Heliothis zea* (Boddie) were lowest at the full moon, as were the numbers of these moths in light traps (Nemec 1971). It is possible that artificial lighting, by simulating moonlight, could disturb this synchronization. In field trials, artificial lighting appeared to suppress *Heliothis zea* oviposition synchronized to lunar cycles, but these results contradicted the findings of earlier studies (Nemec 1969).

Trauma

Lamps desiccate or incinerate moths that fly into lamp housings or land on heated lamp surfaces. Thousands of moths have been found in open or broken housings (Hausmann 1992). Overheating and dehydration caused by electric lighting resembles effects of solar radiation, a selective force that may have given rise to the evolution of nocturnality in moths (Daily and Ehrlich 1996).

Moths flying against lamps and illuminated surfaces tear wings and lose scales. Trapped inside lamp housings with other insects such as beetles, they can lose legs and parts of antennae. Streetlamps divert moths into oncoming traffic (Hessel 1976). Moths may drown flying around lamps over bodies of water.

Number of Species of Moths That Fly to Light

Light traps used for inventories typically capture hundreds of species of moths (Table 13.1). Over a four-year period in the 1930s, a single stationary light trap equipped with a 200-W incandescent (tungsten filament) lamp in Rothamsted, England captured 256 species, which represented more than one-third of species of larger moths known at the time for Britain and the majority for the county of Hertfordshire (Williams 1939). Two decades later, the number of species of larger moths caught in two mercury vapor traps in Rothamsted was twice as high (Hosny 1959), equaling two-thirds of the total for the United Kingdom. In a forest in West Virginia, light traps captured more species of moths than did Malaise traps (tentlike structures designed to capture all insects that fly into their open ends; Butler et al. 1999; Table 13.2).

The number of species of moths that fly to light is greater than the number reported in the light trap surveys cited here. All the studies cited in Table 13.1 are limited to larger moths, known as macrolepidoptera. In the United Kingdom, most species of moths are categorized as microlepidoptera (Young 1997). Although microlepidoptera fly to light, they are more difficult than macrolepidoptera to identify and trap, and most surveys of moths exclude them.

Identifying all moths that fly to an artificial light source may not be

Table 13.2. Comparison of macrolepidoptera moths caught in light and Malaise traps.*

Moths	*Black Light Traps*	*Malaise Traps*
Species	343	250
Individuals	36,160	28,082
Species captured exclusively in trap type	135	42

*Four 8-W black light traps were operated once weekly, and 20 Malaise traps continuously, from May to August for three years in a deciduous hardwood forest in West Virginia. Malaise traps are constructed like tents with one end open; they use no attractant. Some of the species caught exclusively in Malaise traps are described as diurnal (Butler et al. 1999).

feasible, as exemplified in a recent inventory of Lepidoptera at Marine Corps Air Station Miramar in southwestern San Diego County, California. Brown and Bash (1997 [2000]) surveyed both microlepidoptera and macrolepidoptera using several sampling methods, including black light trapping. They estimated that their black light samples included 5–10% more species than they identified, despite consulting with ten authorities at seven institutions. They cite several factors to account for this. Twenty or more species of moths in the black light samples were undescribed species; lack of names for species complicated the task of compiling the inventory. The number of specimens in the samples was so large that the investigators and their collaborators could not make species determinations for all of them. Many microlepidoptera, especially Gelechioidea, were exceedingly small and difficult to prepare and distinguish. Taxonomic expertise in microlepidoptera is scarce (Brown and Bash 1997 [2000]).

Number of Species of Moths That Live Near Artificial Lighting

Although the number of species that fly to light is high in natural habitats such as forests, artificial lighting is concentrated around metropolitan areas, as shown in nocturnal images taken from satellites (see Figure 1.2). How many species of moths live in these areas?

Sixty years ago, Frank Lutz, curator of insects at the American Museum of Natural History in New York City, described the habits and kinds of insects in his own suburban backyard, a lot 23 × 61 m (0.14 ha, 0.35 acre). His inventory included 26 orders and 1,401 species, a third of which were moths—not counting many moth specimens he could not identify (Lutz 1941).

In England, private gardens occupy an area about ten times as large as national nature reserves. Gardens contain exotic species of plants that have become hosts to a rich diversity of native species of moths (Owen 1978). Emmet (1991:64) classified 305 species of British moths as exclusively or predominantly inhabiting "suburban habitats, gardens, parks, orchards, the outside walls of buildings." Bradley and Mere (1966) reported 362 species of moths from the garden of Buckingham Palace in London.

At a larger scale, 1,479 species of moths were recorded within a radius of 20 miles (32 km) from St. Paul's Cathedral in London ("London area,"

3,215 km^2 [1,256 mi^2]). This number of moth species represents almost two-thirds of the total for the United Kingdom. These species were about evenly divided between microlepidoptera and macrolepidoptera (Plant 1999). More than 1.3 million records were used to map the distribution of species of larger moths in a grid containing 856 squares (tetrads) of 4 km^2. These data were used to create an atlas containing distribution maps for each resident species, with names of local host plants, number of annual broods, and seasonal flight periods (Plant 1993). Almost 80% (n = 511) of the 715 species of larger moths recorded in the London area from 1980 to 1991 are considered resident species (Figure 13.1).

Although some of these species may not breed in tetrads where they have been recorded, they all breed in the London area. The atlas defines "resident species" as "one which breeds annually and which survives the winter. Its population may sometimes be reinforced by immigration, but these influxes are not essential to its continued existence here" (Plant 1993:xiv). Thirty-eight percent (n = 193) of the resident species are

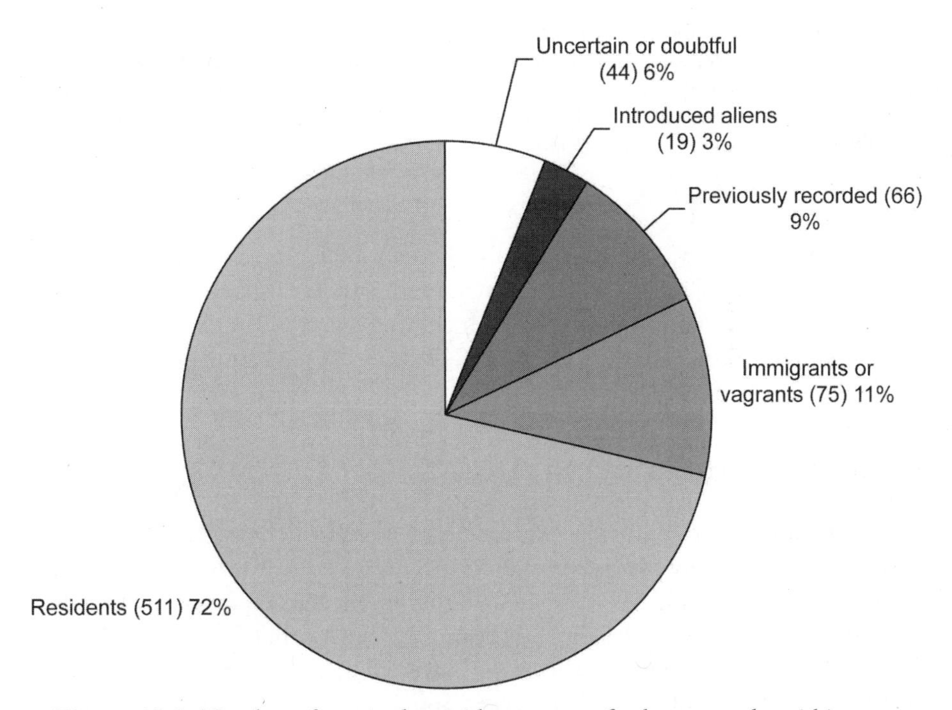

Figure 13.1. Number of species by residency status for larger moths within 32 km of St. Paul's Cathedral, London, from January 1, 1980 to December 31, 1991 (Plant 1993).

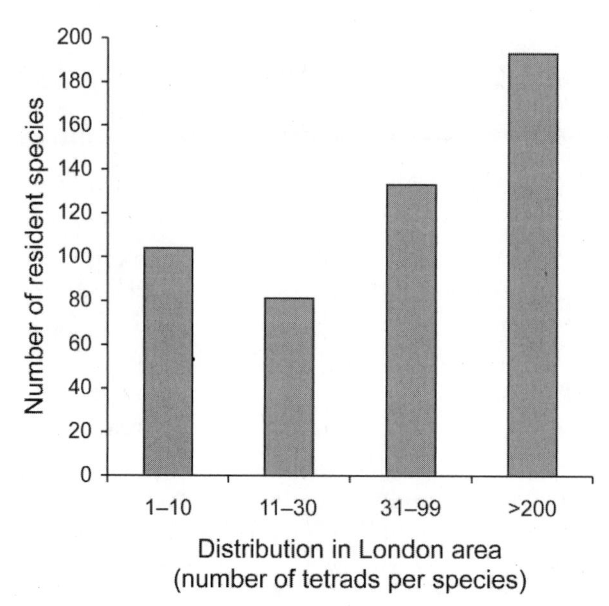

Figure 13.2. Number of resident moth species by distribution in the London area. For example, each of 193 species was recorded in more than 200 tetrads. Tetrads are 2-km squares contained in a grid overlying the London area, defined as the area within 32 km of St. Paul's Cathedral. Data from January 1, 1980 to December 31, 1991 (Plant 1993).

distributed widely, recorded in more than 200 tetrads (Figure 13.2). Eighty-seven of these widely distributed resident species have been recorded in all kinds of habitats, including central London. Resident species have been recorded in about 84% (n = 716) of tetrads in the London area (Plant 1993).

Flight to Light in Urban Areas

Urban and suburban moths exhibit flight-to-light behavior. In the London area, volunteers have conducted species inventories with light traps. The numbers of species trapped at lights in gardens commonly exceeds 100 and sometimes exceeds 300 (Plant 1999). In other European cities, moth inventories with light traps have yielded hundreds of species (Table 13.3). In suburban Kiel, Germany, 71 species of moths were captured in one year at a high pressure sodium vapor streetlight (Kolligs 2000).

Table 13.3. Number of species of moths collected in light traps in urban sites in Europe (collection methods vary).

Country	*City*	*Species*	*References*
Czech Republic	Brno	522	Wolda et al. 1994
Czech Republic	Prague	134	Wolda et al. 1994
Czech Republic	Cernis	407	Wolda et al. 1994
Czech Republic	Ceské Budejovice	314	Wolda et al. 1994
Germany	Kiel	263	Kolligs 2000
England	Portsmouth	219	Langmaid 1959

Types of Lamps That Elicit Flight-to-Light Behavior by Moths

Except for low pressure sodium vapor lamps, spectral emissions of all types of lamps used in outdoor lighting elicit flight-to-light behavior by moths. Short-wavelength light elicits this behavior most strongly for most but not all species of moths (Mikkola 1972). The greatest numbers of individuals and species fly to mercury vapor lamps, which emit substantial amounts of energy in the ultraviolet region of the spectrum. Nearly four times more moths were trapped at mercury vapor streetlamps than at high pressure sodium vapor streetlamps in a rural area of Rheinhessen, Germany (Eisenbeis and Hassel 2000; Chapter 12, this volume). In suburban Kiel, Germany, more than 100 times more individuals of some species were trapped at mercury vapor lamps than at high pressure sodium vapor lamps, but for a few species, such as *Idaea dimidiata* (Hufnagel) (Geometridae), numbers were higher at high pressure sodium vapor lamps (Kolligs 2000). The white color of mercury vapor light resembles that from other sources, such as metal halide vapor lamps, which emit much less ultraviolet energy.

Among light sources used for artificial illumination, only low pressure sodium vapor lamps rarely or never elicit flight-to-light behavior by moths (Robinson 1952, Rydell 1992), although moths' retinal sensitivity extends into the spectral range emitted by these lamps (MacFarlane and Eaton 1973, Mikkola 1972, Mitchell and Agee 1981). These lamps emit deep yellow-orange light that is practically monochromatic at 589 nm (Frank 1988). In most urban areas they have become much less common than high pressure sodium vapor lamps, which produce golden yellow or pinkish light with a broad spectral content.

Factors That Reduce Vulnerability of Moth Populations to Outdoor Lighting

Most artificial light sources have the capacity directly or indirectly to kill moths or to interfere with their reproduction. Most species of moths exhibit flight-to-light behavior. What allows a high diversity of moths to survive near light sources?

Patchy Distribution of Artificial Lighting

Nocturnal illumination in urban and suburban areas is uneven. Viewed from an airplane, outdoor lighting in urban and suburban areas creates an illuminated matrix. Vegetated habitats such as parks and gardens typically are located in dark areas in the matrix. In densely illuminated urban centers, buildings or trees may shield vegetated habitats from direct exposure to electric lamps. Microhabitats in the shadows of herbaceous shrubs may protect moths from light. The London area as defined earlier includes rural woodlands and fields and an expanse of grazing marsh of more than 160 ha (395 acres; Plant 1993). Even when moths live in habitats directly exposed to artificial light, lights may be turned off during moth flight periods, which may occur after midnight (Williams 1939). Short dispersal distances (less than 100 m [328 ft]; Kozlov 1996) may protect some urban microlepidoptera (Nepticulidae) from exposure to artificial lighting. Moths that live in environments that appear to be illuminated may actually spend most of their nocturnal time in places that are still relatively dark.

Inverse Relationship of Lamp Density to Flight-to-Light Behavior

An inverse relationship exists between flight-to-light behavior and lamp density. Robinson and Robinson (1950) observed that few moths flew to concentrations of powerful lamps in urban areas that abut rural habitats with large populations of moths, whereas many more moths flew to isolated, weak lamps in nearby rural phone booths. In experiments in the field, Robinson and Robinson found that numbers of moths flying to 1,500-lumen incandescent lamps decreased as the lamps were moved toward each other. The number of moths began to decrease as the distance separating the lamps fell to 46 m (150 ft) and continued to decrease until the distance dropped to 15 m (50 ft) apart, at which point the numbers of moths at the lamps became negligible (Robinson and Robinson

1950). These results indicate that high lamp densities can suppress flight-to-light behavior when the distance separating the lamps drops below a certain threshold.

The inverse relationship of lamp density and flight-to-light behavior fundamentally limits effects on moth populations. Because the radius over which lamps elicit flight-to-light behavior ranges from 3 m to 130 m (10–427 ft), lamps would have to be spaced relatively close together to influence all moths in an area. Such close spacing, however, can interfere with the flight-to-light response.

Background lighting may explain why two lamps spaced near each other attract so few moths compared with the same lamps spaced far apart. Ambient light from a nearby lamp reduces the contrast between a lamp and its background. At some threshold, the contrast may become insufficient to trigger flight to the lamp. In urban areas, artificial light reflected off the atmosphere increases background lighting and may be expected to reduce flight to lamps.

Moonlight, Clear Nights, and Wind

Moonlight decreases flight to artificial light sources for most species of moths in temperate regions (McGeachie 1989, Williams et al. 1956, Yela and Holyoak 1997). Moonlight may reduce flight-to-light behavior by functioning as a navigational reference point that competes with artificial light sources (Sotthibandhu and Baker 1979) or by simply increasing background lighting. Moths may synchronize their life cycles so that their adult stages do not occur during the full moon (Nemec 1971), or moonlight may suppress flight; flight activity measured by methods other than light trapping, however, usually has not been shown to decrease with the full moon. These methods include suction traps (Danthanarayana 1986, Williams et al. 1956), pheromone traps (Janzen 1984, Saario et al. 1970), radar (Schaefer 1976), and bait (Yela and Holyoak 1997). In one study, collections of noctuids at light did not decline with increased moonlight (McGeachie 1989).

Fewer moths fly to artificial light sources on clear nights than on overcast nights (Dirks 1937), especially during the full moon (Butler et al. 1999). This relationship has been reported for moths collected at light but not at bait (Yela and Holyoak 1997).

Wind can suppress flight (McGeachie 1988), blow moths away from lamps (Muirhead-Thomson 1991), and deliver olfactory and aerodynamic navigational cues that compete with light (Janzen 1984). Radar observations reinforced by captures in nets attached to balloons have revealed dense masses of migrant moths flying at altitudes as high as 0.5 km (1,650 ft), where prevailing winds carry them hundreds of kilometers (Riley et al. 1995). Shifts in wind have determined when and where long-distance migrants fly to lamps (Mikkola 1986, Pedgley and Yathom 1993).

Species, Age, and Activity

Although most species of moths fly to light, some rarely do. Comparing collections in suction and light traps, Taylor and Carter (1961) calculated that the probability of flight to light for individuals of *Xestia c-nigrum* L. (Noctuidae) is 5,000 times greater than for individuals of *Amphipyra tragopoginis* (Clerck) (Noctuidae), which flew to light so rarely that the investigators considered the capture of this species in the light trap to be a chance event. In Kiel, Germany, the red underwing (*Catocala nupta* L. [Noctuidae]) and the copper underwing (*Amphipyra pyramidea* L. [Noctuidae]) were found primarily at natural food sources or bait rather than at light sources that attracted other species in abundance (Kolligs 2000).

Moths may learn to ignore artificial light sources. Janzen (1984) observed that freshly emerged Costa Rican sphingids are more likely to fly to light than are older sphingids; he hypothesized that moths learn to navigate using alternative cues such as memory, wind, odors, and landscape. In a mark–recapture experiment in Finland, some individual moths flew for unknown reasons into light traps more often than did others of the same species (Väisänen and Hublin 1983).

Larval host plants reduce flight to electric lamps by luring gravid females away from these light sources, as noted above. Males of two Costa Rican saturniids, *Rothschildia lebeau* (Guérin-Méneville) and *Rothschildia erycina* (Shaw), fly to light only during the second of their two nocturnal flight periods (Janzen 1984), but the reason for the difference in the response to artificial light during the two flight periods is unclear.

Links to Favorable Ecological Conditions

Environmental change associated with artificial lighting may favor the survival of some populations of moths and counterbalance harmful effects of outdoor lighting. In the city of Brno, Czech Republic, most species light-trapped in numbers of five or more individuals were found to be "tramp" species that thrive in urban habitats (Wolda et al. 1994).

The ailanthus silk moth (*Samia cynthia* [Drury]), a large and colorful saturniid, was introduced into eastern North America more than 100 years ago. It established itself exclusively in cities, even though its host plant, *Ailanthus altissima* (Miller) Swingle (Simaroubaceae), was found in both urban and rural areas. Urban conditions likely protected the moth from rural predators and parasitoids until it disappeared from many locations in North America (Frank 1986, Tuskes et al. 1996). Around Philadelphia, I have found cecropia moths predominantly in inner city areas, probably for similar reasons. Distribution in urban refuges has led to classification of *Hyalophora cecropia* as a "fugitive species" (Sternburg et al. 1981).

Migration and Dispersal

Migrant species may appear at light sources in vast swarms (Howe 1959). Two light traps in Hants, England caught more than 10,000 individuals of two species, *Xestia c-nigrum* and *Autographa gamma* (L.) (Noctuidae) in one year. One of these traps collected about 50,000 *Xestia c-nigrum* in a single night. Both of these species are migrants that feed on a wide variety of economically important crops, and both are widely distributed in Europe, Africa, and Asia (Zhang 1994). In England, populations of *Autographa gamma* are maintained exclusively by annual immigration of moths from southern Europe. No comparable reverse migration occurs (Young 1997). Consequently, the destruction of vast numbers of these moths at lamps in England does not affect the southern European populations that annually supply immigrants to Britain. Populations of *Xestia c-nigrum* are resident, but their populations often are boosted by immigration in the fall (Plant 1993). Migration and dispersal account in part for the failure of light trapping to eradicate populations of moths that are agricultural pests (Hienton 1974).

In Rothamsted, England, numbers of moths and species collected in a light trap in a 1.3-ha (3.2-acre) woodlot were tracked for ten years (Taylor et al. 1978). Although year-to-year variation occurred, the numbers show no trend (Figure 13.3). These investigators concluded that the traps captured only a small fraction of the populations they were sampling. Similar results were reported for a stationary light trap over a four-year period (Williams 1939). For the light trap collection depicted in Figure 13.4, only a few species were captured in numbers exceeding 100 individuals, and the two species trapped in greatest numbers were the two migrant pest species *Xestia c-nigrum* and *Autographa gamma*.

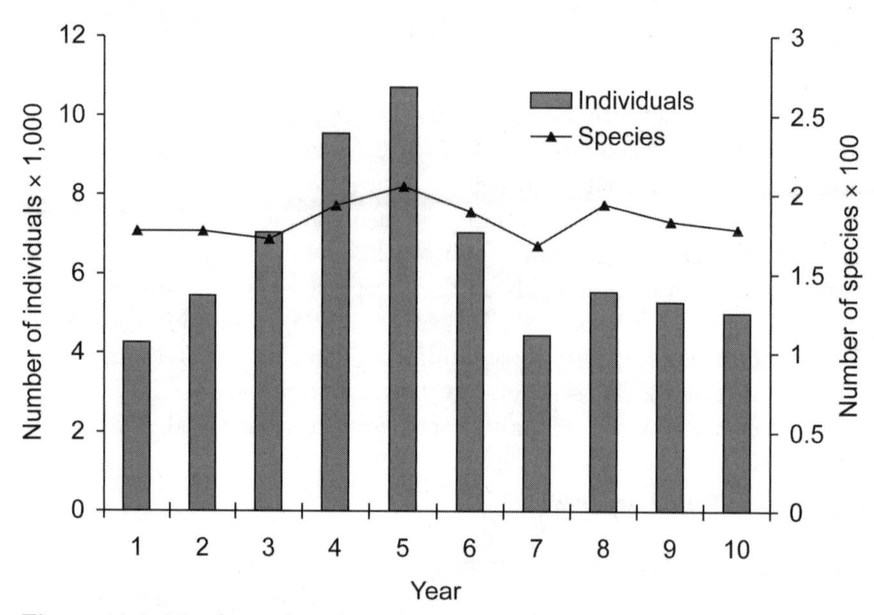

Figure 13.3. Numbers of moths and species trapped each year with a 200-W incandescent lamp in a 1.3-ha (3.2-acre) woodland from 1966 to 1975 at Rothamsted, England. Data from Taylor et al. (1978).

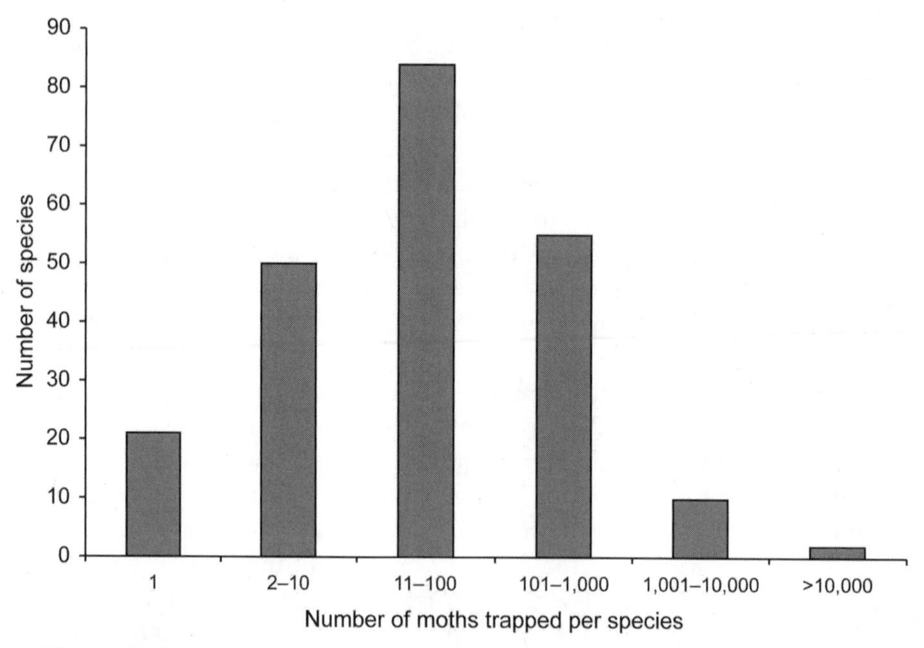

Figure 13.4. Number of species by frequency of capture during light trapping in Hants, England in 1949. For example, two to ten moths were captured for each of 50 species. Data are for families Sphingidae and Noctuidae. Two species, *Xestia c-nigrum* L. and *Autographa gamma* L., were collected in numbers over 10,000 (Robinson and Robinson 1950).

Environmental Factors That Increase Vulnerability of Moth Populations to Outdoor Lighting

Lighting between habitat fragments may disrupt dispersal of moths and decrease colonization or recolonization of habitats, and lights within habitats may threaten populations already at risk.

Lighting Between Habitat Fragments

In the London area, 60 species of larger moths that were once resident are now considered locally extinct. The 104 resident species confined to ten or fewer tetrads are considered to be at risk of local extirpation, and recommendations have been made to protect them (Plant 1993). Thus, more than a quarter of species once resident in this area may be considered either locally extinct or endangered.

Although many species of moths thrive in urban habitats and are attracted to urban light sources, the Rothamsted Insect Survey revealed that diversity and abundance of moths in urban areas in Britain are lower than in the surrounding countryside. Numbers and species of moths are more temporally variable in urban areas than in rural areas, probably because of anthropogenic changes in habitat. Results of the survey provided evidence that immigration from the countryside can reestablish populations that disappear in urban areas (Taylor et al. 1978).

Outdoor lighting, however, can interfere with immigration. In Kiel, Germany, eight moth species considered endangered were caught by traps fitted to suburban streetlights. The host plants for these species were not present in the sites illuminated by the lamps (Kolligs 2000). Although only single individuals of each endangered species were found at the lights (Kolligs 2000), each of these individuals might have established a new population in a new patch of habitat or increased genetic diversity of an existing population. The number of individuals of endangered species attracted to all artificial sources of light in the region would be expected to be much greater than the number caught at the few streetlamps sampled.

For endangered species of moths, the most widespread and serious effect of outdoor lighting probably is disruption of dispersal. Lighting typically is located between habitat patches, where it acts as a barrier to disrupt the movement of moths. By impeding movement of moths between habitat fragments, artificial night lighting compounds the adverse effects of habitat fragmentation produced by urban sprawl.

Daily and Ehrlich (1996) reported that habitat fragmentation reduces species diversity less for moths than for butterflies. They concluded that dispersal at night confers a survival benefit that accounts for the relatively high tolerance of moth species for habitat fragmentation. Although their work was performed in fragmented forest in Costa Rica, the importance of dispersal for species survival has been corroborated for moths in other locations with fragmented habitats, such as an island archipelago in Finland (Nieminen 1996, Nieminen and Hanski 1998), farm woodlands in Britain (Usher and Keiller 1998), and parks and cemeteries in St. Petersburg, Russia (Kozlov 1996). For butterflies in Britain, vulnerability of species to extinction caused by fragmentation of landscapes is linked to dispersal ability (Thomas 2000). In British farm woodlands, species richness of woodland moths correlated inversely with degree of geographic isolation (Usher and Keiller 1998). By interrupting dispersal at night, flight-to-light behavior undermines the ability of moth species to survive habitat fragmentation.

Flight-to-light behavior probably exerts the most pressure on moth populations with particular patterns of dispersal. Lighting would not be expected to prevent immigration into an area by pest species arriving in vast swarms or by species so rare in a particular location that they produce no immigrants. Lighting probably would exert the most influence on populations dispersing at low levels between a few small habitat fragments.

Lepidopterists have observed that the numbers of moths flying to particular lamps decline over years. Janzen (1983) investigated such declines at several locations in Costa Rica and attributed the cause to extensive habitat destruction related to agriculture. Urban lighting, like artificial lighting in agricultural areas, probably is not a primary cause of declines in populations of moths. Urban lighting, however, may undermine shaky populations of moths that have survived urbanization. It may interrupt dispersal of these populations between patches of urban habitat such as parks, gardens, vacant lots, and roadside vegetation.

Lighting Within Habitat Fragments

Within a habitat fragment, light trapping or loss of individuals to lights may directly threaten moth populations. For example, *Hydraecia petasitis* (Doubleday) (Noctuidae) reaches the northern limit of its range in Finland, where it is restricted to just a few colonies located in patches of its food plant, butterbur (*Petasites hybridus* [L.] Gaertner [Asteraceae]). A light trap containing an 80-W mercury vapor lamp was placed in each of two butterbur patches covering 700 m^2 (0.17 acre) and 800 m^2 (0.19 acre) and located 150 m (492 ft) apart (Väisänen and Hublin 1983). Over 48 days, a mark–recapture experiment provided data to estimate the moth's population size as 145 individuals. Traps caught about half of the males and one-third of the females at least once. The investigators concluded that collecting these moths by light trapping could eliminate this population and suggested that such light trapping could pose an even greater danger to colonies of species with a greater tendency to fly to light (Väisänen and Hublin 1983). Because noctuids that approach light traps often are not caught (Muirhead-Thomson 1991), the proportion of the colony that flew to the light probably was higher than that caught.

Light trapping suppressed the hickory shuckworm (*Cydia caryana* [Fitch] [Olethreutidae]) in a 3.2-ha (7.9-acre) pecan orchard in Georgia (Tedders et al. 1972). During three years of trapping, the proportion of infested pecans in the orchard progressively fell to 1.2%, in contrast to 40% infested pecans on nearby control trees. Geographic isolation of the plantation and relatively low dispersal rates for this species may have contributed to the success of this trial (Tedders et al. 1972) compared with other pest control trials with light traps (Hienton 1974). In no agricultural trial, however, has artificial lighting eradicated a pest species from a particular area.

Outdoor lighting is unlikely to eliminate large populations of moths, even when populations are isolated. To assess the effect of light trapping on isolated populations, an extensive light trapping trial was conducted on St. Croix, U.S. Virgin Islands. The trial included 250 light traps distributed at a density of 1/km^2 (2.6/mi^2) and operated for 43 consecutive months. Numbers of catches of 11 species of Lepidoptera were monitored. The volume of insects trapped was enormous: 40 liters (10.5 gallons) in one or two nights at the beginning of the trial. The investigators concluded that the traps reduced populations of most of the target species, in some instances to the brink of extinction (Cantelo et al. 1974). Toward the end of the trial, however, populations of most species were increasing, which the authors attributed to increased rainfall (Cantelo et al. 1972).

The results of light trapping studies may seem to have little relevance to artificial lighting. Light trapping may be expected to affect moth populations more than does outdoor lighting, but the reverse sometimes may be true. For example, in the St. Croix study, populations of *Heliothis virescens* (Fabricius) (Noctuidae) initially declined to near extinction, although the number of individuals trapped was far too low to have depleted the island's population. The investigators attributed the population crash to predators consuming moths on the ground around traps (Cantelo et al. 1973). They noted that moths of this species usually land near traps instead of flying into them.

Possible Evolutionary Responses to Lighting

Industrial melanism in moths demonstrates that populations of moths can evolve rapidly in response to environmental change created by humans (Majerus 1998). Selective pressure from artificial lighting favors individual moths that are less inclined to fly to light than are other individuals of the same species. Variability in flight-to-light behavior between individuals of a single species was observed in the mark–recapture study of *Hydraecia petasitis* in Finland (Väisänen and Hublin 1983). Because some species of moths do not fly to light, or do so only rarely, it is possible that evolutionary modification of flight-to-light behavior has already occurred, although the reasons are unknown.

A century ago, city lights were regarded as among the best places for collecting moths (Denton 1900). Today, the low numbers of moths at city lights can be attributed to many causes: decreased numbers of moths, dilution of moths among many lamps, use of high pressure and low pressure sodium vapor lamps, and background lighting suppressing flight-to-light behavior. The decline in numbers of moths around city lights is consistent with evolutionary loss of flight-to-light behavior, but no evidence has supported this hypothesis.

Moth populations may resist selective pressure to modify flight-to-light behavior. The diversity of insects, and especially moths, that fly to light suggests that this behavior has been deeply conserved. In some instances, Lepidoptera have not evolved adaptations that would seem to be advantageous (Ehrlich 1984). Barriers to evolutionary loss of flight-to-light behavior, in some cases, may be insurmountable. For example, a potential evolutionary response may be a circadian shift to an earlier or later flight time when ambient light exists at levels high enough to prevent flight-to-light behavior. Diurnal or crepuscular flight, however, could expose moths to attacks by birds, dehydrate them, desynchronize flight periods with periods of nectar production and pheromone release, and disrupt species isolation based on allochronic flight periods.

One possible effect of outdoor lighting could be a shift in moth fauna in favor of species that are pre-adapted to survive in illuminated environments. The mouse moth (*Amphipyra tragopoginis* [Clerck]), which flies to light only rarely, if at all (Taylor and Carter 1961), is common throughout metropolitan London, including at the very center of the city (Plant 1993). Urban pests include clearwings (Sesiidae), which are diurnal, and clothes moths (Tineinae), which do not fly to artificial light (Ebeling 1978). Many North American urban pests have flightless females and therefore cannot fly to light, such as the bagworm moth (*Thyridopteryx ephemeraeformis* [Haworth] [Psychidae]), gypsy moth (*Lymantria dispar* [L.] [Lymantriidae]), white-marked tussock moth (*Orgyia leucostigma* [J.E. Smith] [Lymantriidae]), and fall cankerworm (*Alsophila pometaria* [Harris] [Geometridae]).

Potential Disruption of Moth Parasitoids

Parasitoids are parasites that kill their hosts. Moth parasitoids include insects that attack eggs, larvae, or pupae. Many parasitoids fly to artificial light and are minute enough to enter small openings in lamp housings, where thermal and radiant injury can destroy them. Two flashes from an ordinary photographer's flash (100 J) at a distance of 3 mm (0.13 in) killed or sterilized 100% of *Dahlbominus fuscipennis* (Zetterstedt) (Hymenoptera: Eulophidae), a parasitoid of sawfly pupa (Diprionidae; Riordan 1964). Parasitoids generally are more prone than their hosts to extinction because they are positioned at higher trophic levels and live at lower population densities (LaSalle 1993).

Parasitoids co-occur with moths in illuminated suburban habitats. During three years of Malaise trapping in a suburban garden, Owen (1978) collected 529 species of ichneumonid wasps (Hymenoptera). The 0.07-ha (0.16-acre) garden was located on the corner of a busy road 3.8 km (2.4 mi) from the center of the city of Leicester, England. Single specimens represented about one-third of the species, and the average number of individuals per species was fewer than 12. The ichneumonids in this garden parasitize larvae belonging to several of the largest moth families (Pyralidae, Geometridae, and Noctuidae; Owen 1978).

In suburban Newark, Delaware, an inventory of insects killed in bugzappers (black light traps that destroy insects by electrocution) revealed that predators and parasitoids of insects outnumbered Lepidoptera (Frick and Tallamy 1996). Predators and parasitoids included 13.5% of the total number of insects killed, representing seven orders and 35 families (Frick and Tallamy 1996; Table 13.4). Eggs and larvae of moths are known to be attacked by insects that belong to five of these orders (Dermaptera, Coleoptera, Hemiptera, Diptera, and Hymenoptera; Young 1997). Saturniids are attacked by parasitoids that belong to three

Table 13.4. Orders and families of insect predators and parasitoids killed in UV-light bugzappers in yards of six homes in suburban Newark, Delaware.

Orders	*Families (numbers of individuals killed)*
Dermaptera	Labiidae (2)
Hemiptera	Hebridae (2)
	Nabidae (2)
Neuroptera	Chrysopidae (8)
	Corydalidae (1)
Coleoptera	Cantharidae (104)
	Carabidae (661)
	Cleridae (4)
	Coccinellidae (15)
	Dytiscidae (21)
	Lampyridae (12)
	Mordellidae (10)
	Staphylinidae (306)
Diptera	Asilidae (1)
	Dolichopodidae (70)
	Empididae (58)
	Pipunculidae (1)
	Rhagionidae (2)
	Scenopinidae (1)
	Sciomyzidae (1)
	Stratiomyidae (5)
	Tachinidae (16)
	Xylophagidae (1)
Lepidoptera	Epipyropidae (5)
Hymenoptera	Braconidae (377)
	Chrysididae (3)
	Encyrtidae (1)
	Eulophidae (1)
	Formicidae (84)
	Ichneumonidae (77)
	Mymaridae (1)
	Perilampidae (1)
	Pteromalidae (1)
	Torymidae (2)
	Vespidae (1)

Six devices were run for at least two hours during each of six nights. Numbers of individuals captured are listed in parentheses after each family. The total number of insects killed was 13,789, including 1,868 (13.5%) predators and parasitoids, 1,586 (11.5%) Lepidoptera, and 31 (0.22%) biting flies (Frick and Tallamy 1996).

of the families: Tachinidae (Diptera), Braconidae (Hymenoptera), and Ichneumonidae (Hymenoptera; Tuskes et al. 1996). Wasps of the ichneumonid subfamily Ophioninae fly to light (Kolligs 2000) and parasitize saturniids (Tuskes et al. 1996). Parasitoids and other entomophagous insects have been collected at high pressure sodium vapor light and white fluorescent light (Kolligs 2000, Nabli et al. 1999).

Because parasitoids function as important regulators of their host populations (Hawkins 1993), artificial lighting could, at least in theory, disturb natural control of populations of Lepidoptera. Artificial light could enhance populations of moths if it protected their eggs or larvae from attack by parasitoids and predators; the reverse could occur if artificial lighting protected parasitoids from attack by hyperparasites. Identifying parasitoids (Gauld 1986, LaSalle and Gauld 1993) and their host ranges (Memmott and Godfray 1993, Young 1997) is difficult, and no studies have investigated the effect of outdoor lighting on populations of entomophagous insects.

Methods for Restricting Outdoor Lighting to Protect Moths

The most effective method for protecting Lepidoptera from artificial lighting is to simply turn off lights. If lights are to be used, they should be operated only when needed. Owners of billboards can save money and moths by limiting illumination to the hours of the night when people are most likely to view outdoor advertising. Light sources should be located away from structures that are likely to trap moths. Lamp housings should have reflectors that direct illumination only to areas where needed. Light fixtures should be tightly sealed to prevent entrance of insects.

If lighting is required in areas where Lepidoptera need to be protected, low pressure sodium vapor lamps should be used, and mercury vapor lamps and other lamps with high ultraviolet emissions should be avoided or equipped with filters to block ultraviolet light. High pressure sodium vapor and incandescent lamps attract fewer moths than do mercury vapor lamps, but they attract more moths than do low pressure sodium vapor lamps.

Low pressure sodium vapor lamps present both advantages and disadvantages when evaluated for municipal use. On the negative side, the deep monochromatic yellow-orange of low pressure sodium light does not allow accurate color rendition. Faces appear gray, and colors of vehicles cannot be distinguished. Total operating costs of these lamps have been calculated to be about 50% higher than that for high pressure sodium vapor lamps. Light from low pressure sodium lamps is harder to direct than light from high pressure sodium lamps, which are smaller. On the positive side, low pressure sodium vapor lamps do not attract moths indoors when they are used to illuminate entrances to buildings. Remains of moths do not foul lamps and block light emission. Compared with the broad spectral emission of other light sources, the narrow spectral output of low pressure sodium vapor light causes less interference with astronomical observation. The energy efficiency of a 135-W low pressure sodium vapor lamp is 148 lumens/W, more than six times higher than that of a 1,000-W incandescent (tungsten filament) lamp (Frank 1988). Low pressure sodium vapor lamps do not need orange filters such as those coating incandescent bulbs (e.g., "Bug-Lites") to block short wavelengths that attract moths.

Conclusion

Most species of larger moths can breed near urban lighting. Moth populations already threatened by habitat loss and fragmentation may be vulnerable to artificial lighting. Artificial lighting will exert its greatest force on populations when it is located in or near fragments of habitat suitable for breeding, but even when located at a distance remote from these sites, it may interrupt dispersal. In rare native habitats, assemblages of endemic or regionally restricted species may warrant protection even when no particular member of the assemblage is itself endangered (Brown and Bash 1997 [2000]).

In the future, controlled trials may clarify effects of artificial lighting on populations of moths. Such studies could measure the extent to which lighting disrupts dispersal of moths between habitat fragments. In the meantime, a reasonable strategy would be to incorporate restrictions on artificial lighting into plans for conservation of important ecological habitats.

Acknowledgments

Robert L. Edwards gave me insight into the relation of lighting to habitat fragmentation. Detlef Kolligs reviewed with me his research and other work written in German. Dale Schweitzer and Timothy McCabe shared their observations and assessments of moth behavior around lights. John W. Brown provided information about taxonomic problems in the inventory of moths at Marine Corps Air Station Miramar. Jason Weintraub directed me to recent literature on urban moths. This chapter updates an earlier literature review of this subject (Frank 1988).

Literature Cited

Acharya, L., and M. B. Fenton. 1999. Bat attacks and moth defensive behaviour around street lights. *Canadian Journal of Zoology* 77:27–33.

Agee, H. R. 1969. Mating behavior of bollworm moths. *Annals of the Entomological Society of America* 62:1120–1122.

Agee, H. R. 1972. Sensory response of the compound eye of adult *Heliothis zea* and *H. virescens* to ultraviolet stimuli. *Annals of the Entomological Society of America* 65:701–705.

Allen, N., and C. R. Hodge. 1955. Mating habits of the tobacco hornworm. *Journal of Economic Entomology* 48:526–528.

Baker, R. R., and Y. Sadovy. 1978. The distance and nature of the light-trap response of moths. *Nature* 276:818–821.

Barth, F. G. 1985. *Insects and flowers: the biology of a partnership*. Princeton University Press, Princeton, New Jersey.

Beaty, H. H., J. H. Lilly, and D. L. Calderwood. 1951. Use of radiant energy for corn borer control. *Agricultural Engineering* 32:421–422, 426, 429.

Bebas, P., B. Cymborowski, and J. M. Giebultowicz. 2001. Circadian rhythm of sperm release in males of the cotton leafworm, *Spodoptera littoralis*: in vivo and in vitro studies. *Journal of Insect Physiology* 47:859–866.

Beebe, W., and H. Fleming. 1951. Migration of day-flying moths through Portachuelo Pass, Rancho Grande, north-central Venezuela. *Zoologica* 36:243–254.

Berlinger, M. J., and G. W. Ankersmit. 1976. Manipulation with the photoperiod as a method of control of *Adoxophyes orana* (Lepidoptera, Tortricidae). *Entomologia Experimentalis et Applicata* 19:96–107.

Bernhard, C. G., and D. Ottoson. 1960. Studies on the relation between the pigment migration and the sensitivity changes during dark adaptation in diurnal and nocturnal Lepidoptera. *Journal of General Physiology* 44:205–215.

Blest, A. D. 1963. Longevity, palatability and natural selection in five species of New World saturniid moth. *Nature* 197:1183–1186.

Bowden, J., and M. G. Morris. 1975. The influence of moonlight on catches of insects in light-traps in Africa. III. The effective radius of a mercury-vapour light-trap and the analysis of catches using effective radius. *Bulletin of Entomological Research* 65:303–348.

Bradley, J. D., and R. M. Mere. 1966. Natural history of the garden of Buckingham Palace. Further records and observations, 1964–5. Lepidoptera. *Proceedings of the South London Entomological and Natural History Society* 1966:15–17.

Bristowe, W. S. 1941. *The comity of spiders*. Volume II. The Ray Society, London.

Brown, J. W., and K. Bash. 1997 (2000). The Lepidoptera of Marine Corps Air Station Miramar: calculating faunal similarity among sampling sites and estimating total species richness. *Journal of Research on the Lepidoptera* 36:45–78.

Brown, L. N. 1984. Population outbreak of pandora moths (*Coloradia pandora* Blake) on the Kaibab Plateau, Arizona (Saturniidae). *Journal of the Lepidopterists' Society* 38:65.

Butler, L., V. Kondo, E. M. Barrows, and E. C. Townsend. 1999. Effects of weather conditions and trap types on sampling for richness and abundance of forest macrolepidoptera. *Environmental Entomology* 28:795–811.

Cantelo, W. W., J. L. Goodenough, A. H. Baumhover, J. S. Smith Jr., J. M. Stanley, and T. J. Henneberry. 1974. Mass trapping with blacklight: effects on isolated populations of insects. *Environmental Entomology* 3:389–395.

Cantelo, W. W., J. S. Smith Jr., A. H. Baumhover, J. M. Stanley, T. J. Henneberry, and M. B. Peace. 1972. The suppression of isolated populations of sphingids by blacklight traps. *Environmental Entomology* 1:753–759.

Cantelo, W. W., J. S. Smith Jr., A. H. Baumhover, J. M. Stanley, T. J. Henneberry, and M. B. Peace. 1973. Changes in the population levels of 17 insect species during a 3 1/2-year blacklight trapping program. *Environmental Entomology* 2:1033–1038.

Chey, V. K., J. D. Holloway, and M. R. Speight. 1997. Diversity of moths in forest plantations and natural forests in Sabah. *Bulletin of Entomological Research* 87:371–385.

Collins, C. T., and A. Watson. 1983. Field observations of bird predation on neotropical moths. *Biotropica* 15:53–60.

Daily, G. C., and P. R. Ehrlich. 1996. Nocturnality and species survival. *Proceedings of the National Academy of Sciences of the United States of America* 93:11709–11712.

Danthanarayana, W. 1986. Lunar periodicity of insect flight and migration. Pages 88–119 in W. Danthanarayana (ed.), *Insect flight: dispersal and migration*. Springer-Verlag, Berlin.

Delisle, J., R. J. West, and W. W. Bowers. 1998. The relative performance of pheromone and light traps in monitoring the seasonal activity of both sexes of the eastern hemlock looper, *Lambdina fiscellaria fiscellaria*. *Entomologia Experimentalis et Applicata* 89:87–98.

Denton, S. F. 1900. *As nature shows them: moths and butterflies of the United States, east of the Rocky Mountains*. Volume 1. *The moths*. J.B. Millet Company, Boston.

Dirks, C. O. 1937. Biological studies of Maine moths by light trap methods. *Maine Agricultural Experiment Station, Orono: Bulletin* 389:33–162.

Dunning, D. C., and M. Krüger. 1996. Predation upon moths by free-foraging *Hipposideros caffer*. *Journal of Mammalogy* 77:708–715.

Ebeling, W. 1978. *Urban entomology*. University of California, Berkeley.

Eguchi, E., K. Watanabe, T. Hariyama, and K. Yamamoto. 1982. A comparison of electrophysiologically determined spectral responses in 35 species of Lepidoptera. *Journal of Insect Physiology* 28:675–682.

Ehrlich, P. R. 1984. The structure and dynamics of butterfly populations. Pages 24–40 in R. I. Vane-Wright and P. R. Ackery (eds.), *The biology of butterflies*. Academic Press, London.

Eisenbeis, G., and F. Hassel. 2000. Zur Anziehung nachtaktiver Insekten durch Straßenlaternen: eine Studie kommunaler Beleuchtungseinrichtungen in der Agrarlandschaft Rheinhessens [Attraction of nocturnal insects to street lights: a study of municipal lighting systems in a rural area of Rheinhessen (Germany)]. *Natur und Landschaft* 75(4):145–156.

Emmet, A. M. 1991. Chart showing the life history and habits of the British Lepidoptera. Pages 61–303 in A. M. Emmet and J. Heath (eds.), *The moths and butterflies of Great Britain and Ireland*. Volume 7, Part 2. *Lasiocampidae: Thyatiridae with life history chart of the British Lepidoptera*. Harley Books, Colchester, Essex, UK.

Fatzinger, C. W. 1973. Circadian rhythmicity of sex pheromone release by *Dioryctria abietella* (Lepidoptera: Pyralidae (Phycitinae)) and the effect of a diel light cycle on its precopulatory behavior. *Annals of the Entomological Society of America* 66:1147–1153.

Ficht, G. A., T. E. Hienton, and J. M. Fore. 1940. The use of electric light traps in the control of the European corn borer. *Agricultural Engineering* 21:87–89.

Frank, K. D. 1986. History of the ailanthus silk moth (Lepidoptera: Saturniidae) in Philadelphia: a case study in urban ecology. *Entomological News* 97:41–51.

Frank, K. D. 1988. Impact of outdoor lighting on moths: an assessment. *Journal of the Lepidopterists' Society* 42:63–93.

Frick, T. B., and D. W. Tallamy. 1996. Density and diversity of nontarget insects killed by suburban electric insect traps. *Entomological News* 107:77–82.

Fullard, J. H. 1990. The sensory ecology of moths and bats: global lessons in staying alive. Pages 203–226 in D. L. Evans and J. O. Schmidt (eds.), *Insect defenses: adaptive mechanisms and strategies of prey and predators*. State University of New York Press, Albany.

Gauld, I. D. 1986. Taxonomy, its limitations and its role in understanding parasitoid biology. Pages 1–21 in J. Waage and D. Greathead (eds.), *Insect parasitoids*. Academic Press, London.

Gentry, C. R., W. A. Dickerson Jr., and J. M. Stanley. 1971. Populations and mating of adult tobacco budworms and corn earworms in northwest Florida indicated by traps. *Journal of Economic Entomology* 64:335–338.

Giebultowicz, J. M. 2000. Molecular mechanism and cellular distribution of insect circadian clocks. *Annual Review of Entomology* 45:769–793.

Graham, H. M., P. A. Glick, and J. P. Hollingsworth. 1961. Effective range of argon glow lamp survey traps for pink bollworm adults. *Journal of Economic Entomology* 54:788–789.

Graham, H. M., P. A. Glick, and D. F. Martin. 1964. Nocturnal activity of adults of six lepidopterous pests of cotton as indicated by light-trap collections. *Annals of the Entomological Society of America* 57:328–332.

Grimble, D. G., R. C. Beckwith, and P. C. Hammond. 1992. A survey of the Lepidoptera fauna from the Blue Mountains of eastern Oregon. *Journal of Research on the Lepidoptera* 31:83–102.

Grünewald, B., and H. Wunderer. 1996. The ocelli of arctiid moths: ultrastructure of the retina during light and dark adaptation. *Tissue & Cell* 28:267–277.

Hamdorf, K., and G. Höglund. 1981. Light induced retinal screening pigment migration independent of visual cell activity. *Journal of Comparative Physiology A* 143:305–309.

Hamilton, D. W., and L. F. Steiner. 1939. Light traps and codling moth control. *Journal of Economic Entomology* 32:867–872.

Hartstack, A. W. Jr., J. P. Hollingsworth, and D. A. Lindquist. 1968. A technique for measuring trapping efficiency of electric insect traps. *Journal of Economic Entomology* 61:546–552.

Hartstack, A. W. Jr., J. P. Hollingsworth, R. L. Ridgway, and H. H. Hunt. 1971. Determination of trap spacings required to control an insect population. *Journal of Economic Entomology* 64:1090–1100.

Hausmann, A. 1992. Untersuchungen zum Massensterben von Nachtfaltern an Industriebeleuchtungen (Lepidoptera, Macroheterocera) [Studies of the mass mortality of moths near municipal lights (Lepidoptera, Macroheterocera)]. *Atalanta* 23:411–416.

Hawkins, B. A. 1993. Refuges, host population dynamics and the genesis of parasitoid diversity. Pages 235–256 in J. LaSalle and I. D. Gauld (eds.), *Hymenoptera and biodiversity*. CAB International, Wallingford, UK.

Heiling, A. M. 1999. Why do nocturnal orb-web spiders (Araneidae) search for light? *Behavioral Ecology and Sociobiology* 46:43–49.

Henderson, R. W., and R. Powell. 2001. Responses by the West Indian herpetofauna to human-influenced resources. *Caribbean Journal of Science* 37:41–54.

Herczig, B., and Z. Mészáros. 1994. A study of dispersion and migration of *Lepidoptera* by light trap pairs. *Acta Phytopathologica et Entomologica Hungarica* 29:187–193.

Herms, W. B. 1929. A field test of the effect of artificial light on the behavior of the codling moth *Carpocapsa pomonella* Linn. *Journal of Economic Entomology* 22:78–88.

Herms, W. B. 1932. Deterrent effect of artificial light on the codling moth. *Hilgardia* 7:263–280.

Hessel, S. A. 1976. A preliminary scan of rare and endangered Nearctic moths. *Atala* 4:19–21.

Hickey, M. B. C., L. Acharya, and S. Pennington. 1996. Resource partitioning by two species of vespertilionid bats (*Lasiurus cinereus* and *Lasiurus borealis*) feeding around street lights. *Journal of Mammalogy* 77:325–334.

Hienton, T. E. 1974. *Summary of investigations of electric insect traps*. Technical Bulletin No. 1498 of the Agricultural Research Service. United States Department of Agriculture, Washington, D.C.

Hoffman, J. D., F. R. Lawson, and B. Peace. 1966. Attraction of blacklight traps baited with virgin female tobacco hornworm moths. *Journal of Economic Entomology* 59:809–811.

Hosny, M. M. 1959. A review of results and a complete list of macrolepidoptera caught in two ultra-violet light traps during 24 months, at Rothamsted, Hertfordshire. *Entomologist's Monthly Magazine* 95:226–236.

Howe, W. H. 1959. A swarm of noctuid moths in southeastern Kansas. *Journal of the Lepidopterists' Society* 13:26.

Hsiao, H. S. 1972. *Attraction of moths to light and to infrared radiation.* San Francisco Press, San Francisco.

Hulse, A. C., C. J. McCoy, and E. J. Censky. 2001. *Amphibians and reptiles of Pennsylvania and the Northeast.* Cornell University Press, Ithaca, New York.

Janzen, D. H. 1983. [Insects] Introduction. Pages 619–645 in D. H. Janzen (ed.), *Costa Rican natural history.* University of Chicago Press, Chicago.

Janzen, D. H. 1984. Two ways to be a tropical big moth: Santa Rosa saturniids and sphingids. Pages 85–140 in R. Dawkins and M. Ridley (eds.), *Oxford surveys in evolutionary biology.* Volume 1. Oxford University Press, Oxford.

Johnson, C. G. 1969. *Migration and dispersal of insects by flight.* Methuen, London.

Johnson, S. D., A. Ellis, P. Carrick, A. Swift, N. Horner, S. Janse van Rensburg, and W. J. Bond. 1993. Moth pollination and rhythms of advertisement and reward in *Crassula fascicularis* (Crassulaceae). *South African Journal of Botany* 59:511–513.

Kaston, B. J. 1981. *Spiders of Connecticut.* State Geological and Natural History Survey of Connecticut, Bulletin 70. Revised edition. Connecticut Department of Environmental Protection, Hartford.

Kitching, R. L., A. G. Orr, L. Thalib, H. Mitchell, M. S. Hopkins, and A. W. Graham. 2000. Moth assemblages as indicators of environmental quality in remnants of upland Australian rain forest. *Journal of Applied Ecology* 37:284–297.

Kolligs, D. 2000. Ökologische Auswirkungen künstlicher Lichtquellen auf nachtaktive Insekten, insbesondere Schmetterlinge (Lepidoptera) [Ecological effects of artificial light sources on nocturnally active insects, in particular on moths (Lepidoptera)]. *Faunistisch–Ökologische Mitteilungen* Supplement 28:1–136.

Kozlov, M. V. 1996. Patterns of forest insect distribution within a large city: microlepidoptera in St Peterburg, Russia. *Journal of Biogeography* 23:95–103.

Langmaid, J. R. 1959. Moths of a Portsmouth garden: a four-year appreciation. *Entomologist's Gazette* 10:159–164.

LaSalle, J. 1993. Parasitic hymenoptera, biological control and biodiversity. Pages 197–215 in J. LaSalle and I. D. Gauld (eds.), *Hymenoptera and biodiversity.* CAB International, Wallingford, UK.

LaSalle, J., and I. D. Gauld. 1993. Hymenoptera: their diversity, and their impact on the diversity of other organisms. Pages 1–26 in J. LaSalle and I. D. Gauld (eds.), *Hymenoptera and biodiversity.* CAB International, Wallingford, UK.

Lutz, F. E. 1941. *A lot of insects: entomology in a suburban garden.* G. P. Putnam's Sons, New York.

MacFarlane, J. H., and J. L. Eaton. 1973. Comparison of electroretinogram and electromyogram responses to radiant energy stimulation in the moth, *Trichoplusia ni. Journal of Insect Physiology* 19:811–822.

Majerus, M. E. N. 1998. *Melanism: evolution in action.* Oxford University Press, Oxford.

Martin, C. H., and J. S. Houser. 1941. Numbers of *Heliothis armigera* (Hbn.) and two other moths captured at light traps. *Journal of Economic Entomology* 34:555–559.

McGeachie, W. J. 1988. A remote sensing method for the estimation of light-trap efficiency. *Bulletin of Entomological Research* 78:379–385.

McGeachie, W. J. 1989. The effects of moonlight illuminance, temperature and wind speed on light-trap catches of moths. *Bulletin of Entomological Research* 79:185–192.

Memmott, J., and H. C. J. Godfray. 1993. Parasitoid webs. Pages 217–234 in J. LaSalle and I. D. Gauld (eds.), *Hymenoptera and biodiversity*. CAB International, Wallingford, UK.

Mikkola, K. 1972. Behavioural and electrophysiological responses of night-flying insects, especially Lepidoptera, to near-ultraviolet and visible light. *Annales Zoologici Fennici* 9:225–254.

Mikkola, K. 1986. Direction of insect migrations in relation to the wind. Pages 152–171 in W. Danthanarayana (ed.), *Insect flight: dispersal and migration*. Springer-Verlag, Berlin.

Mitchell, E. R., and H. R. Agee. 1981. Response of beet and fall armyworm moths to different colored lamps in the laboratory and field. *Journal of Environmental Science and Health, Part A: Environmental Science and Engineering* A16:387–396.

Mizunami, M. 1995. Functional diversity of neural organization in insect ocellar systems. *Vision Research* 35:443–452.

Muirhead-Thomson, R. C. 1991. *Trap responses of flying insects: the influence of trap design on capture efficiency*. Academic Press, London.

Nabli, H., W. C. Bailey, and S. Necibi. 1999. Responses of Lepidoptera in central Missouri to traps with different light sources. *Journal of the Kansas Entomological Society* 72:82–90.

Nässig, W. A., and R. S. Peigler. 1984. The life-history of *Actias maenas* (Saturniidae). *Journal of the Lepidopterists' Society* 38:114–123.

Nemec, S. J. 1969. Use of artificial lighting to reduce *Heliothis* spp. populations in cotton fields. *Journal of Economic Entomology* 62:1138–1140.

Nemec, S. J. 1971. Effects of lunar phases on light-trap collections and populations of bollworm moths. *Journal of Economic Entomology* 64:860–864.

Nieminen, M. 1996. Migration of moth species in a network of small islands. *Oecologia* 108:643–651.

Nieminen, M., and I. Hanski. 1998. Metapopulations of moths on islands: a test of two contrasting models. *Journal of Animal Ecology* 67:149–160.

Nomura, K. 1969. Studies on orchard illumination against fruit piercing moths. *Review of Plant Protection Research* 2:122–124.

Norris, M. J. 1936. The feeding-habits of the adult Lepidoptera Heteroneura. *Transactions of the Royal Entomological Society of London* 85:61–90.

Opell, B. D. 1994. The ability of spider cribellar prey capture thread to hold insects with different surface features. *Functional Ecology* 8:145–150.

Owen, D. F. 1978. Insect diversity in an English suburban garden. Pages 13–29 in G. W. Frankie and C. S. Koehler (eds.), *Perspectives in urban entomology*. Academic Press, New York.

Pedgley, D. E., and S. Yathom. 1993. Windborne moth migration over the Middle East. *Ecological Entomology* 18:67–72.

Pfrimmer, T. R., M. J. Lukefahr, and J. P. Hollingsworth. 1955. *Experiments with light traps for control of the pink bollworm*. U.S. Department of Agriculture, Agricultural Research Service, ARS-33-6, Washington, D.C.

Pittendrigh, C. S., and D. H. Minis. 1971. The photoperiodic time measurement in *Pectinophora gossypiella* and its relation to the circadian system in that species. Pages 212–250 in M. Menaker (ed.), *Biochronometry*. National Academy of Sciences, Washington, D.C.

Plant, C. W. 1993. *Larger moths of the London area*. London Natural History Society, London.

Plant, C. W. 1999. A review of the butterflies and moths (Lepidoptera) of the London Area for 1997 and 1998. *London Naturalist* 78:147–171.

Plaut, H. N. 1971. Distance of attraction of moths of *Spodoptera littoralis* to BL radiation, and recapture of moths released at different distances of an ESA blacklight standard trap. *Journal of Economic Entomology* 64:1402–1404.

Powell, J. A., and J. W. Brown. 1990. Concentrations of lowland sphingid and noctuid moths at high mountain passes in eastern Mexico. *Biotropica* 22:316–319.

Riley, J. R., D. R. Reynolds, A. D. Smith, A. S. Edwards, Zhang X.-X., Cheng X.-N., Wang H.-K., Cheng J.-Y., and Zhai B.-P. 1995. Observations of the autumn migration of the rice leaf roller *Cnaphalocrocis medinalis* (Lepidoptera: Pyralidae) and other moths in eastern China. *Bulletin of Entomological Research* 85:397–414.

Riordan, D. F. 1964. High-intensity flash discharge as a source of radiant energy for sterilizing insects. *Nature* 204:1332.

Robinson, H. S. 1952. On the behaviour of night-flying insects in the neighbourhood of a bright source of light. *Proceedings of the Royal Entomological Society of London, Series A. General Entomology* 27:13–21.

Robinson, H. S., and P. J. M. Robinson. 1950. Some notes on the observed behaviour of Lepidoptera in flight in the vicinity of light-sources together with a description of a light-trap designed to take entomological samples. *Entomologist's Gazette* 1:3–20.

Robinson, P. J. M. 1960. An experiment with moths on the effectiveness of a mercury vapour light trap. *Entomologist's Gazette* 11:121–132.

Rydell, J. 1992. Exploitation of insects around streetlamps by bats in Sweden. *Functional Ecology* 6:744–750.

Saario, C. A., H. H. Shorey, and L. K. Gaston. 1970. Sex pheromones of noctuid moths. XIX. Effect of environmental and seasonal factors on captures of males of *Trichoplusia ni* in pheromone-baited traps. *Annals of the Entomological Society of America* 63:667–672.

Sargent, T. D. 1990. Startle as an anti-predator mechanism, with special reference to the underwing moths (*Catocala*). Pages 229–249 in D. L. Evans and J. O. Schmidt (eds.), *Insect defenses: adaptive mechanisms and strategies of prey and predators*. State University of New York Press, Albany.

Saunders, D. S. 1982. *Insect clocks*. Pergamon Press, Oxford.

Schaefer, G. W. 1976. Radar observations of insect flight. Pages 157–197 in R. C. Rainey (ed.), *Insect flight*. Symposia of the Royal Entomological Society of London: Number Seven. John Wiley & Sons, New York.

Shorey, H. H., and L. K. Gaston. 1964. Sex pheromones of noctuid moths. III. Inhibition of male responses to the sex pheromone in *Trichoplusia ni* (Lepidoptera: Noctuidae). *Annals of the Entomological Society of America* 57:775–779.
Shorey, H. H., and L. K. Gaston. 1965. Sex pheromones of noctuid moths. VIII. Orientation to light by pheromone-stimulated males of *Trichoplusia ni* (Lepidoptera: Noctuidae). *Annals of the Entomological Society of America* 58:833–836.
Sierro, A., and R. Arlettaz. 1997. Barbastelle bats (*Barbastella* spp.) specialize in the predation of moths: implications for foraging tactics and conservation. *Acta Œcologica* 18:91–106.
Sotthibandhu, S., and R. R. Baker. 1979. Celestial orientation by the large yellow underwing moth, *Noctua pronuba* L. *Animal Behaviour* 27:786–800.
Sower, L. L., H. H. Shorey, and L. K. Gaston. 1970. Sex pheromones of noctuid moths. XXI. Light:dark cycle regulation and light inhibition of sex pheromone release by females of *Trichoplusia ni*. *Annals of the Entomological Society of America* 63:1090–1092.
Spangler, H. G. 1988. Moth hearing, defense, and communication. *Annual Review of Entomology* 33:59–81.
Spencer, J. L., L. J. Gewax, J. E. Keller, and J. R. Miller. 1997. Chemiluminescent tags for tracking insect movement in darkness: application to moth photo-orientation. *Great Lakes Entomologist* 30:33–43.
Stanley, W. W. 1932. Observations on the flight of noctuid moths. *Annals of the Entomological Society of America* 25:366–368.
Sternburg, J. G., G. P. Waldbauer, and A. G. Scarbrough. 1981. Distribution of cecropia moth (Saturniidae) in central Illinois: a study in urban ecology. *Journal of the Lepidopterists' Society* 35:304–320.
Stewart, P. A., J. J. Lam, and J. D. Hoffman. 1967. Activity of tobacco hornworm and corn earworm moths as determined by traps equipped with blacklight lamps. *Journal of Economic Entomology* 60:1520–1522.
Svensson, A. M., and J. Rydell. 1998. Mercury vapour lamps interfere with the bat defence of tympanate moths (*Operophtera* spp.; Geometridae). *Animal Behaviour* 55:223–226.
Taylor, L. R., and C. I. Carter. 1961. The analysis of numbers and distribution in an aerial population of macrolepidoptera. *Transactions of the Royal Entomological Society of London* 113:369–386.
Taylor, L. R., R. A. French, and I. P. Woiwod. 1978. The Rothamsted insect survey and the urbanization of land in Great Britain. Pages 31–65 in G. W. Frankie and C. S. Koehler (eds.), *Perspectives in urban entomology*. Academic Press, London.
Tedders, W. L. Jr., J. G. Hartsock, and M. Osburn. 1972. Suppression of hickory shuckworm in a pecan orchard with blacklight traps. *Journal of Economic Entomology* 65:148–155.
Teraguchi, S. E., and K. J. Lublin. 1999. Checklist of the moths of Pallister State Nature Preserve, Ashtabula County, Ohio (1988–1992) with analyses of abundance. *Kirtlandia* 51:3–18.
Thomas, A. W. 1996. Light-trap catches of moths within and above the canopy of a northeastern forest. *Journal of the Lepidopterists' Society* 50:21–45.

Thomas, C. D. 2000. Dispersal and extinction in fragmented landscapes. *Proceedings of the Royal Society of London. Series B, Biological Sciences* 267:139–145.

Tigar, B. J., and P. E. Osborne. 1999. Patterns of biomass and diversity of aerial insects in Abu Dhabi's sandy deserts. *Journal of Arid Environments* 43:159–170.

Turnbull, A. L. 1964. The search for prey by a web-building spider *Achaearanea tepidariorum* (C. L. Koch) (Araneae, Theridiidae). *Canadian Entomologist* 96:568–579.

Tuskes, P. M., J. P. Tuttle, and M. M. Collins. 1996. *The wild silk moths of North America: a natural history of the Saturniidae of the United States and Canada.* Cornell University Press, Ithaca, New York.

Usher, M. B., and S. W. J. Keiller. 1998. The macrolepidoptera of farm woodlands: determinants of diversity and community structure. *Biodiversity and Conservation* 7:725–748.

Vail, P. V., A. F. Howland, and T. J. Henneberry. 1968. Seasonal distribution, sex ratios, and mating of female noctuid moths in blacklight trapping studies. *Annals of the Entomological Society of America* 61:405–411.

Väisänen, R., and C. Hublin. 1983. The effect of continuous light trapping on moth populations: a mark–recapture experiment on *Hydraecia petasitis* (Lepidoptera, Noctuidae). *Notulae Entomologicae* 63:187–191.

Walcott, B. 1975. Anatomical changes during light-adaptation in insect compound eyes. Pages 20–33 in G. A. Horridge (ed.), *The compound eye and vision of insects*. Clarendon Press, Oxford.

Waldbauer, G. P., and J. G. Sternburg. 1979. Inbreeding depression and a behavioral mechanism for its avoidance in *Hyalophora cecropia*. *American Midland Naturalist* 102:204–208.

Waldbauer, G. P., and J. G. Sternburg. 1982. Long mating flights by male *Hyalophora cecropia* (L.) (Saturniidae). *Journal of the Lepidopterists' Society* 36:154–155.

Wehner, R. 1998. Navigation in context: grand theories and basic mechanisms. *Journal of Avian Biology* 29:370–386.

Williams, C. B. 1939. An analysis of four years captures of insects in a light trap. Part I. General survey; sex proportion; phenology; and time of flight. *Transactions of the Royal Entomological Society of London* 89:79–131.

Williams, C. B., B. P. Singh, and S. el Ziady. 1956. An investigation into the possible effects of moonlight on the activity of insects in the field. *Proceedings of the Royal Entomological Society of London, Series A. General Entomology* 31:135–144.

Winter, W. D. Jr. 2000. *Basic techniques for observing and studying moths and butterflies*. The Lepidopterists' Society Memoir Number 5. The Lepidopterists' Society, Los Angeles, California.

Wolda, H., J. Marek, K. Spitzer, and I. Novák. 1994. Diversity and variability of Lepidoptera populations in urban Brno, Czech Republic. *European Journal of Entomology* 91:213–226.

Wolf, W. W., A. N. Sparks, S. D. Pair, J. K. Westbrook, and F. M. Truesdale. 1986. Radar observations and collections of insects in the Gulf of Mexico.

Pages 221–234 in W. Danthanarayana (ed.), *Insect flight: dispersal and migration*. Springer-Verlag, Berlin.
Worth, C. B., and J. Muller. 1979. Captures of large moths by an ultraviolet light trap. *Journal of the Lepidopterists' Society* 33:261–264.
Yela, J. L., and M. Holyoak. 1997. Effects of moonlight and meteorological factors on light and bait trap catches of noctuid moths (Lepidoptera: Noctuidae). *Environmental Entomology* 26:1283–1290.
Young, M. 1997. *The natural history of moths*. T & A D Poyser, London.
Zhang, B.-C. 1994. *Index of economically important Lepidoptera*. CAB International, Wallingford, UK.

Chapter 14

Stray Light, Fireflies, and Fireflyers

James E. Lloyd

Fireflies (Coleoptera: Lampyridae) use self-generated chemical luminescence for sexual communication and have a number of attributes that make them appropriate and unique subjects for examining the consequences of artificial light entering natural environments. They are also important in many cultures, which may make them useful icons for the conservation of the nighttime environment. Their bioluminescence provides a magic to many a backyard that would be sorely missed.

In this chapter I discuss light and life, reviewing first the organisms that produce light and introducing the ecology of fireflies. I recount the methods of studying fireflies, then expand on the potential effects of stray light interfering with fireflies in their environments. These effects are exacerbated by the cumulative effects of other ecological insults; hope for the amelioration of these effects may lie in part in the universal appeal of these small beetles.

Light and Life

One of my favorite productions of the film industry is *Doc Hollywood*, and my favorite character is the cantankerous town doctor because I identify with him. He was a general practitioner who had seen to the births, illnesses, injuries, health education, and deaths of the natives in his domain across the several decades of his watch. Through intimate knowledge of his people and general understanding of medicine and health, when conditions warranted he could bring his patients to the attention of specialists. I identify with the Old Doc because I am a "GP" of sorts, a mere naturalist and taxonomist in a world of specialization who has chased fireflies and other terrestrial bioluminescence for several decades, mostly in the United States but in a few other places as well. I can bring information about fireflies and their possibilities as research organisms and models to the attention of specialists who can use them. My studies on fireflies have always been those of a provincial biotaxonomist, focused primarily on knowing and understanding the nature of the species of North America (Lloyd 1966, 1969, 2001, 2003), but after all of this chasing and watching I have been forced to the realization that species in nature are not what they appear or are often purported to be. Indeed, the more I learn and think about firefly species the less I understand them, and I now view "them" (that is, those things we want to call species) as transcendent entities (Lloyd 2003:100).

Likewise, I suspect that what we are seeking in this book may be of the same nature, and regardless of how much detail we learn and how comprehensive and insightful our theories, it is beyond our capacity to truly experience, comprehend, and, in particular, understand the living world's relations with light, so complex is the fabric. Yet can anything be more important to us than the maintenance of the very wellspring of our livelihood, spirit, and humanity? "Light and life"—a simple and elegant phrase but so profound and subtle in its reality that only poetry can satisfyingly deal with it. "What is life? It is the flash of a firefly in the night . . . the little shadow which runs across the grass and loses itself in the sunset" (Chief Crowfoot, Blackfoot Confederation, circa 1890).

Fireflies are the most common terrestrial, light-emitting organisms in the world; others can be mentioned, though they seem to be even less consequential elements in natural ecosystems (Lloyd and Gentry 2003). These include luminous mushrooms (Figure 14.1a), whose luminosity may attract flies that transport spores on their "hairy" legs to other suitable sites (Lloyd 1974); luminescent fungus gnats such as those responsible for the pinpoints of blue-green light one sees along streams and in impatiens sloughs in the Appalachians, whose larvae glow and attract minute prey to tiny webs (Figure 14.1b; Fulton 1941, Sivinski 1982); and certain click beetles, their adults being the illuminated, hard-headed bullets that streak through rainforest canopies at dusk in tropical America and open fields in Florida and Texas (Figure 14.1c; Lloyd 1983) and whose larvae may use their light to capture prey, as is known of a Brazilian species that lives with and specializes on termites that live in carton towers (Redford 1982). There are even luminescent collembolans, little-known and primitive insects that occur broadly in damp forests under logs and leaves and flash brightly when disturbed (Figure 14.1b inset; Barber 1913).

To find evidence that light plays an important part in the lives of organisms, lab- and library-bound biologists need go no further than a student dictionary: *photosynthesis*, *photoperiod*, *photomorphogenesis*, *photophore*, *photophosphorylation*, *photopia*, *phototaxis*, *phototropism*, and *skototaxis*, the last being the term for orientation to darkness. In the days before study of the adaptive significance of animal behavior became respectable (via European ethology and thence behavioral ecology), zoologists were permitted to look at the mechanics and machine-like responses of behavior, and the vocabulary they generated from their studies gives further evidence of the intimate connection between light and life. A now-classic summation of this evidence is the book by Fraenkel and Gunn, *The Orientation of Animals* (1961, a republication in English of the 1940 German text), where one finds *photophil*, *photophobe*, *photo-horo-taxis*, and *photokinesis*, as well as the two-light experiment, dorsal-light reaction, and light-compass reaction.

What can be more convincing of the likely importance of artificial light sources near natural environments than the millions of individuals of thousands of species of insects, pollinators, and pest parasites among them, that die under street and gas station lights and are hunted there by experienced predators, including toads and domestic cats? Under the lights at a repair shop in the highlands of New Guinea I found scarab beetles the size of a child's fist and Atlas moths nearly a foot in wingspan. Piles like snowdrifts of mayflies and midges sometimes accumulate under the lights on bridges over the upper Mississippi River. Entomologists have long been aware of the importance of light in the lives of their subjects and its fatal attraction for them, and taxonomists and students still use light traps to make their collections. Curiously, the explanations for the lure of light and for the misreckoning and misguidance that occur in insect brains and servo systems remain mostly a mystery.

Other evidence of the subtleties in the relation of insects with light can be adduced from the work of practical entomologists over the past century in their studies and attempts to manipulate and eradicate pest insects. When I joined the entomology department at the University of Florida in

(a)

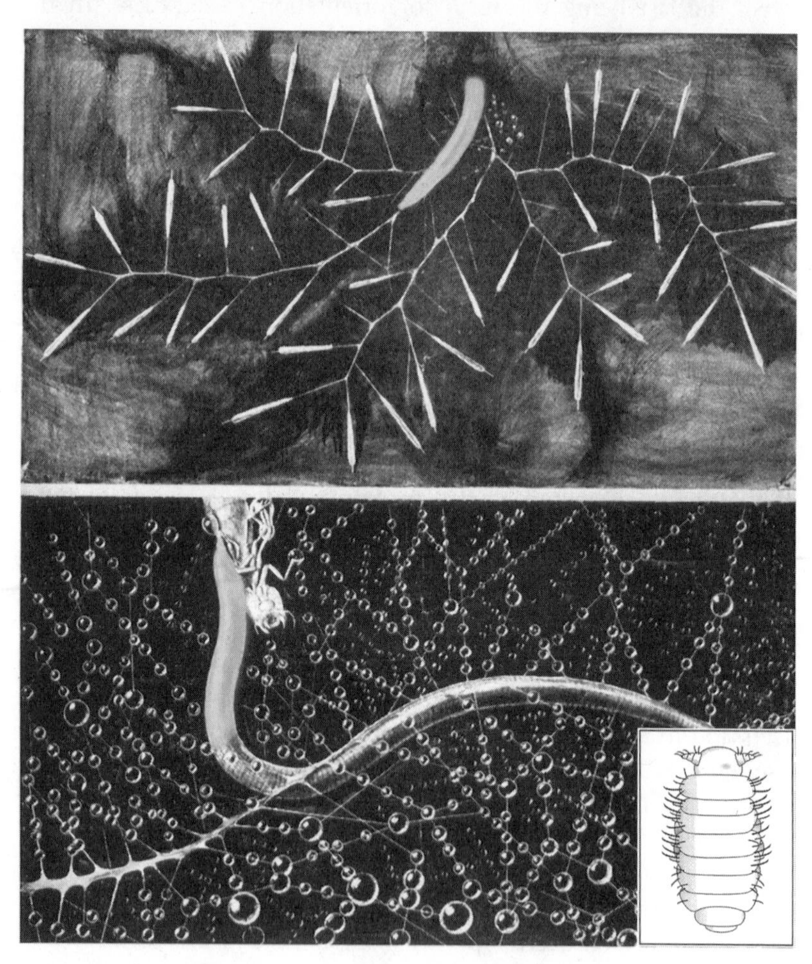

(b)

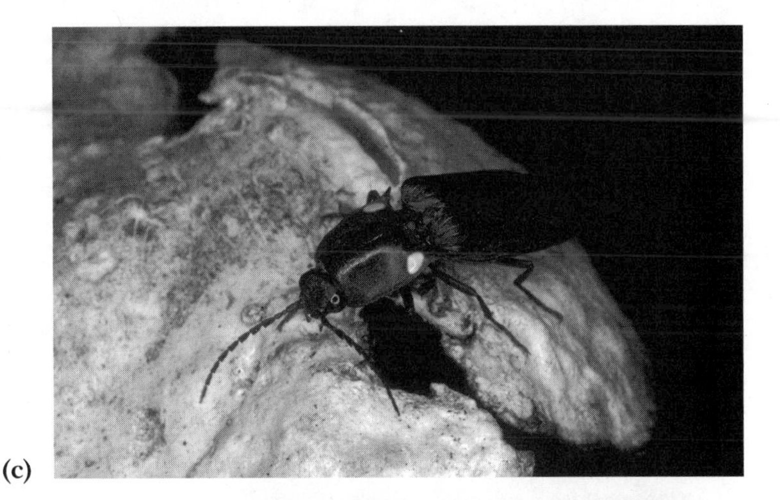
(c)

(d)

Figure 14.1. Examples of bioluminescent organisms. *(a)* Glow of luminous mushrooms from the island of Espiritu Santo in the New Hebrides in the southwestern Pacific. *(b)* Luminous fungus gnat larva on its spider-like web with prey that it has attracted with its glow. A full-grown larva is about 15 mm (0.6 in) long (after Fulton 1941). Inset, sketch of a luminescent Collembola, *Neanura barberi*, length 1–2 mm (0.04–0.08 in). The luminescence of this insect has not been studied, and the intended receiver of its flashes and the tissue or organ that emits the light remain unknown (from Lloyd, unpublished instruction manual). *(c)* A luminescent click beetle, family Elateridae. The pale spot visible on the pronotum is a lantern. There is one like it on the other side, and both are visible from beneath. There is a third lantern near the junction between the abdomen and thorax that shines downward when the beetle flies. This example is from south Florida and is probably *Deilelater physoderus* (Germar). *(d)* A male firefly, *Photinus ignitus* Fall, with simulated luminescence shining from his lantern and in his airspace above.

1966 I found an insect-rearing chamber (for mosquitos?) with a crude experimental setup that used fluorescent and incandescent lights on the ceiling and walls to simulate a shift in intensity, wavelength, and direction, such as that which occurs at sunset in nature. Entomologists have also used artificial illumination in crop fields to prevent diapause and pupation and UV-reflective material around fields to deter approaching pests. In the 1970s, when the love-bug (a fly, Bibionidae) moved around the Gulf into Florida (Buschman 1976) and began splattering the fast automobiles of tourists bringing money, one remedy suggested was that reflective foil along the highways might deter the flies dallying there *in copulo*.

Fireflies 101

Fireflies are beetles (Coleoptera; Figure 14.1d) of the family Lampyridae, not flies (Diptera) or bugs (Hemiptera), as common American colloquial names suggest. Worldwide there are about 2,000 "officially" named species (i.e., following the required procedure of the *International Code of Zoological Nomenclature*), but more than twice that number remain to be discovered, identified, and named, mostly from tropical rainforests. Firefly life history is similar to that of butterflies and includes egg, larva, pupa, and adult stages (so-called development with complete metamorphosis). A complete cycle takes from a few months in the south to two or three years in the northern United States, depending on species, temperature, and probably also larval hunting success. During the pupal stage the snail- and/or earthworm-eating larva transforms into the reproductive stage; pupation takes about two weeks, but adults of most North American species probably live less than two weeks because it is a high-risk life stage. The photochemistry that produces firefly light (bioluminescence) uses molecules of substrates called luciferin and enzymes known as luciferase and is used in science for research (e.g., quantifying enzyme kinetics and monitoring gene activation and manipulation) and in medicine for disease diagnosis. The gene that codes for the enzyme luciferase was taken from a firefly and put into the genome of a bacterium, making it possible to produce this protein "by the ton" without killing millions of fireflies for the market each year, although a "natural harvest" continues; one parent involved with her children in collecting specimens for luciferase commerce commented that every time she sees a firefly flash she sees a penny!

About 120 species of North American fireflies have been given scientific names, and about 40 more are referred to by nicknames or letter codes (e.g., "Whistler's Mother" and "LIV"). Fireflies emit their light in

Figure 14.2. A graphic scene of firefly emissions as they might be recorded in a time-lapse photograph. Alphanumeric coordinates index the text discussion. Several of the illustrated situations indicate the importance of light in the lives of these beetles; other situations, such as light emission when trapped in a water puddle (N7), probably have more value to investigating fireflyers than to the fireflies themselves. Shown are male sexual (advertising and identifying) signals (B2–B4, C2–C3, B5–B7, B3–B9, E3–D7, F2–K11, F9–F11), a female response flash (L12), a warning flash (N3), an attack flash (E7), illumination landing and take-off flashes (J2–L6 and K7–H6), a flash pattern switch (D9), and some (probably?) meaningless flashes of "stressed" fireflies including one trapped in a water puddle (N7), one in a spider web (J6), one in a snarl of Spanish moss (F9), and one being grasped by a hitchhiking, tree-living pseudoscorpion under a tree at H9. Note the larval glow (L8–L10) and crashing adult (F12).

what a compulsive–obsessive organizer could list in twenty or more situations. As examples, referring to Figure 14.2 at alphanumeric coordinates, larvae glow as they walk on the ground (L8–L10); pupae glow when disturbed in their earthen chambers (N11); adults blink and flicker when trapped in water (N7) and spider webs (J6), when walking through tangles of Spanish moss (F9), and when grasped by hitchhiking pseudoscorpions (G9); adults of some species use their lights when landing and taking off (J2–L6 and K7–H9); and when flying males crash into twigs or stems in the dark, their fall and sometimes aerial recovery are illuminated by a streak of glow (F12). Adults especially use their light for sexual signaling (B2–B4, C2–C3, B5–B7, C3–C9, E3–D7, etc.); at E7 a predator firefly turns on its light as it attacks a flashing male in the air; at L12 a female flashes to answer a male *Photinus pyralis* that has flashed overhead; near D9 a *Photuris tremulans* male that has been advertising for a mate with his flicker pattern is answered, and he defaults (switches) to his identification pattern, a short flash in his species.

Fireflies as Research Subjects and Models

The mate-seeking flash patterns of males (Figure 14.2) account for more than 99% of the flashes seen in the field on most nights. Because these mate-seeking flashes of males are signature patterns unique to species, they are convenient for human observers and researchers: to find firefly sites by driving slowly and looking along suburban and country roads, along streams, and near marshes; to correctly and quickly identify each species being studied and not mix or confuse samples; to determine when each species is sexually active, count how many individuals are active in a population during the adult season, pinpoint local hotspots in time and space (i.e., detect mating sprees and leks), and quantify how populations change through each night and season and from year to year; and to determine what kind of breeding synchrony, and possible gene exchange, exists between local populations. This is to say that populations of fireflies are more easily monitored and quantified than those of most other kinds of organisms, facilitating the assessment of the size, vigor, longevity, and interconnection of local populations, as well as distinctly revealing their demise. Amateur firefly watchers have been doing this informally for a long time, as evidenced by letters received from them over the past several years asking, "Where have all the fireflies gone—we don't see them anymore?" (Lloyd 1994, Monchamp and Lewis 1994).

In summary, because of the distinctive and species-identifying mating signals of male fireflies, these species may be useful as model systems for studying conservation ecology, especially the significance of intrusions of foreign illumination as it affects basic biological phenomena (e.g., diapause and photoperiodism). Foreign illumination may be particularly detrimental because it intrudes on the signal channel of organisms that use costly, self-generated light for communication. Fireflies may have special utility for assessing the effects of a variety of human activities such as lowered water tables, introduced chemicals, and reduced natural areas.

Effects of Stray Light on Fireflies

Observers have long noted that light can wash out the glow of fireflies. The Japanese poet Busenji wrote, "After leaving bushes / the light of fireflies / disappears into moonlight." In the modern world, intrusive illumination from sky glow and from local light emitters as "point" sources such as streetlights will have various harmful effects on nocturnally active and especially light-emitting insects. A clear distinction cannot always be made between these two categories, thus the following examples are an assorted sampler and invite further examination.

Intrusive illumination as signal noise makes bioluminescent emissions less efficient as mating signals to potential mates, less attractive for predator–trappers that use them, and less effective as aposematic warnings to attackers. The levels of light intensity that organisms normally emit have been tuned to operate against darker backgrounds, and although some emitters may facultatively increase their luminous output to compensate, such emissions decrease stored energy supplies, one molecule of adenosine triphosphate per emitted photon (Herring 1978), that would otherwise go directly into reproduction. For example, when females of the twilight firefly *Photinus collustrans* (LeConte) are delayed in mating and oviposition it costs them about seven eggs per day of delay (Wing 1989). Foreign light also illuminates and reveals targets to predators, potentially shifting the balance and outcomes of predator–prey interactions (Figure 14.3).

Figure 14.3. A wolf spider (*Lycosa* sp.), a visual hunter, grasping a female *Photuris* firefly that it probably located via its flashes, aided by ambient illumination; wasted photons from nearby streetlights make it easier for the hunter.

Mating activity of twilight fireflies is triggered by diminishing light intensity. In shady places and on cloudy days males of twilight fireflies (e.g., *Photinus pyralis* [L.]) become active several minutes earlier than otherwise. For example, Allard (1931) observed the onset of *Photinus pyralis* flight activity during one summer in Arlington, Virginia as males tracked the time change of sunset through their season; his chart portrays their response to late, average, and early twilight conditions (Figure 14.4). When individual males of many species, and even late-twilight species, become active in the evening, those that begin in darker shady spaces, such as those within shrubs and under herbaceous canopies, land when their flight brings them out of shade into the open. Adults of such species may be confused by intrusive light and therefore inappropriately initiate, continue, or terminate their evening mate search.

Photinus collustrans reveals another response to ambient light level that may be influenced by intrusions of stray light. Early in their 20-minute

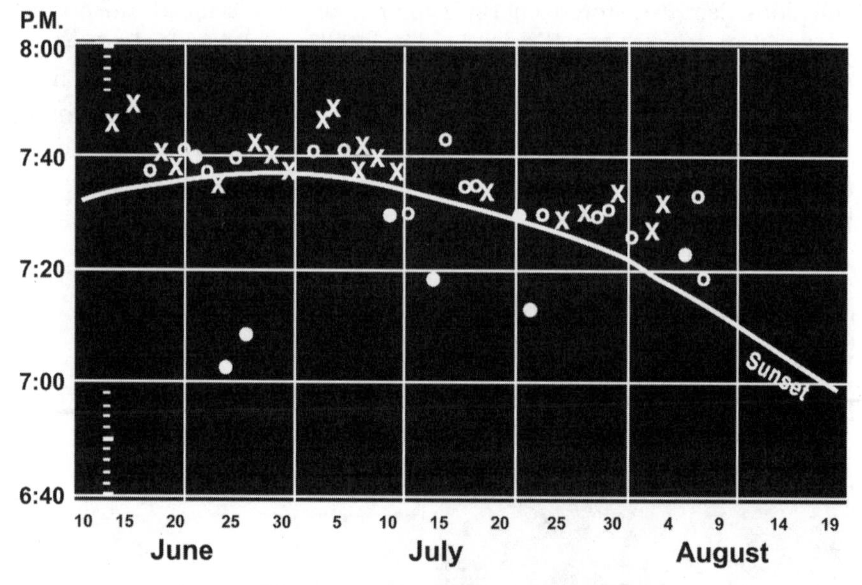

Figure 14.4. Observed onset of *Photinus pyralis* (L.) flight activity during one summer in Arlington, Virginia as males cued on twilight through their season. The line shows sunset time; Xs show later appearance on clear, bright days; circles, appearance with hazy or partly cloudy skies; and dots, time of appearance on darkly overcast days. Now overcast skies greatly increase sky glow, and males may start later. After Allard (1931); his original chart also showed temperature and relative humidity at his firefly study site.

window of evening flight, which begins about 20 minutes after sunset, males fly less than one meter above the ground, but as light level declines they gradually fly higher (Lloyd 1979, 2000). When early-flying males pass through the shade of trees they adjust their altitude accordingly, perhaps suggesting that males also respond to subtle changes in ambient light.

One dimension of light as noise in firefly activity space is color, that is, the wavelengths that are present and dominant. Although there are natural causes for the conspicuous red shift that sometimes occurs at evening twilight (i.e., from particles in the atmosphere arising from dust storms, volcanoes, forest fires, aboriginal cooking fires, and industry), artificial light from various nearby sources may be more intense, invasive, and long-lasting. Whether such color alteration is significant for fireflies whose window of sexual opportunity occurs shortly after sunset is not known, but the color of their bioluminescence is significant for them and has been specially tuned (Biggley et al. 1967). The spectra of the two most common American genera, *Photinus* and *Photuris*, differ in their peaks and half-maximum-energy wavelengths. To generalize, twilight fireflies emit yellow light and late-evening fireflies emit green light, albeit with significant and yet unexplained exceptions. The evolutionary direction of firefly color change probably was from green to yellow—daytime insect vision has its peak sensitivity in the green and matches the spectrum of green-flashing fireflies (Seliger et al. 1982a, 1982b). Twilight fireflies have filters in their eyes that allow them to see yellow better by filtering out the green wavelengths that reflect from vegetation (Lall et al. 1980). This tuning improves the signal-to-noise ratio in their communication channel, but presumably fireflies are colorblind and do not differentiate between colors. Higher levels of long wavelengths, such as from sodium vapor streetlights, may have special significance for yellow-flashing fireflies.

The shift to twilight activity by certain *Photinus* species may have been brought about through predation by other fireflies. Key predators are the females of several species in the subfamily Photurinae (Lloyd 1984, Eisner et al. 1997). In North America *Photuris* females have two known tactics. In the first, females take perches at sites where prey species are active, mimic the mating signals of females of the prey species, attract the males, and eat them (Figure 14.5). In the second tactic, females attack males by aiming at their light emissions and striking them in the air (Lloyd and Wing 1983). In an intermediate tactic, mimicking females leave their perches and strike approaching, but perhaps hesitating, males. Increases in ambient light may reveal mimics to approaching males but negate one of the few defenses that flying males have, that of hiding in the dark when

Figure 14.5. A female *Photuris* of an unnamed Florida species (temporarily called "B") eating a male *Pyractomena angulata* (Say). One of his eyes is detached and is near her right front foot.

not flashing. Intrusive light may thus alter the dynamics of predator–prey interactions in two ways: revealing mimics to potential prey and revealing signaling males to aerial attackers.

The flashed mating signals of lightningbug fireflies, those of the genus *Photuris* in particular, are far more complex than could have been suspected (Barber 1951, Lloyd 1998). Whether the subtleties observed in the flash patterns of males of some species are to fool mate competitors or predators, or to persuade females to answer and mate with them because the signals reflect particularly good health or genes, or all of the above (and more) remains to be discovered. The many flash patterns of an unnamed New England firefly, *Photuris* "LIV" (Figure 14.6a), include a flicker that, except for color, is a copy of that of another resident species, *Pyractomena angulata* (Say), which is probably preyed upon by their females, and a graded series of pulsing patterns (Figure 14.6b). Furthermore, these patterns change through the evening, with the flicker pattern appearing shortly after activity has begun, then rising to a peak an hour or so later and falling around midnight and beyond (Lloyd 1990). The occurrence of patterns in the pulse series may also change through the evening or with competition levels, as noted in related *Photuris* species (Forrest and Eubanks 1995). Subtle variations occur in the signals of many (especially *Photuris*) fireflies, and they may be obscured, inappropriately triggered, or rendered ineffective for their evolved significance by

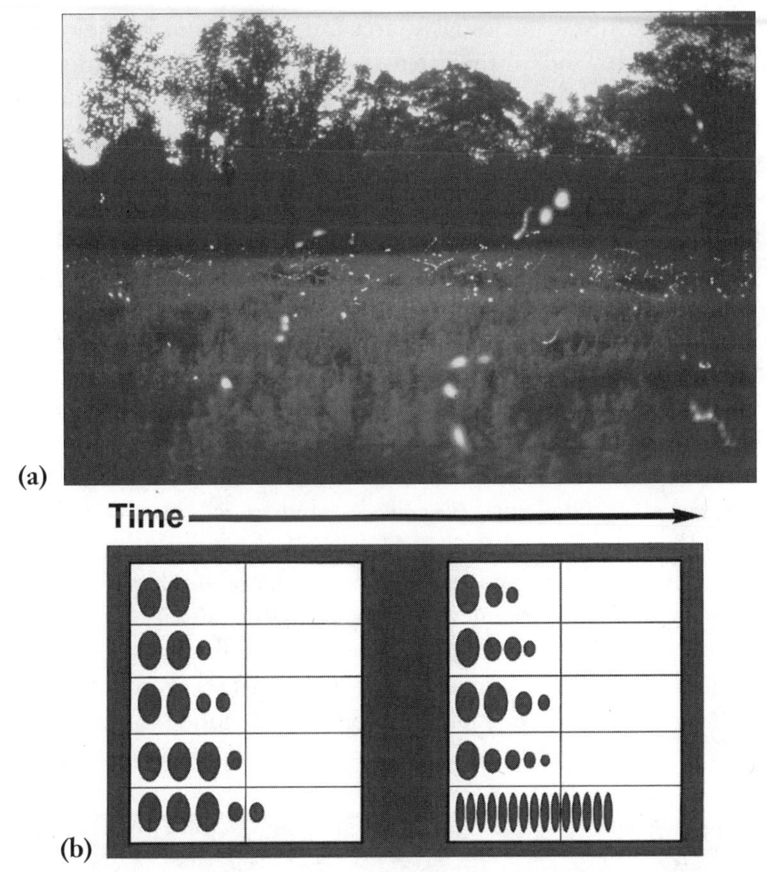

Figure 14.6. *(a)* A Connecticut old-field with *Photuris* "LIV" males showing many of the pulsed variations and the flicker pattern of a complex flash pattern repertoire. *(b)* Diagrams of the flash pattern repertoire of *Photuris* "LIV" males showing the apparent graded series of pulse patterns and the flicker pattern, some of which are visible in *(a)*. The significance of this variation probably is to be found in the contexts of female choice and male competition, the two basic themes of sexual selection. Time is on the horizontal axis and relative intensity on the vertical axis. Vertical lines indicate half-second intervals, but note that flashing rates and duration change predictably with temperature.

intrusive light. Note that such alterations in the occurrence and timing of flash patterns within a species may be useful as bioassays in the study of the effects of intrusive light in natural ecosystems.

Light that enters natural areas is not only noise but also may be a source of misinformation as it affects the triggering, timing, and orientation of critical activities. Light travels in straight lines, so animals use light from "infinitely" distant celestial sources for navigation in their movement over the Earth and for the orientation of postures and positions they take within their habitats. Intrusive light from local sources can misinform them in both situations. By maintaining a fixed angle (Figure 14.7a) on parallel light rays arriving from remote sources, night flyers are able to steer and maintain straight flight over long distances. For example, as an adaptation explanation (behavioral ecology), males may maximize their opportunity to intercept downwind-streaming female pheromone plumes by flying perpendicular to the wind on star-oriented straight lines. Without a remote orientation reference, nocturnal flyers would tend to stray off course, perhaps even fly in circles. Intrusive light, such as from sky glow, will not only hide remote navigation markers (stars), but if night flyers take a fixed bearing on a local light source that has diverging rays, they will spiral into the light (Figure 14.7b). Males of glowworm fireflies (Lampyridae, e.g., *Pleotomus*, *Pleotomodes*) and giant glowworm beetles (Phengodidae, e.g., *Phengodes*) probably all track pheromone plumes, and males sometimes are captured in light traps in large numbers.

Another case for disorientation is suggested by fireflies of the genus *Pyractomena*. Unlike larvae of other North American fireflies, which pupate underground or in dead logs, those of this genus do it above ground on vegetation. *Pyractomena borealis* (Randall) occurs from Florida to Canada, and in late January in north central Florida larvae attach to, and pupate on, tree

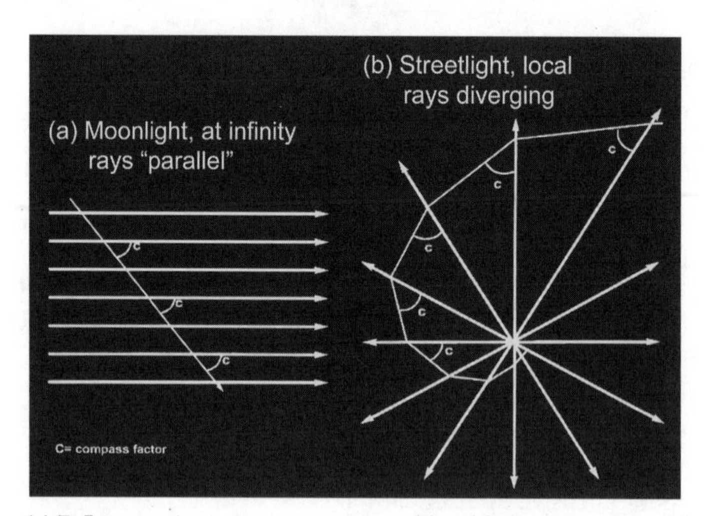

Figure 14.7. Insect orientation to remote and local light sources. *(a)* The insect flies a straight course over time by maintaining a constant compass angle ("C") to the parallel rays from a celestial source at "infinity." *(b)* If an insect uses the same method but with the diverging rays from a local source, the navigator will spiral into the light.

trunks in damp forests and emerge as adults a week or two later, depending on temperature (Lloyd 1997). By eclosing at this time of year adults avoid predaceous fireflies, and their larvae may avoid competition from larvae of other species during a critical early stage. In north central Florida temperatures in January sometimes drop below freezing at night, but days are usually sunny, and insolation warms the trees and the water within their woody tissues. *Pyractomena borealis* larvae have a strong tendency to pupate on the sun-exposed southern aspect of live trees and thus receive the warmth of the sun during the day, and on cold nights warmed tree water continues to support their metabolism and development (Figure 14.8a; Lloyd 1997). If these larvae use rays from the solar disk (say, instead of using the warm bark of the tree) to determine the southern exposure on trees, nearby streetlights may disorient them. A related species, *Pyractomena limbicollis* (Green), occurs in many of the same forests but pupates with a north-facing orientation and ecloses a few weeks later (Figure 14.8b).

Artificial light sources may attract colonizing and site-changing individual fireflies if they interpret them as the glow of the collective emissions of fireflies in suitable habitats, with *suitable* defined as "good for young mate seeking" by males and "suitable for oviposition" by female emigrants. This may be most consequential and deadly for Asian species

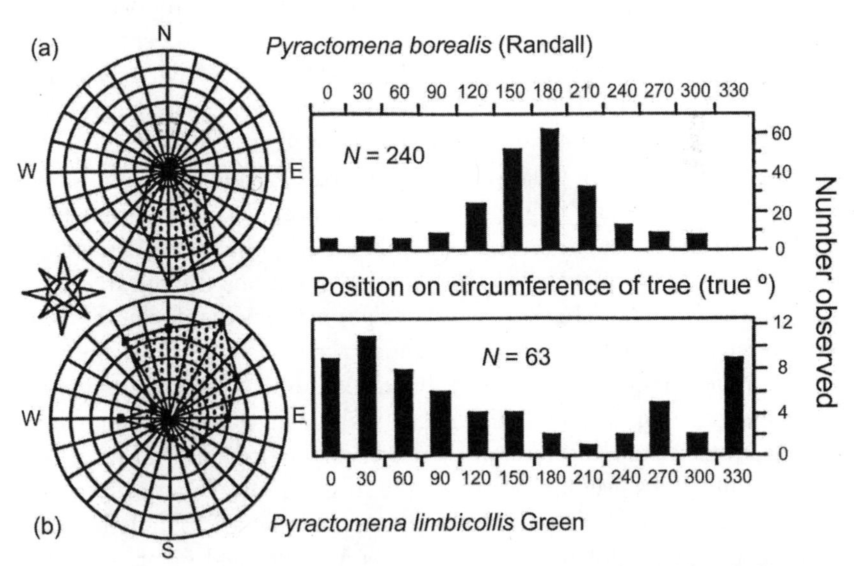

Figure 14.8. Directional orientations selected by larvae of two Florida *Pyractomena* species on tree trunks, in preparation for pupation, shown in both radar and histogram presentations. *(a)* Southern orientation of *Pyractomena borealis* (Randall) and *(b)* northern orientation of *Pyractomena limbicollis* (Green).

of *Pteroptyx* fireflies, which gather in arboreal mating swarms and are attracted to the collective emissions of "firefly trees." Along intertidal rivers in Southeast Asia some of these species once were reported to swarm by the thousands, or perhaps even millions, but pollution of their habitats apparently has reduced their numbers (Lloyd et al. 1989).

Another potential source of photic misinformation for *Pyractomena borealis* was suggested when larvae were brought into the lab in early December and, by chance, were exposed to continuous room light associated with other research. The larvae climbed and pupated a few days after they were brought into the lab and soon eclosed as adults, two to three months too early. If continuous light triggered their metamorphosis (i.e., caused them to break diapause), then field populations exposed to continuous artificial illumination from sky glow or local lights may likewise eclose early or make other developmental miscalculations.

Stray Light in Concert with Other Ecological Insults

Two of the major ecological changes that have had an adverse effect on firefly populations are the decrease of natural areas, including woodlands, meadows, and bottomland near streams and rivers, and the lowering of water tables, resulting in fewer creeks and their floodplain forests, ponds and pond edges, marshes, and wet meadows. These are important habitats of fireflies. There are more than a dozen species that occurred in the Gainesville, Florida area in the years 1966–1980 that I have not seen for more than a decade. Marshes and rivulets that once occurred at the sites where these species were found have shrunk or vanished completely. The normal patchwork of populations of species that existed across the area in former times has many more holes in it. As more and more populations are lost, the holes get larger, and remaining local populations become less well connected by migrants. When local populations disappear, as they naturally do from time to time, neighbor populations that formerly would have reseeded now-empty sites are much farther away. Thus there are fewer and fewer suitable sites, and these remaining sites are separated by large and growing distances. The inhospitable spaces between remaining sites are punctuated and bathed by noisy, distracting, and misinforming bright lights. This presents a formidable formula for extinction. A mathematical ecologist has modeled the extinction of local populations and concluded that total extinction of species was retarded or prevented by the chaotic (in the formal, mathematical sense) nature of local extinction and the repopulation of suitable sites (Allen et al. 1993). I expect that formal chaos models for fireflies will now predict less optimistic outcomes.

Fireflies in Education: Everyone Is a Fireflyer

There are a number of projects with luminescent insects and the effects of intrusive light upon them that students can perform with little supervision. Many will be suggested to the reader from the preceding discussion, such as spotting trips to find and map local populations of various species of fireflies and woodlands with luminescent collembola or a Harry Allard–like study of alterations of starting times of *Photinus pyralis* in unlit and streetlight-lit populations (note that sky glow now should make a significant difference from what he found for overcast twilights). Here are a few other possible projects.

- Track local populations through time to determine their seasonal variation in occurrence and whether populations are changing in size, or if local populations are more likely to be out of synchrony if they experience intrusive illumination.
- Identify existing populations near sites where there are plans for development to see whether luminescent species are present and whether these species occur elsewhere in the region.
- Compare activities such as start and stop time, flash pattern changes, and mate-finding success in populations of a species that occur in dark and artificially lit sites (*Photinus pyralis* may be especially well suited for this project).
- Determine height of flight during twilight and post-twilight flight in dark and artificially lit sites.
- Simulate bioluminescent fungi with chemiluminescent tubes to determine what insects are attracted (should luminescent mushrooms appear in their area). Smeared with sticky paste, these will capture potential spore transporters. If condoms are placed over a tube and smeared, the same glow-tube can be used during successive time periods in an evening, and separate catches can be preserved in vials of alcohol.
- Place shields on offending lights to protect sensitive firefly populations nearby.
- Quantify the orientation and other characteristics of *Pyractomena borealis* pupation sites (tree species, bark roughness, pupation height, water content, thermal features). Students can make physical models of trees and fireflies and the thermal consequences of various tree characteristics; sand-filled (wet or dry) plastic jars can be used for model trees and clay balls with thermocouples around the circumference for model larvae (see Lloyd 1997).
- Compare populations of *Pyractomena borealis* to determine whether flashing adults appear earlier in woods near intrusive light sources.

- Observe aggressive mimicry and the aerial attack of *Photuris* females. Students can find female fireflies by walking about a firefly site flashing a penlight, simulating the male flash pattern. Responses of females of prey species (commonly *Photinus*) are identifiable by their yellow light; predator females emit green light (dark adaptation of the observer and the ambient light may cause some error). When a fishpole with a small hanging light at the tip is substituted for the more efficient penlight and the "lure" flashes but does not approach the predator, she may leave her perch and attack. A similar target flown around bushes and along hedgerows where fireflies are active will elicit attacks. Hanging targets are observed via their silhouettes against skylight (Lloyd and Wing 1983).

Conclusion

Fireflies may be expected to have inappropriate "innate" responses to foreign light, similar to those that occur in other organisms; because of their conspicuous luminescent signals, such alterations may be more easily monitored and quantified with fireflies, as individuals and populations, than with other organisms. Fireflies therefore may be useful as subjects and model systems for the study of the long- and short-term consequences of ecological insults that occur in combination. Furthermore, because much of firefly life is mediated through their own pinpoints of light in otherwise dimly lit or dark environments, the firefly's relation to light is virtually unique in the terrestrial world; foreign light will have even more serious consequences for them, and they provide a special case for study.

Fireflies have long held a special place in human cultures. The Japanese have used fireflies as metaphors for experiences of the human condition and to punctuate subtle points of a reverent natural philosophy. In North America fireflies also have cultural significance, and even today urban Americans often have fond feelings and memories for the fireflies they once met in the park or at a summer camp, or even in a book. It is a culturally impoverished American who has not heard of Wah-wah-taysee, the firefly in a passage from Longfellow's *Song of Hiawatha*, or does not know of the glowworm in *Hamlet*, or what a glowworm did for the Mills Brothers, or who has not seen the golden flicker of a firefly and recalled the 1940s trio "The Three Sons." This is to say that fireflies offer a way to present the intrusive light problem to an attentive public. A burned bear cub became the icon of a well-known cause in conservation, and Smokey was recognized and understood by millions of children and adults. Perhaps a "Blinky the Firefly," who might say, "Remember, be enlightened, keep us in the dark," or "Keep us, not yourselves, in the dark about light," would be useful now?

Acknowledgments

I am indebted to Steve Wing and John Sivinski for providing information and making helpful comments and technical corrections on the manuscript and to Pam Howell and Mike Sanford for computer graphic and layout assistance. This chapter has been assigned Florida Agricultural Experiment Station Journal Series number R-08688.

Literature Cited

Allard, H. A. 1931. The photoperiodism of the firefly *Photinus pyralis* Linn. [*sic*]; its relation to the evening twilight and other conditions. *Proceedings of the Entomological Society of Washington* 33:49–58.

Allen, J. C., W. M. Schaffer, and D. Rosko. 1993. Chaos reduces species extinction by amplifying local population noise. *Nature* 364:229–232.

Barber, H. S. 1913. Luminous collembola. *Proceedings of the Entomological Society of Washington* 15:46–50.

Barber, H. S. 1951. North American fireflies of the genus *Photuris*. *Smithsonian Miscellaneous Collections* 117(1):1–58.

Biggley, W. H., J. E. Lloyd, and H. H. Seliger. 1967. The spectral distribution of firefly light. II. *Journal of General Physiology* 50:1681–1692.

Buschman, L. L. 1976. Invasion of Florida by the "lovebug" *Plecia nearctica* (Diptera: Bibionidae). *Florida Entomologist* 59:191–194.

Eisner, T., M. A. Goetz, D. E. Hill, S. R. Smedley, and J. Meinwald. 1997. Firefly "femmes fatales" acquire defensive steroids (lucibufagins) from their firefly prey. *Proceedings of the National Academy of Sciences of the United States of America* 94:9723–9728.

Forrest, T. G., and M. D. Eubanks. 1995. Variation in the flash pattern of the firefly, *Photuris versicolor quadrifulgens* (Coleoptera: Lampyridae). *Journal of Insect Behavior* 8:33–45.

Fraenkel, G. S., and D. L. Gunn. 1961. *The orientation of animals: kineses, taxes and compass reactions*. Dover Publications, New York.

Fulton, B. B. 1941. A luminous fly larva with spider traits (Diptera, Mycetophilidae). *Annals of the Entomological Society of America* 34:289–302.

Herring, P. J. (ed.). 1978. *Bioluminescence in action*. Academic Press, London.

Lall, A. B., H. H. Seliger, W. H. Biggley, and J. E. Lloyd. 1980. Ecology of colors of firefly bioluminescence. *Science* 210:560–562.

Lloyd, J. E. 1966. Studies on the flash communication system in *Photinus* fireflies. *University of Michigan Museum of Zoology, Miscellaneous Publication* 130:1–95.

Lloyd, J. E. 1969. Flashes of *Photuris* fireflies: their value and use in recognizing species. *Florida Entomologist* 52:29–35.

Lloyd, J. E. 1974. Bioluminescent communication between fungi and insects. *Florida Entomologist* 57:90.

Lloyd, J. E. 1979. Sexual selection in luminescent beetles. Pages 293–342 in

M. S. Blum and N. A. Blum (eds.), *Sexual selection and reproductive competition in insects*. Academic Press, New York.

Lloyd, J. E. 1983. Bioluminescence and communication in insects. *Annual Review of Entomology* 28:131–160.

Lloyd, J. E. 1984. Occurrence of aggressive mimicry in fireflies. *Florida Entomologist* 67:368–376.

Lloyd, J. E. 1990. Firefly semiosystematics and predation: a history. *Florida Entomologist* 73:51–66.

Lloyd, J. E. 1994. Where are the lightningbugs? *Fireflyer Companion* 1(1):1–2, 5, 10.

Lloyd, J. E. 1997. On research and entomological education, and a different light in the lives of fireflies (Coleoptera: Lampyridae; *Pyractomena*). *Florida Entomologist* 80:120–131.

Lloyd, J. E. 1998. Firefly mating ecology, selection and evolution. Pages 184–192 in J. C. Choe and B. J. Crespi (eds.), *The evolution of mating systems in insects and arachnids*. Cambridge University Press, Cambridge.

Lloyd, J. E. 2000. On research and entomological education IV: quantifying mate search in a perfect insect—seeking true facts and insight (Coleoptera: Lampyridae, *Photinus*). *Florida Entomologist* 83:211–228.

Lloyd, J. E. 2001. On research and entomological education V: a species (c)oncept for fireflyers, at the bench and in old fields, and back to the Wisconsian glacier. *Florida Entomologist* 84:587–601.

Lloyd, J. E. 2003. On research and entomological education VI: firefly species and lists, old and now. *Florida Entomologist* 86:99–113.

Lloyd, J. E., and E. C. Gentry. 2003. Bioluminescence. Pages 115–120 in V. H. Resh and R. T. Cardé (eds.), *Encyclopedia of insects*. Academic Press, New York.

Lloyd, J. E., and S. R. Wing. 1983. Nocturnal aerial predation of fireflies by light-seeking fireflies. *Science* 222:634–635.

Lloyd, J. E., S. R. Wing, and T. Hongtrakul. 1989. Ecology, flashes, and behavior of congregating Thai fireflies. *Biotropica* 21:373–376.

Monchamp, J., and S. Lewis. 1994. Fireflies at risk. *Fireflyer Companion* 1(1):3.

Redford, K. H. 1982. Prey attraction as a possible function of bioluminescence in the larvae of *Pyrearinus termitilluminans* (Coleoptera: Elateridae). *Revista Brasileira de Zoologia* 1:31–34.

Seliger, H. H., A. B. Lall, J. E. Lloyd, and W. H. Biggley. 1982a. The colors of firefly bioluminescence—I. Optimization model. *Photochemistry and Photobiology* 36:673–680.

Seliger, H. H., A. B. Lall, J. E. Lloyd, and W. H. Biggley. 1982b. The colors of firefly bioluminescence—II. Experimental evidence for the optimization model. *Photochemistry and Photobiology* 36:681–688.

Sivinski, J. 1982. Prey attraction by luminous larvae of the fungus gnat *Orfelia fultoni*. *Ecological Entomology* 7:443–446.

Wing, S. R. 1989. Energetic costs of mating in a flightless female firefly, *Photinus collustrans* (Coleoptera: Lampyridae). *Journal of Insect Behavior* 2:841–847.

Chapter 15

Artificial Light at Night in Freshwater Habitats and Its Potential Ecological Effects

Marianne V. Moore, Susan J. Kohler, and Melani S. Cheers

The effects of artificial night lighting should be particularly intriguing to freshwater ecologists because artificial lighting illuminates a vast array of freshwater habitats in both urban and rural areas. Major cities and their lights often flank or surround aquatic ecosystems including lakes, large rivers, and coastal areas. In the United States, Chicago borders Lake Michigan, Saint Louis abuts the Mississippi River, and New York City encircles Long Island Sound. Furthermore, lakes, large rivers, and coastal waters, in contrast to small pools or streams, are sufficiently large that their open waters away from shore (i.e., pelagic zone) are exposed to artificial night lighting and usually not shaded by trees or large buildings. Organisms in the pelagic zone, including fish and their zooplankton prey, may be exposed nightly to artificial lighting and respond strongly to it. Importantly, however, artificial nighttime illumination is not limited to urban aquatic environments. The nearshore areas (i.e., littoral zone) of lakes, streams, and wetlands in rural areas are also vulnerable. Security lights associated with waterfront property or vacation homes often illuminate portions of rural shorelines.

Artificial night lighting includes both glare and sky glow, and their relative intensities differ in urban and rural areas. Glare is direct light shining from a fixture into the eye of an observer; sky glow is the composite illumination of the nighttime sky by lights (Wilson 1998). Sky glow results from fixtures lacking shielding that cast light upward above the horizontal plane (Wilson 1998) or light that reflects off particles in the atmosphere between a fixture and the ground. When this light is reflected back from Earth, it creates an aura above metropolitan regions that can sometimes be viewed more than 160 km (100 miles) from the city center (Crawford and Hunter 1990). Sky glow is associated with major cities or other large agglomerations of outdoor lighting (e.g., prisons, hydrocarbon platforms, power plants, greenhouses), whereas direct glare can occur in any area with outdoor lighting.

Both glare and sky glow must be quantified to predict how they might affect freshwater organisms and ecological processes. Until recently, this was not possible because intensities of artificial night lighting (particularly that of sky glow) were below the detection limit of commercially available scientific instruments. Recently, however, we modified or developed two instruments—a spectrometer and a radiometer—to quantify attributes of artificial night lighting at the surface of lakes along an urban-to-suburban gradient in New England. We compared these measurements and existing data for natural equivalents of nighttime light (e.g., full moonlight) with light thresholds of freshwater organisms to begin predicting the types of organisms and processes that might be affected by artificial night lighting.

To this end, the objectives of this chapter are to present preliminary measurements of artificial night lighting at the surface of lakes; to predict the types of freshwater organisms, habitats, and ecological processes likely to be affected by this light; and to highlight future research needs regarding artificial night lighting in freshwater habitats.

Measurement of Artificial Night Lighting in Freshwater Lakes

We quantified the level of illumination at the surface of freshwater lakes across an urban-to-suburban gradient to determine whether this light reached biologically relevant thresholds and to develop instruments appropriate for such measurements.

Site Description and Methods

Measurements of artificial night lighting, predominantly from sky glow, were made during the new moon period in June–August 2000 at four lakes located along an urban-to-suburban gradient in the northeastern United States. All study lakes are small glacial lakes (10–50 ha [25–124 acres]) located in or within 31 km (19 mi) of Boston, Massachusetts. Each of the lakes experiences both sky glow and glare. Sky glow was visible to the human eye, and streetlamps or security lights lacking full cutoff fixtures and generating glare were within 30 m (98 ft) of shore at all lakes. Measurements reported here, however, describe predominantly sky glow because measurements were taken in the center of each lake away from the influence of isolated lights near the shoreline.

Two different instruments, a modified spectrometer and a custom-built radiometer, were used to quantify levels of illumination striking the surface of the study lakes at night. Two instruments were used because each quantified a different property of light. The spectrometer (Ocean Optics USB 2000 with 200- to 850-nm spectral range) measured the wavelength or color distribution, whereas the radiometer (a photomultiplier tube–based instrument) provided a single measurement of relative light intensity over a broad range of wavelengths (275–700 nm).

Measurements of light intensity and spectral composition were made with the instruments at least two hours after sunset from a boat on new moon nights (i.e., no moonlight) in the summer of 2000 in the center of each lake. Typically, a two-minute spectrum composed of 40 averages was acquired per lake with the spectrometer. Using the radiometer, five to ten measurements at approximately five-second intervals were obtained at each lake and averaged. To quantify how light intensity varied over the course of a single night, readings every five minutes were obtained at the center of each lake using the radiometer and a datalogger between approximately 2200 h and 0400 h.

Color Distribution

The spectra of light striking the surface of urban and suburban lakes were nearly identical (Figure 15.1), with yellow light (585 nm) predominating at all lakes. These lake spectra matched closely the emission spectrum of high pressure sodium vapor lamps (Figure 15.1), which are the most common streetlamps in the United States (Crawford and Hunter 1990). Maximum emission from these lights occurs in the yellow-orange portion of the

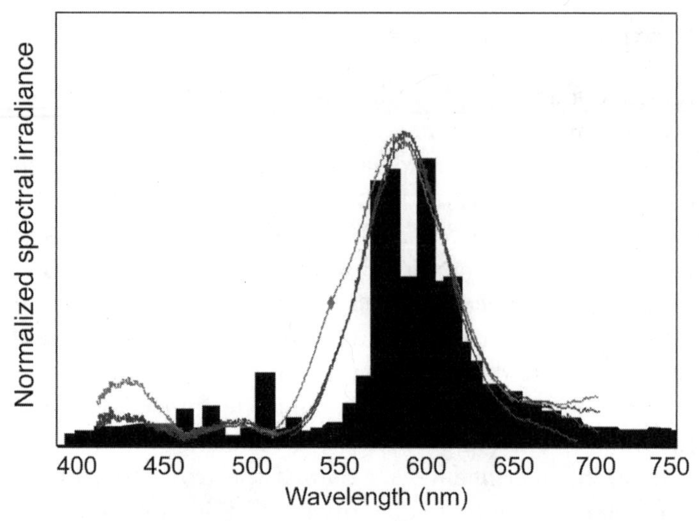

Figure 15.1. Black lines indicate normalized spectra of artificial night lighting measured immediately above the surface of four urban or suburban lakes in the Boston, Massachusetts area in summer 2000 on cloudy, new moon nights when no moonlight was present. The diamond indicates the spectrum for the most urban lake located in Boston, Massachusetts, and spectra for the remaining suburban lakes are so similar that they are indistinguishable. The histogram represents the emission spectrum of a high pressure sodium vapor lamp (McGowan 2000), the most common streetlamp in the United States. Maximum emission for the high pressure sodium vapor lamp occurs in the yellow-orange portion of the spectrum, from approximately 540 to 630 nm, coinciding exactly with the prominent peak for the lakes' spectra.

spectrum from approximately 540 to 630 nm (Cinzano et al. 2000), exactly coinciding with the prominent peak for the lakes' spectra. Yellow light (585-nm wavelength) is also the color of light that penetrates most deeply in many lakes (Wetzel 2001). Lake water typically contains measurable quantities of dissolved organic carbon, algae, and silt, and when none of these three substances dominates, yellow light is transmitted farthest (Moss 1988).

The spectra of nighttime illumination reported here differed from that recorded for a full moon (*sensu* Gal et al. 1999). Because moonlight is reflected sunlight, it is white and consequently composed of more wavelengths than the sky glow we measured. Specifically, the spectrum of full moonlight is broad, extending from 380 nm to well beyond 700 nm, so it contains not only orange light but also shorter, blue wavelengths (about 380–500 nm) as well as longer, red wavelengths (about 650–750 nm). Organisms responding to full moonlight may be sensing its broader variety of wavelengths, whereas organisms reacting to sky glow probably are sensing yellow-orange light. Although few spectral response curves exist for freshwater organisms, those for *Daphnia* and yellow perch (*Perca flavescens*) confirm their sensitivity to yellow wavelengths (e.g., Buchanan and Goldberg 1981, Gromov 1993, Loew et al. 1993). Importantly, however, color vision may not be possible at the low light intensities typical of artificial night lighting because the stimulation of cone cells necessary for color vision usually requires higher light intensities comparable to that of daylight and twilight (i.e., 10 to 100,000 lux).

Relative Light Intensity

Because the suburban and urban lakes all shared the same spectrum of light at night (Figure 15.1), the relative intensity of nighttime illumination at the water surface could be compared between lakes using the radiometer. This comparison shows that the relative intensity of illumination increased along the suburban-to-urban gradient at night under either clear or cloudy

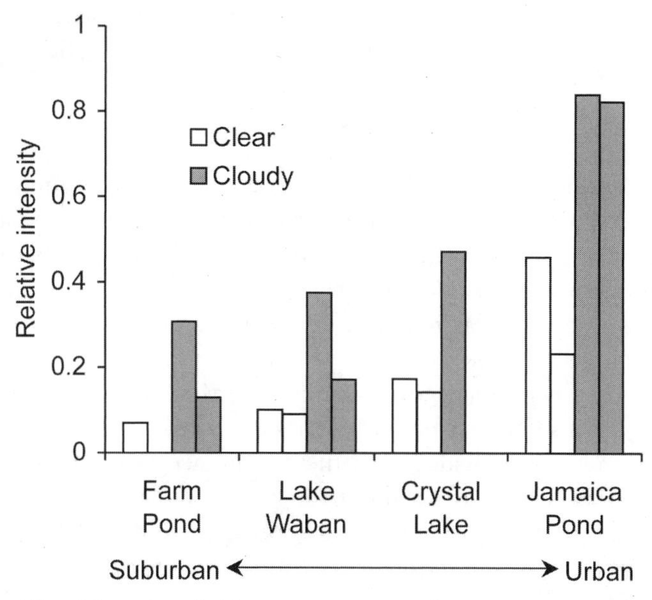

Figure 15.2. Mean (n = 5–10 measurements) relative intensities of artificial night lighting on clear and cloudy nights immediately above the surface of urban and suburban lakes in eastern Massachusetts. The most urban lake, Jamaica Pond, is located in Boston, and the remaining suburban lakes—Crystal Lake, Lake Waban, and Farm Pond—are located at increasing distances from Boston. Farm Pond, the most distant suburban lake, is 31 km (19 mi) from Boston. Measurements were obtained with a custom-built radiometer at a central location in each lake between 2200 h and 0100 h on new moon nights.

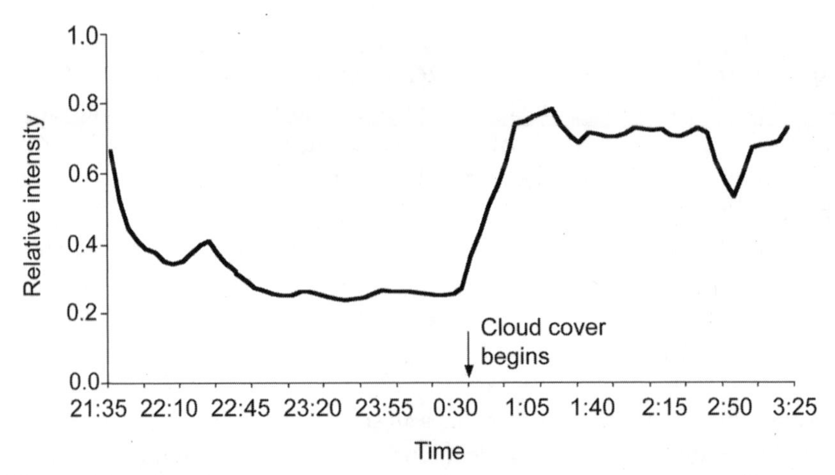

Figure 15.3. Continuous measurements (i.e., one measurement every five minutes) of relative light intensity of artificial night lighting at the surface of Lake Waban, a suburban Massachusetts lake, on the night of June 1–2, 2000. Light intensity was measured with a custom-built radiometer connected to a datalogger over the course of the night. The decline in intensity at the onset of the recording resulted from a gradual darkening of the sky after sunset at 2014 h. The sharp increase in intensity beginning at 0030 h was caused by the onset of cloud cover, which reflects additional artificial light to Earth (i.e., sky glow).

weather conditions (Figure 15.2). On new moon nights, the most urban lake in Boston (Jamaica Pond) experienced levels of illumination that were three to six times higher than those at the most distant suburban lake (Farm Pond). On cloudy nights, intensity of illumination on the suburban and urban lakes was as much as four times higher than that on clear nights. Continuous measurements of the relative intensity of illumination during a single night in which cloud cover changed corroborated these findings (Figure 15.3). Measurements on one of the lakes doubled between 0030 h and 0105 h, and this coincided with the appearance of cloud cover. Sky glow is enhanced by cloudy conditions because clouds reflect light back toward Earth. The frequency and intensity of sky glow therefore are highest in and near cities with the cloudiest weather (e.g., Seattle, London).

Underwater Light Estimates

Although we have not yet quantified the amount of artificial illumination transmitted underwater, our preliminary results show that lakes with clear, uncolored water and located close to a city center are most vulnerable to effects from artificial night lighting. Such lakes are not only exposed to higher intensities of artificial illumination because of their urban location, but the water of these lakes appears to transmit light to greater depths than does water in rural or suburban lakes. Water uncolored by dissolved organic carbon is common in urban lakes because these lakes often are disconnected from wetlands by development. Wetlands are a source of dissolved organic carbon that typically colors water brown and attenuates light underwater. Rural and suburban lakes are more likely to maintain some or all of their connections to wetlands because of less development within their watersheds. Dissolved organic carbon levels in these lakes are likely to be higher and light transmission lower. Our preliminary measurements in the Boston area suggest this pattern, but additional data are needed to confirm it.

Artificial night lighting cannot be measured directly below the water surface because light levels are below the detection limit of commercial instrumentation; nevertheless, light levels can be estimated once absolute intensities of artificial illumination are quantified at the lake's surface. These latter measurements are then used in conjunction with the lake's diffuse attenuation coefficient to estimate the amount of light per wavelength transmitted underwater at night. Diffuse attenuation coefficients describe the percentage of light extinguished per meter as the light travels down through the water column, and these coefficients are routinely determined during the day using existing instrumentation. Once absolute intensities of artificial illumination are quantified at night above water, the intensity and depth of penetration of light underwater can be estimated easily.

Challenges of Measuring Artificial Night Lighting

We were unable to measure the absolute intensity of artificial night lighting above water; however, it is possible to do so using customized fore optics in conjunction with the spectrometer (S. Johnsen, personal communication, 2003). Customized fore optics are necessary because the requisite cosine corrector, which collects light coming from all angles and delivers it to the detector, cannot be used when measuring the dim light characteristic of artificial night lighting, particularly sky glow. Cosine correctors reduce the signal intensity of the light by as much as a factor of

1,000, all but extinguishing the dim light of artificial night lighting at the locations we measured. This problem can be overcome, however, by measuring illumination that is diffused and reflected from a diffuse reflectance standard mounted such that the sensor of the spectrometer collects only reflected light. In this manner, it is possible to make measurements of absolute intensities of artificial nocturnal illumination that are cosine-corrected and accurate.

Predicted Responses of Aquatic Animals to Artificial Night Lighting

The lower limit of light detection for most aquatic organisms is unknown (E. Loew, personal communication, 2003). Light thresholds eliciting particular behaviors (e.g., avoidance or foraging), however, have been documented in the laboratory and, in some instances, in nature. According to these published thresholds, a variety of aquatic animals respond behaviorally to light levels as low as 10^{-7} lux (Table 15.1), which includes light intensities ranging from less than starlight (0.0005–0.001 lux) to that approaching full moonlight (0.05–0.1 lux). Although absolute intensities of artificial night lighting have yet to be recorded, our preliminary measurements with the spectrometer at the most urban lake were of the same order of magnitude as that of full moonlight. Specifically, intensities of nighttime artificial illumination at this lake approached two-thirds the value of full moonlight. Consequently, artificial night lighting is likely to produce biological effects in such lakes.

Aquatic organisms most likely to be affected by the estimated levels of artificial illumination underwater include invertebrates and fish that move, forage, or reproduce at night. Most, if not all, of these nocturnal organisms respond to light from a full moon (0.05–0.1 lux), and this level of light intensity is assumed, for purposes of this review, to represent maximal levels of artificial night lighting at the water surface. We will discuss in more detail how selected invertebrates and fish respond to moonlight or to artificial light in scientific experiments and how these same organisms may respond to artificial night lighting in nature.

Zooplankton

Many pelagic organisms in both lakes and the ocean are exquisitely sensitive to light. For example, zooplankton exhibit diel vertical migration in response to small changes in light intensities (see reviews by Haney 1993,

Table 15.1. Lower light thresholds for a variety of behaviors performed by freshwater invertebrates and fish.

Organism	*Behavior*	*Threshold (lux)**	*References*
INVERTEBRATES			
Lake zooplankton			
Daphnia	Positive phototaxis	0.01–3.00	Flik et al. 1997
Mysid shrimp	Negative phototaxis	≥0.00006–0.0004	Gal et al. 1999
Phantom midge	Negative phototaxis	>0.0000004	Forward 1988
Stream organisms			
Stream insects	Initiation of drift	2.4–31.5	Haney et al. 1983
Stream inverts	Maximum stream drift	0.001–1.0	Holt and Waters 1967, Bishop 1969
FISH			
Fish	Schooling	0.1	Blaxter 1975
Kokanee salmon**	Foraging (*Daphnia*)	0.1	Koski and Johnson 2002
Pike, coho salmon, carp, and minnow	Foraging (zooplankton)	0.00001–0.01	Blaxter 1975
European perch	Foraging (*Daphnia* and phantom midge)	0.02	Bergman 1988
Rainbow trout	Foraging (drifting stream insects)	0.03–0.1	Tanaka 1970, Jenkins et al. 1970
Bream	Foraging (*Daphnia*)	0.005	Townsend and Risebrow 1982
Brown trout	Foraging (brine shrimp)	0.001	Robinson and Tash 1979
Rainbow smelt	Negative phototaxis	>0.2	Appenzeller and Leggett 1995
Bonneville cisco	Negative phototaxis	≥0.01	Luecke and Wurtsbaugh 1993
Rainbow trout	Emergence from substrate in winter	<1.0	Contor and Griffith 1995

*Light units converted to lux using conversions in Table 2.1 in Wetzel and Likens (2000). For comparative purposes, the surface light intensity of a full moon ranges from 0.05–0.1 lux, and starlight only is 0.0005–0.001 lux (Contor and Griffith 1995). Names of organisms in parentheses refer to prey of foraging fish.
**Young-of-the-year (fingerlings).

Ringelberg 1999). The typical "nocturnal" pattern of diel vertical migration refers to zooplankton residing deep in the water column during the day, ascending at dusk to shallower depths where they feed, and returning at dawn to deeper depths (Hutchinson 1967). Although avoidance of visual predators and harmful ultraviolet radiation probably are the ultimate causes of this phenomenon (Lampert 1989, 1993, Ringelberg 1999), light is the fundamental proximate factor controlling it in zooplankton (Haney 1993, Ringelberg 1999). For example, the rate of change of light intensity is the proximate cue triggering the ascent of zooplankton at dusk and their descent at dawn (Ringelberg 1964, 1999). Also, light alters the amplitude of migration, which is defined as the depth range of the population over a 24-hour period (Dodson 1990). In 14 north temperate lakes, the light of a full moon reduced the amplitude of *Daphnia* migrations by about 2 m (6.6 ft; Dodson 1990).

Recent field experiments demonstrate that artificial night lighting can suppress zooplankton vertical migration in lakes in the Boston area (Moore et al. 2000, Moore and Pierce, unpublished data). In underwater enclosures that blocked artificial night lighting, zooplankton taxa (e.g., *Daphnia* and *Bosmina*) ascended 2–3 m (6.6–9.8 ft) higher than in enclosures that transmitted artificial lighting or in the lake itself. Although only 10–20% of the individuals of responding taxa ascended when artificial lighting was removed, this effect was significant statistically. Furthermore, in the most urban lake, most taxa exhibited no vertical migration, suggesting that the presence of artificial night lighting curtailed or eliminated this expected behavior. Reduction of either the amplitude (depth range of the population) or magnitude (percentage of population migrating) of zooplankton diel vertical migration by artificial lighting is potentially important because such light, unlike that from a full moon, occurs every night. Artificial night lighting therefore represents a sustained perturbation that may have long-term cumulative effects. For example, if zooplankton are chronically confined by artificial lighting to deeper depths than normal, then the effects of their grazing in surface waters could be reduced (Moore and Pierce, unpublished data).

Among members of the zooplankton community, however, the larger predatory species such as mysid shrimp and vertically migrating species of the phantom midge (e.g., *Chaoborus punctipennis*) are likely to be most vulnerable to artificial night lighting. These large taxa, unlike the smaller grazing zooplankton, exhibit pronounced avoidance of even very dim light (Table 15.1; Forward 1988, Smith et al. 1992, Gal et al. 1999) that is comparable to starlight intensity (i.e., 0.0005–0.001 lux) or lower. Their avoidance of these low light levels is presumably necessary to protect these larger zooplankters from planktivorous fish who select for larger prey as light levels decrease (O'Brien 1979, Gal et al. 1999). It is possible that artificial night lighting could result in the elimination of these large zooplankton taxa or indirectly reduce their population sizes by confining them to deep, cold waters where food is less abundant and where growth is slow. In stratified lakes or reservoirs with clear water, artificial night lighting may penetrate relatively deeply and at a sufficient intensity to illuminate the entire upper layer of warm water. Consequently, continuous confinement of these large taxa to deep, cold, dark waters 24 hours per day could potentially slow rates of individual growth, prevent completion of the life cycle, and limit the size of the population.

Stream Macroinvertebrates

The aquatic life stage of stream insects (e.g., mayfly nymphs, blackfly larvae, chironomid larvae, caddisfly larvae) and benthic crustaceans (amphipods) in temperate and tropical streams exhibit a nocturnal pattern of movement called stream drift (Waters 1972, Giller and Malmqvist 1998). During the day, stream insects and crustaceans remain cryptic and attached to the stream substrate, but at night, usually while foraging, they detach, enter the water column, and drift downstream, where they reattach to the substrate. The selective advantage of this nocturnal behavior is avoidance of visual predators (e.g., fish) while moving to new locations where there are fewer competitors or improved foraging areas.

Importantly, drift is not simply a passive behavior resulting from stream turbulence but an active behavior cued by low light conditions (Giller and Malmqvist 1998). For example, the light of a full moon dramatically suppresses drift density (number of individuals drifting per cubic meter; Anderson 1966, Haney et al. 1983), and similarly, the constant light in polar areas during summer eliminates drift (Giller and Malmqvist 1998). In many streams, a peak in drift density occurs shortly after sunset, and the stimuli eliciting this peak include not only the low light intensities of twilight but also the decrease in light intensity associated with dusk (Haney et al. 1983). Because sky glow can approach the intensity of full moonlight in urban areas, and light levels from glare will likely exceed moonlight intensity, both could suppress drift and possibly delay its initiation at night. If artificial lights cause a steady increase in light intensity at dusk that offsets the decrease in natural light, the decrease in illumination identified by Haney et al. (1983) as a proximate signal necessary for initiating the evening peak in stream drift would be eliminated.

Shallow streams with clear water are more vulnerable to artificial night lighting than large, turbid rivers because light is more likely to penetrate to the substrate where drift-dwelling organisms reside. Ecological consequences of suppressing or extinguishing stream drift with artificial night lighting are unknown, but they may include depressing the productivity of both stream insects and their fish consumers. A decline or elimination of drift may increase interspecific competition for food between stream insects, which could slow their growth and lower stream productivity. Likewise, a severe reduction in drift densities could slow growth rates of fish (e.g., selected salmonid taxa) capable of feeding on drifting organisms at low light intensities (Table 15.1; Jenkins et al. 1970, Tanaka 1970).

Fish

Light similar in intensity to that of a full moon or lower can affect dramatically the behavior and spatial distributions of freshwater and estuarine fishes (Table 15.1). For example, fish foraging (Blaxter 1980, Gliwicz 1986), schooling (Blaxter 1975), spawning (Robertson et al. 1988), vertical movement in the pelagic zone (Blaxter 1975, Luecke and Wurtsbaugh 1993, Appenzeller and Leggett 1995), and even nocturnal emergence from winter concealment (Contor and Griffith 1995) are all affected by natural diel and lunar cycles of light. The lower light threshold for fish schooling is approximately 0.1 lux (Blaxter 1975), which is similar to that of a full moon (0.05–0.1 lux). The threshold for foraging is often considerably lower, particularly for adult fishes. Light thresholds for many freshwater fish (Table 15.1; pike, minnow, coho salmon, carp, bream, and perch) feeding on zooplankton range between 0.00001 and 0.01 lux (Blaxter 1975, Townsend and Risebrow 1982, Bergman 1988), and these intensities are much less than that of full moonlight at the water surface. In general, adult piscivorous fish (e.g., pike, salmonids) detect their fish prey by the contrast of the prey with the background (Vogel and Beauchamp 1999), and piscivores exhibit greater visual sensitivity under low light conditions than do smaller fish feeding on invertebrate prey (Blaxter 1980). Piscivore foraging therefore may be more sensitive to artificial night lighting than that of planktivorous fish. It is also likely that the vertical distribution of some fishes will be altered by artificial night lighting, especially in clear water habitats. For example, some fishes (e.g., Bonneville cisco [*Prosopium gemmifer*]) are so negatively phototactic that light from a full moon causes them to congregate at the lake bottom, where they are difficult to detect acoustically (Luecke and Wurtsbaugh 1993). Consequently, fishery biologists are urged not to make hydroacoustic estimates of population size of these taxa on such nights, and the same recommendation could be extended to freshwater habitats exposed to high levels of artificial night lighting. Finally, in winter, artificial night lighting may prevent emergence of stream-dwelling salmonids from their benthic refugia and consequently impair both wintertime feeding and overwintering success. During the winter, juvenile salmonids hide in the stream substrate during the day and emerge only at night to feed, but this nocturnal emergence is inhibited by the light of a full moon or artificial lighting (Contor and Griffith 1995). Using the light of a large commercial billboard near the riverbank, Contor and Griffith (1995) generated an illumination intensity of 0.003 lux, and this relatively low level caused 30% of the fish to seek cover. This instance illustrates that artificial light from a single source and equivalent to the intensity of only starlight is sufficient to alter the behavior of some freshwater fishes.

Future Research Needs

Many intriguing avenues for future research exist, and they can be grouped into at least two categories: measuring artificial night lighting and probing its organismal and ecological effects.

Measuring Artificial Night Lighting

Comparative measurements of artificial night lighting in open-water and nearshore habitats are essential to gauge the potential for ecological effects in freshwater systems. Importantly, measurements of artificial lighting reported here represent predominantly sky glow rather than glare because measurements were taken at the middle of each lake away from point sources of light on shore. Light intensity measurements near shore, however, are needed because this is where glare predominates, and these light intensities will be greater than those resulting from sky glow. Also, biological effects could be more pronounced near shore because the stronger light intensities are more likely to penetrate the entire water column in these shallower waters and to illuminate the substrate where light-sensitive, benthic animals dwell. Furthermore, glare probably affects more lakes and streams than does sky glow because security lights associated with waterfront homes, summer cottages, and other shoreline development illuminate nonurban freshwater habitats as well as those in urban areas.

Measurements of seasonal variation in exposure to artificial lighting are also essential because lighting increases in duration, and possibly intensity, in winter; the duration of exposure to artificial night lighting in winter increases because of longer nights and the longer need for illumination. But it is also possible that the intensity of illumination from artificial lights, particularly in nearshore habitats in the temperate zone, may increase in winter because of loss of shading from tree leaves. Also, in geographic regions and habitats receiving snow, the reflection of artificial lights from snow on land into the sky should increase the intensity of sky glow reaching the pelagic zones of lakes, large rivers, and coastal waters. This phenomenon should be most pronounced in and near cities with relatively mild winters, resulting in snow but no ice cover, because the latter attenuates light underwater.

Exploring Organismal and Ecological Effects

Artificial night lighting poses interesting biological, ecological, and evolutionary questions for aquatic ecologists. Biological investigations of the visual capabilities of freshwater invertebrates and fishes are needed at light intensities typical of those caused by artificial night lighting, both above and below water. Specifically, action spectra recorded at the wavelengths and intensities typical of artificial night lighting for a wide variety of aquatic taxa will determine what light characteristics alter the behavior of aquatic animals and whether animals are responding to spectrum, intensity, contrast, or combinations of these characteristics of light in particular situations. Any biological process responsive to low light (i.e., up to 10^{15} photons/m^2/s) may be altered by artificial lighting, and these processes warrant investigation. For example, zooplankton resting eggs accumulate in lake and estuarine sediments, and they need dim light to trigger hatching. Could glare from artificial lights be sufficiently bright in nearshore areas to trigger hatching of resting eggs, and, if so, what are the consequences? Although artificial night lighting is much too dim to influence photosynthesis underwater, it may have an effect on cellular or physiological processes.

Ecologically, effects of glare in the nearshore area of lakes, streams, rivers, and wetlands warrant investigation. For example, in streams and lakes where predation pressure from visually oriented fish is strong, crayfish grazers and macroinvertebrate grazers often remain concealed during the day and emerge at night to feed. Glare at night, however, may limit the emergence of these grazers or facilitate their consumption by fish predators to such a degree that grazing is dramatically reduced. Therefore, an algae-covered substrate may result from the indirect effects of artificial night lighting.

Fish spawning is another activity that frequently occurs in the littoral zone, and it is often influenced by lunar cycles. Given that artificial lighting can be as bright as the full moon, an assessment of the effects of glare on fish spawning and its demographic consequences for fish populations may yield interesting results. Enclosure experiments incorporating treatments that block or allow transmission of glare into nearshore waters could evaluate these possibilities. Such experiments, particularly if performed in multiple lakes across a gradient of artificial lighting intensity, would be especially appropriate given the finding in northern Wisconsin that growth rates of some fish are negatively correlated with the amount of lakeshore residential development (Schindler et al. 2000).

At the ecological level, it would be useful to know how much artificial lighting mutes or eliminates diel vertical migration of zooplankton and drift of stream organisms because the magnitude of these effects will determine whether adjacent trophic levels are also affected. If artificial lighting prevents a substantial proportion of the zooplankton grazers from ascending into the epilimnion (warm surface waters) of lakes at night, then artificial lighting could indirectly increase the biomass of algae by reducing nocturnal consumption by zooplankton. In streams, if artificial lighting suppresses drift substantially, then nocturnal fish predators may experience lower rates of feeding and growth, or the drifting stream organisms may experience enhanced competition and reduced growth.

Finally, artificial night lighting poses fascinating evolutionary questions; it has illuminated suburban and urban waterways for the last 50–100 years in North America, and this is more than sufficient time for the evolution of short-lived invertebrates (Hairston et al. 1999, Fischer et al. 2001). Assuming that this increased illumination has facilitated intense fish predation at night over the last century, then persistent, ongoing selection against vertical migration of zooplankton and drift of stream invertebrates may have occurred. In other words, nonmigrating and nondrifting genotypes of zooplankton and stream invertebrates, respectively, may have been favored evolutionarily in urban waterways that have been lighted at night for decades. Experimental designs using reciprocal transplants of invertebrates between urban lakes (or streams) with artificial lighting and remote lakes (or streams) lacking a history of artificial lighting could help establish whether such evolution has occurred. Importantly, the possibility of evolutionary adaptation is predicated on the assumption that artificial lighting is detected underwater by visual predators and that it enables them to locate and capture prey at night. This assumption must be verified. In addition, field experiments performed on remote, undeveloped waterways targeting native communities that lack an exposure history to artificial night lighting would be fruitful. Such experiments not only would simulate present situations in which development and its associated lights suddenly illuminate "pristine" habitats but also would allow comparison of biological responses of native communities to those from urban areas with a long exposure history.

Conclusion

Lakes, large rivers, and coastal waters in or near cities may experience high levels of artificial night lighting because trees or buildings generally do not shade their open waters. Our measurements of artificial lighting, predominantly sky glow, at the surface of lakes in the northeastern United States revealed that wavelengths striking the surface of urban and suburban lakes are dominated by yellow light, and these wavelengths match those emitted from the most common streetlamp in the United States. The relative intensity of artificial illumination at the surface of lakes at night along a suburban-to-urban gradient increased three- to six-fold, and cloud cover increased the intensity of artificial illumination by a factor of three to four. Aquatic organisms most likely to be affected by artificial night lighting include zooplankton and fish that migrate vertically at night, stream insects and crustaceans that drift at night, nocturnal predators that visually locate their prey, and organisms using lunar cues for reproduction. From our initial findings, coupled with the existing literature, we conclude that artificial night lighting may alter the spatial distribution, diel movements (i.e., drift of stream insects, vertical migration of zooplankton and fish), demography, and overwintering success of some freshwater organisms. The sensitivity and magnitude of these responses and their effects on aquatic ecosystems, however, remain to be determined. Future research should include expanding existing measurements of artificial lighting, probing ecological effects of glare in nearshore areas, testing for cascading of ecological effects, and exploring the potential for evolutionary responses to artificial lighting in waterways with different exposure histories.

Acknowledgments

We are grateful to Georgina Scarlata for measuring artificial and natural light in lakes at night, Sonke Johnsen for advising us on how to measure artificial light at night, Jim Haney for discussing some of the ideas presented here, Ellis Loew and Jon Moore for supplying references regarding fish and invertebrate vision, and Jeannie Benton and Laurie Spitler for assisting with figures. Awards from Wellesley College funded the purchase of equipment, and grants from the National Science Foundation and the Howard Hughes Medical Institute supported student assistants. We also thank The Urban Wildlands Group and the UCLA Institute of the Environment for sponsoring the first North American symposium on the ecological consequences of artificial night lighting.

Literature Cited

Anderson, N. H. 1966. Depressant effect of moonlight on activity of aquatic insects. *Nature* 209:319–320.

Appenzeller, A. R., and W. C. Leggett. 1995. An evaluation of light-mediated vertical migration of fish based on hydroacoustic analysis of the diel vertical movements of rainbow smelt (*Osmerus mordax*). *Canadian Journal of Fisheries and Aquatic Sciences* 52:504–511.

Bergman, E. 1988. Foraging abilities and niche breadths of two percids, *Perca fluviatilis* and *Gymnocephalus cernua*, under different environmental conditions. *Journal of Animal Ecology* 57:443–453.

Bishop, J. E. 1969. Light control of aquatic insect activity and drift. *Ecology* 50:371–380.

Blaxter, J. H. S. 1975. Fish vision and applied research. Pages 757–773 in M. A. Ali (ed.), *Vision in fishes: new approaches in research*. Plenum Press, New York.

Blaxter, J. H. S. 1980. Vision and the feeding of fishes. Pages 32–56 in J. E. Bardach, J. J. Magnuson, R. C. May, and J. M. Reinhart (eds.), *Fish behavior and its use in the capture and culture of fishes*. International Center for Living Aquatic Resources Management, Manila.

Buchanan, C., and B. Goldberg. 1981. The action spectrum of *Daphnia magna* (Crustacea) phototaxis in a simulated natural environment. *Photochemistry and Photobiology* 34:711–717.

Cinzano, P., F. Falchi, C. D. Elvidge, and K. E. Baugh. 2000. The artificial night sky brightness mapped from DMSP satellite Operational Linescan System measurements. *Monthly Notices of the Royal Astronomical Society* 318: 641–657.

Contor, C. R., and J. S. Griffith. 1995. Nocturnal emergence of juvenile rainbow trout from winter concealment relative to light intensity. *Hydrobiologia* 299:179–183.

Crawford, D. L., and T. B. Hunter. 1990. The battle against light pollution. *Sky & Telescope* 80:23–29.

Dodson, S. 1990. Predicting diel vertical migration of zooplankton. *Limnology and Oceanography* 35:1195–1200.

Fischer, J. M., J. L. Klug, A. R. Ives, and T. M. Frost. 2001. Ecological history affects zooplankton community responses to acidification. *Ecology* 82:2984–3000.

Flik, B. J. G., D. K. Aanen, and J. Ringelberg. 1997. The extent of predation by juvenile perch during diel vertical migration of *Daphnia. Archiv für Hydrobiologie. Beihefte. Ergebnisse der Limnologie* 49:51–58.

Forward, R. B. Jr. 1988. Diel vertical migration: zooplankton photobiology and behaviour. *Oceanography and Marine Biology: An Annual Review* 26:361–393.

Gal, G., E. R. Loew, L. G. Rudstam, and A. M. Mohammadian. 1999. Light and diel vertical migration: spectral sensitivity and light avoidance by *Mysis relicta. Canadian Journal of Fisheries and Aquatic Sciences* 56:311–322.

Giller, P. S., and B. Malmqvist. 1998. *The biology of streams and rivers*. Oxford University Press, Oxford.

Gliwicz, Z. M. 1986. A lunar cycle in zooplankton. *Ecology* 67:883–897.

Gromov, A. Y. 1993. Effect of wavelength of light on phototaxis of *Daphnia. Hydrobiological Journal* 29:75–81.

Hairston, N. G. Jr., W. Lampert, C. E. Cáceres, C. L. Holtmeier, L. J. Weider, U. Gaedke, J. M. Fischer, J. A. Fox, and D. M. Post. 1999. Rapid evolution revealed by dormant eggs. *Nature* 401:446.

Haney, J. F. 1993. Environmental control of diel vertical migration behaviour. *Archiv für Hydrobiologie. Beihefte. Ergebnisse der Limnologie* 39:1–17.

Haney, J. F., T. R. Beaulieu, R. P. Berry, D. P. Mason, C. R. Miner, E. S. McLean, K. L. Price, M. A. Trout, R. A. Vinton, and S. J. Weiss. 1983. Light intensity and relative light change as factors regulating stream drift. *Archiv für Hydrobiologie* 97:73–88.

Holt, C. S., and T. F. Waters. 1967. Effects of light intensity on the drift of stream invertebrates. *Ecology* 48:225–234.

Hutchinson, G. E. 1967. *A treatise on limnology. Volume II: Introduction to lake biology and the limnoplankton*. John Wiley & Sons, New York.

Jenkins, T. M. Jr., C. R. Feldmeth, and G. V. Elliott. 1970. Feeding of rainbow trout (*Salmo gairdneri*) in relation to abundance of drifting invertebrates in a mountain stream. *Journal of the Fisheries Research Board of Canada* 27:2356–2361.

Koski, M. L., and B. M. Johnson. 2002. Functional response of kokanee salmon (*Oncorhynchus nerka*) to *Daphnia* at different light levels. *Canadian Journal of Fisheries and Aquatic Sciences* 59:707–716.

Lampert, W. 1989. The adaptive significance of diel vertical migration of zooplankton. *Functional Ecology* 3:21–27.

Lampert, W. 1993. Ultimate causes of diel vertical migration of zooplankton: new evidence for the predator-avoidance hypothesis. *Archiv für Hydrobiologie. Beihefte. Ergebnisse der Limnologie* 39:79–88.

Loew, E. R., W. N. McFarland, E. L. Mills, and D. Hunter. 1993. A chromatic action spectrum for planktonic predation by juvenile yellow perch, *Perca flavescens*. *Canadian Journal of Zoology* 71:384–386.

Luecke, C., and W. A. Wurtsbaugh. 1993. Effects of moonlight and daylight on hydroacoustic estimates of pelagic fish abundance. *Transactions of the American Fisheries Society* 122:112–120.

McGowan, T. 2000. *Specifying light and color*; *GE Lighting Application Bulletin*. Publication 70143. General Electric Lighting, Nela Park, Cleveland, Ohio.

Moore, M. V., S. M. Pierce, H. M. Walsh, S. K. Kvalvik, and J. D. Lim. 2000. Urban light pollution alters the diel vertical migration of *Daphnia*. *Verhandlungen der Internationalen Vereinigung für Theoretische und Angewandte Limnologie* 27:779–782.

Moss, B. 1988. *Ecology of fresh waters: man and medium*. Second edition. Blackwell Scientific Publications, Oxford.

O'Brien, W. J. 1979. The predator–prey interaction of planktivorous fish and zooplankton. *American Scientist* 67:572–581.

Ringelberg, J. 1964. The positively phototactic reaction of *Daphnia magna* Straus: a contribution to the understanding of diurnal vertical migration. *Netherlands Journal of Sea Research* 2:319–334.

Ringelberg, J. 1999. The photobehaviour of *Daphnia* spp. as a model to explain diel vertical migration in zooplankton. *Biological Reviews* 74:397–423.

Robertson, D. R., D. G. Green, and B. C. Victor. 1988. Temporal coupling of production and recruitment of larvae of a Caribbean reef fish. *Ecology* 69:370–381.

Robinson, F. W., and J. C. Tash. 1979. Feeding by Arizona trout (*Salmo apache*) and brown trout (*Salmo trutta*) at different light intensities. *Environmental Biology of Fishes* 4:363–368.

Schindler, D. E., S. I. Geib, and M. R. Williams. 2000. Patterns of fish growth along a residential development gradient in north temperate lakes. *Ecosystems* 3:229–237.

Smith, S. L., R. E. Pieper, M. V. Moore, L. G. Rudstam, C. H. Greene, J. E. Zamon, C. N. Flagg, and C. E. Williamson. 1992. Acoustic techniques for the in situ observation of zooplankton. *Archiv für Hydrobiologie. Beihefte. Ergebnisse der Limnologie* 36:23–43.

Tanaka, H. 1970. On the nocturnal feeding activity of rainbow trout (*Salmo gairdnerii*) in streams. *Bulletin of Freshwater Fisheries Research Laboratory* 20:73–82.

Townsend, C. R., and A. J. Risebrow. 1982. The influence of light level on the functional response of a zooplanktonivorous fish. *Oecologia* 53:293–295.

Vogel, J. L., and D. A. Beauchamp. 1999. Effects of light, prey size, and turbidity on reaction distances of lake trout (*Salvelinus namaycush*) to salmonid prey. *Canadian Journal of Fisheries and Aquatic Sciences* 56:1293–1297.

Waters. T. F. 1972. The drift of stream insects. *Annual Review of Entomology* 17:253–272.

Wetzel, R. G. 2001. *Limnology*. Third edition. Academic Press, San Diego, California.

Wetzel, R. G., and G. E. Likens. 2000. *Limnological analyses*. Third edition. Springer-Verlag, New York.

Wilson, A. 1998. Light pollution: efforts to bring back the night sky. *Environmental Building News* 7(8):1, 8–14.

Part VI

Plants

Night, Massachusetts

Many men walk by day; few walk by night. It is a very different season. Take a July night, for instance. About ten o'clock,—when man is asleep, and day fairly forgotten,—the beauty of moonlight is seen over lonely pastures where cattle are silently feeding. On all sides novelties present themselves. Instead of the sun, there are the moon and stars; instead of the wood-thrush, there is the whippoorwill; instead of butterflies in the meadows, fire-flies, winged sparks of fire!—who would have believed it? What kind of cool, deliberate life dwells in those dewy abodes associated with a spark of fire? So man has fire in his eyes, or blood, or brain. Instead of singing birds, the half-throttled note of a cuckoo flying over, the croaking of frogs, and the intenser dream of crickets,—but above all, the wonderful trump of the bull-frog, ringing from Maine to Georgia. The potato-vines stand upright, the corn grows apace, the bushes loom, the grain-fields are boundless. On our open river-terraces, once cultivated by the Indian, they appear to occupy the ground like an army,—their heads nodding in the breeze. Small trees and shrubs are seen in the midst, overwhelmed as by an inundation. The shadows of rocks and trees and shrubs and hills are more conspicuous than the objects themselves. The slightest irregularities in the ground are revealed by the shadows, and what the feet find comparatively smooth appears rough and diversified in consequence. For the same reason the whole landscape is more variegated and picturesque than by day. The smallest recesses in the rocks are dim and cavernous; the ferns in the wood appear of tropical size. The sweet-fern and indigo in overgrown wood-paths wet you with dew up to your middle. The leaves of the shrub-oak are shining as if a liquid were flowing over them. The pools seen though the trees are as full of light as the sky. "The light of the day takes refuge in their bosoms," as the Purana says of the ocean. All white objects are more remarkable than by day. A distant cliff looks like a phosphorescent space on a hill-side. The woods are heavy and dark. Nature slumbers. You see the moonlight reflected from particular stumps in the recesses of the forest, as if she selected what to shine on. These small fractions of her light remind one of the plant called moon-seed,—as if the moon were sowing it in such places.

In the night the eyes are partly closed, or retire into the head. Other senses take the lead. The walker is guided as well by the sense of smell. Every plant and field and forest emits its odor now, swamp-pink in the meadow, and tansy in the road; and there is the peculiar dry scent of corn which has begun to show its tassels. The senses both of hearing and smelling are more alert. We hear the tinkling of rills which we never detected before. From time to time, high up on the sides of hills, you pass through a stratum of warm air: a blast which has come up from the sultry plains of noon. It tells of the day, of sunny noon-tide hours and banks, of the laborer wiping his brow and the bee humming amid flowers. It is an air in which work has been done,—which men have breathed. It circulates about from wood-side to hill-sides like a dog that has lost its master, now that the sun is gone. The rocks retain all night the warmth of the sun which they have absorbed. And so does the sand: if you dig a few inches into it, you find a warm bed.

You lie on your back on a rock in a pasture on the top of some bare hill at midnight, and speculate on the height of the starry canopy. The stars are the jewels of the night, and perchance surpass anything which day has to show.

Henry David Thoreau

From "Night and Moonlight," 1863.

Chapter 16

Physiology of Plant Responses to Artificial Lighting

Winslow R. Briggs

Plants are continuously bombarded by biotic and abiotic signals from their environment. Biotic signals include attacks by insects and pathogens and grazing by larger herbivores. Abiotic signals include temperature changes, changes in water availability, nutrient limitation, osmotic stress, and changes in the light environment. In the long course of evolution, plants have developed exquisite mechanisms for detecting and responding to these many signals. This chapter is concerned with those signals that arise from the light environment, of which artificial night lighting is a part. The focus of this chapter is plant photoreceptors, which are the molecules that detect light signals, and the consequent physiological responses to light. Some possible consequences of the excitation of photoreceptors by artificial lighting are discussed. This chapter does not consider the spillover effects on plants of the disruption of ecological interactions (e.g., herbivory, pollination) that may be caused by artificial night lighting, rather concentrating on the physiology and growth of plants themselves.

Although there is an extensive literature on the effects of light spectral quality, quantity, and duration on plant growth and development, its focus has been almost exclusively on results from laboratory, growth chamber, and greenhouse experiments. Most of the work has focused on understanding basic mechanisms for the regulation of plant growth by light. Indeed, a substantial amount of horticultural research attempts to determine the optimal lighting conditions for plants being grown under artificial lighting. Yet except for two articles (Cathey and Campbell 1975a, 1975b), no rigorous studies have examined effects of artificial night lighting on plants in conditions approaching their natural environment. A lack of clear evidence for such effects, however, should not lead one to conclude that no effects exist or that any particular effects that may be noticed are either beneficial, deleterious, or of no consequence to the plant. As a baseline on which to design a research program to address these issues in the future, it will be useful to review what is known about plant photoreceptors—what they are and what responses they mediate.

Currently, four different families of plant photoreceptors have been identified and characterized from higher plants: phytochromes, cryptochromes, phototropins, and a photoreceptor designated as FKF1. These photoreceptors mediate a bewildering range of physiological and developmental responses in plants. Readers interested in further details regarding the extensive information accumulated for these photoreceptors and the many biochemical and physiological consequences of their photoexcitation should consult the following recent review articles: Smith (1995, 2000), Huq and Quail (2005), and Tu and Lagarias (2005) for phytochromes; Batschauer (2005) and Cashmore (2005) for cryptochromes; Briggs and Christie (2002) and Christie and Briggs (2005) for phototropins; and Briggs and Huala (1999) for both cryptochromes and phototropins.

In this chapter I introduce the four photoreceptor families and recount their mechanisms of operation. I then outline various physiological responses that are governed by photoreceptors: germination, phototropism, flowering, and dormancy. With this background, I review the limited literature that investigates the effects of artificial lighting on the physiology of plants and discuss other potential effects.

The Four Photoreceptor Families

Four families of photoreceptors are found in plant tissues: phytochromes, cryptochromes, phototropins, and the photoreceptor FKF1.

The Phytochromes

The first plant photoreceptor to be isolated and characterized was a phytochrome, which is a plant pigment protein that absorbs light and triggers a physiological response. The earliest evidence of this plant pigment was obtained from experiments with a particular lettuce seed cultivar in the 1930s (Flint 1934, Flint and McAlister 1935, 1937). These authors found that red light would induce germination in a greater percentage of a population of lettuce seeds than a dark control, whereas near-infrared light (later designated far-red) would suppress germination. Later workers determined that the signal induced by a brief exposure to red light could be canceled by a subsequent exposure to far-red light (Borthwick et al. 1952b). Indeed, one could provide a series of alternating red and far-red pulses, and the seeds would germinate or not depending on whether the terminal pulse was red or far-red. Many other aspects of plant development were subsequently shown to be regulated by this red/far-red reversible system, including stem elongation (inhibited), leaf expansion (promoted), development of the entire photosynthetic machinery (promoted), flowering (inhibited or promoted depending on timing of irradiation and plant species), and entry into dormancy (inhibited if red light is given at the wrong time).

More than two decades passed after the Flint and McAlister experiments until Butler et al. (1959) used standard biochemical techniques to isolate one of the responsible photoreceptors, the first of the phytochromes, now designated as phytochrome A or phyA. All of the phytochromes are chromoproteins carrying a bilitriene (i.e., a linear tetrapyrrole related to the cyanobacterial and red-algal photosynthetic antenna pigments phycocyanin and phycoerythrin) as their chromophore (the part of the molecule that provides color). Phytochromes are synthesized in the dark, with the chromophore in a molecular conformation that absorbs red light; in this state the photoreceptor is designated Pr. On exposure to red light, the chromophore undergoes isomerization (changes shape) to a form that absorbs far-red light, and the photoreceptor is then designated Pfr. This isomerization results in a protein conformational change that converts an inactive chromoprotein Pr into what is widely regarded as the biologically active form Pfr. Far-red light returns the photoreceptor to its red-absorbing form. The two spectral forms are illustrated in Figure 16.1. Thus, red light initiates the many responses mediated by phytochrome by converting Pr to Pfr. If far-red light is given sufficiently soon after red, however, the activated

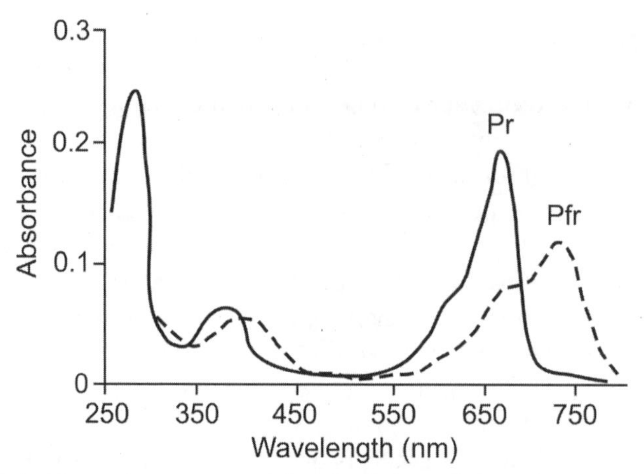

Figure 16.1. Absorption spectra of the two forms of phytochrome, Pr and Pfr. Pfr is widely regarded as the biologically active form. Courtesy of J. Clark Lagarias.

photoreceptor Pfr is returned to the Pr form and turned off before it can complete its assignment, and the response does not occur. Sage (1992) presents a detailed and delightfully readable history of the events leading to the discovery of the phytochromes and the explosion of research that followed their discovery.

We now know that in the model plant *Arabidopsis thaliana* there are five phytochromes, designated phyA through phyE (Sharrock and Quail 1989, Clack et al. 1994). *Arabidopsis* mutants are now available that lack each of these single phytochromes. Mutants lacking two, three, or even four of these phytochromes also have been constructed. Such mutants have been of great assistance in sorting out the various physiological and biochemical roles of the individual phytochromes. PhyA and phyB play the most dominant roles in mediating many of these responses, with phyC through phyE playing largely backup roles (Neff et al. 2000). Responses mediated by phyA typically are far more sensitive and need far less light that those mediated by phyB. This is partly because far higher levels of phyA than phyB exist in dark-grown seedlings and partly because fewer molecules of phyA than of phyB are needed to be transformed to Pfr to bring about a response. Indeed, as a seedling emerges into the bright light above the soil surface, most of the phyA is degraded, and, as a consequence, the remaining highly stable phyB is the phytochrome that predominates in the light-grown plant.

The Cryptochromes

Sachs (1864, 1887), citing his own work and that of much earlier workers (Poggioli 1817; see Briggs in press), had already described a specific effect of blue light on plants long before the beginning of the twentieth century. He and others had observed that plant shoots would grow toward a source of blue light but not toward a source of red light. (The tendency of plant shoots to grow toward a source of light was subsequently designated as phototropism.) Despite this recognition of a dramatic effect of blue light more than a century before Flint and McAlister's experiments establishing an effect of red light, identification of the responsible blue-light receptor remained elusive. Over the years, and before 1993, a large literature accrued devoted to the biological consequences of blue-light irradiation on plants (see Senger and Briggs 1981, Briggs and Iino 1983, and Short and Briggs 1994 for reviews). It was clear that a number of blue-light receptors independent of phytochrome (phytochrome does have small absorption bands in the UV-A in the Pr form and blue in the Pfr form) must exist, but no agreement was reached on how many different receptors might exist or what their blue-light-absorbing chromophore(s) might be. Ultimately studies with *Arabidopsis* mutants (Lascève et al. 1999) indicated the participation of at least four different blue-light-activated signal-transduction pathways in *Arabidopsis* and therefore presumably at least four different blue-light receptors.

A large number of these blue-light-activated responses share a common wavelength sensitivity in both plants and fungi. They are all activated by wavelengths between 330 and 500 nm (UV-A through blue), with a peak of activity in the UV-A and a higher peak of activity in the blue. A typical action spectrum for one of these responses, phototropism in this example, is shown in Figure 16.2. This action spectrum has features that belong to the family of yellow to red pigments known as carotenoids (peaks and shoulders in the blue part of the spectrum). It also shows a peak in the UV-A, characteristic of the compound riboflavin or one of its derivatives. Gressel (1979) gave the unknown photoreceptors exhibiting this action spectrum the name *cryptochrome* partly because they were evidently common in cryptogamic organisms such as fungi and partly because they remained stubbornly hidden or cryptic. Because there are large numbers of carotenoid- and flavin-binding proteins in plants, sorting out which might be a photoreceptor for a particular process proved a daunting task.

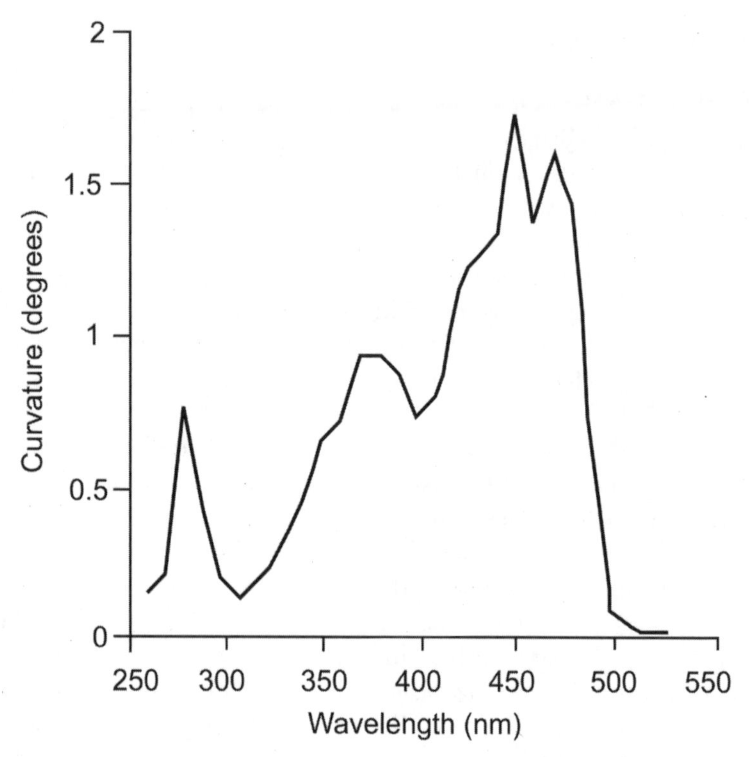

Figure 16.2. Action spectrum for phototropism. Redrawn from Baskin and Iino (1987).

It was through the use of *Arabidopsis* mutants that Ahmad and Cashmore (1993) first isolated a gene encoding a protein that likely serves as a blue-light photoreceptor. Koornneef et al. (1980) had isolated six classes of *Arabidopsis* mutants that lacked the normal light-activated suppression of hypocotyl (seedling) elongation and designated them *hy1* through *hy6*. These mutants all had different lesions related to photosensing and responding to light. For example, *hy3* had a mutation in the *phyB* gene. Because it could not synthesize phyB, its capacity to respond to red light by hypocotyl inhibition was severely impaired. Like red light, blue light suppresses stem elongation. Koornneef et al. (1980) had shown that the *Arabidopsis* mutant *hy4* possessed normal inhibition by red and far-red light but was insensitive to blue light. Ahmad and Cashmore (1993) then determined that the *hy4* mutant lacked the gene encoding a protein with high homology in its amino acid sequence with a group of prokaryotic proteins known as DNA photolyases. These enzymes serve as photoreceptors to repair damaged DNA and are activated by blue and UV-A light. Given this similarity with a known group of photoreceptors, Ahmad and Cashmore (1993) proposed that the protein was the photoreceptor mediating blue-light-induced suppression of stem elongation. Lin et al. (1995) subsequently designated the putative protein cryptochrome 1 or cry1. The cry1 protein has an additional C-terminal domain that is not found in any photolyases but is essential for cryptochrome function (see Cashmore 2005). Furthermore, it fails to show photolyase (DNA repair) activity. Despite its similarity with the photolyases, its biochemical mode of action differs from theirs.

Malhotra et al. (1995) and Lin et al. (1995) both expressed the cry1 protein in a heterologous system and independently demonstrated that it bound the riboflavin derivative flavin adenine dinucleotide (FAD) as a chromophore. Malhotra et al. (1995) also found that cry1 bound a second chromophore, methenyltetrahydrofolate (MTHF). One group of the light-activated prokaryotic DNA photolyases also binds both FAD and MTHF, strengthening the hypothesis that cry1 was indeed a photoreceptor. Later work (Lin et al. 1996, Hoffman et al. 1996) uncovered a second cryptochrome, designated cry2, that also participates in blue-light-induced suppression of stem elongation but is sensitive to lower intensities of light than cry1 (Lin et al. 1998). Thus, the two cryptochromes are partly redundant in the growth suppression response but operate over different light intensity ranges. So, more than four decades after the discovery of the phytochromes, the first true higher-plant blue-light receptors were finally identified. Most recently, Kleine et al. (2003) and Brudler et al. (2003) reported the existence in several organisms, including flies and humans in addition to *Arabidopsis* and a cyanobacterium, of a third cryptochrome, cryDASH, subsequently named cry3. The role of cry3 in plants is not yet known, but efforts to elucidate its role are under way.

The Phototropins

The discovery of the cryptochromes did not identify photoreceptors to account for all blue-light-activated responses in higher plants. Indeed, phototropism, the response that provided the very first "cryptochrome-type" action spectrum (Figure 16.2), was perfectly normal in *Arabidopsis* double mutants lacking both cry1 and cry2 (Lascève et al. 1999). Therefore, there had to be another blue-light receptor in addition to the cryptochromes. As with the cryptochromes, it was the use of mutants that resolved this issue. Liscum and Briggs (1995, 1996) described a series of

Arabidopsis mutants that failed to curve toward low-intensity blue light and designated them *nph1* through *nph4* (*nph* for *nonphototropic hypocotyl*). It had been shown previously that blue light caused a particular plasma membrane protein to become phosphorylated, adding a phosphate group to it (see Briggs et al. 2001b for review). (The addition and removal of phosphate groups to and from proteins is an important mechanism for regulating their activity.) An *Arabidopsis* mutant called JK224 (Khurana and Poff 1989), impaired in phototropism, also showed impaired light-activated phosphorylation (Reymond et al. 1992), indicating that the blue-light-activated phosphorylation reaction played some role in phototropism. *Arabidopsis* mutants identified as belonging to the *nph1* class were also impaired in carrying out the light-activated phosphorylation reaction. JK224 belonged to the *nph1* class.

Note that three components are needed for blue-light-activated phosphorylation: the protein that becomes phosphorylated, the enzyme that carries out the phosphorylation, and a photoreceptor to detect the blue light. Huala et al. (1997) finally isolated and characterized the gene that was mutated in the *nph1* mutant's alleles. It turned out to be a classic serine/threonine protein kinase (a protein kinase is an enzyme that phosphorylates proteins). Hence, two of the three components belonged to a single protein. Christie et al. (1998) then showed that the protein itself bound a flavin (in this case, not FAD but rather the riboflavin derivative flavin mononucleotide [FMN]) and that exposure of the chromophore-bearing protein (chromoprotein) to light in the absence of any other plant proteins drove its own phosphorylation (autophosphorylation). This result demonstrated that the single protein performs all three functions: substrate for phosphorylation, kinase, and photoreceptor for the reaction. The authors concluded that the protein that became phosphorylated on blue-light treatment was itself a photoreceptor for phototropism.

Subsequent work (Christie et al. 1999) demonstrated that this chromoprotein actually bound two molecules of FMN. The two stretches of amino acids binding FMN were highly conserved in a wide range of other proteins in animals and bacteria involved in regulating responses to light, oxygen, or voltage; Huala et al. (1997) thus designated these domains as LOV domains. Salomon et al. (2000) then demonstrated that LOV domains underwent a unique photochemistry, previously undescribed, that we now know forms the basis for their action (Christie and Briggs 2005). Because the name *cryptochrome* had already been preempted, Christie et al. (1999) named the photoreceptor phototropin after its role in phototropism.

Shortly after Huala et al. (1997) characterized the first phototropin, Jarillo et al. (1998) reported that the *Arabidopsis* genome contained a second phototropin. These two photoreceptors are now designated phototropin 1 (phot1) and phototropin 2 (phot2; see Briggs et al. 2001a). A second family of blue-light receptors was thereby identified. Phototropin 2 also mediates phototropism (Sakai et al. 2001) but requires higher intensities of light than phototropin 1. Hence, for all three families of plant photoreceptors, some sense high light levels (phyB, cry1, phot2) and some sense low light levels (phyA, cry2, phot1).

In addition to the role of phototropins as photoreceptors for phototropism, they now have been shown to mediate blue-light-activated stomatal opening, chloroplast movements in response to changes in light intensity, leaf expansion, and a rapid transient inhibition of stem elongation of dark-grown seedlings (Briggs and Christie 2002).

FKF1 and Its Relatives

With the publication of the complete DNA sequence of the model plant *Arabidopsis thaliana*, it became possible to search this database for proteins related to the known photoreceptors. There are now three proteins, FKF1, ZTL, and LKP2, each of which contains a single LOV domain, highly similar to the LOV domains in the phototropins, also binding FMN, and showing the same unique photochemistry. Otherwise, these proteins are entirely different from the phototropins. Imaizumi et al. (2003) recently demonstrated that it was one of these, FKF1, that detected blue-light signals from the long day that would induce *Arabidopsis* to flower. It is not currently known whether the other two, ZTL and LKP2, are genuine photoreceptors. However, the presence of a LOV domain in each that performs the same photochemistry as the LOV domains of phototropin and FKF1 suggests that they might be.

Physiological Responses of Plants to Visible Light

The description of the number and character of the plant photoreceptors leads to a series of questions about their function and the physiological responses of plants to visible light. Why are there red/far-red- and blue-light receptors? Why do plants have so many photoreceptors? These questions lead to another. Why is daylength measurement important to plants?

Why Are There Red/Far-Red- and Blue-Light Receptors?

Full sunlight contains both red and far-red components. Therefore, a plant growing unshaded will have a photostationary mixture of Pr and Pfr. As the forward reaction from Pr to Pfr predominates, perhaps two-thirds of a plant's phytochrome will be in the Pfr form. The chlorophyll in green leaves absorbs large amounts of red and blue light, the energy of which is used for photosynthesis. On the other hand, chlorophyll absorbs very little far-red light. As a consequence, plants in the understory, under a dense canopy of leaves, are left with a minimal amount of light in the very wavelength ranges that they need for their own photosynthesis. By contrast, they are not deprived of far-red light. Hence, their phytochrome (largely phyB) is mostly in the Pr form. As mentioned above, it is Pfr that mediates the inhibition of stem growth. In the absence of Pfr, an understory plant's stem elongates rapidly, and leaf development and expansion are curtailed. In a mixed population of herbaceous plants or seedlings of woody plants, this response can permit a given plant to reach full sunlight and make maximal use of its photosynthetic capacity. This light-induced developmental change is called the shade avoidance response.

Just as plants differentially absorb red and far-red light, they also differentially reflect light in these two spectral regions. The reflected light from a given plant will be somewhat enriched in far-red light and somewhat depleted of red light. Thus, neighboring plants will have some of their Pfr driven back to Pr. These plants can detect and respond to this relatively minor depletion of their Pfr by accelerating their elongation and curtailing their leaf development even though they are not directly shaded. Smith (1995) has written a detailed review of these responses to shading by a leaf canopy or interception of light reflected from neighboring plants. These responses obviously are worthless under a canopy of redwood trees or in the shade of a barn but are nevertheless an important element of plant competition in many vegetation communities.

Much of the light reaching a plant within a plant community may come not from direct sunlight but rather from a patch of open sky. This skylight is enriched in both blue and UV-A light compared with red, a consequence of light-scattering phenomena in the atmosphere. It is perhaps not surprising that phototropism, the response that directs plant growth toward (or in the case of roots away from) a light source, depends on photoreceptors that absorb light in those spectral regions. In guiding a plant's growth toward a blue-light source, phototropism provides a second mechanism to maximize light harvesting and photosynthesis.

As mentioned above, there are several more responses to blue light, and these also play important roles in regulating photosynthesis. When the light intensity is relatively low, the chloroplasts, the subcellular organelles containing all the photosynthetic pigments and machinery, arrange themselves along the surface of the individual chloroplast-containing cells that is at right angles to the direction of incident light (Senn 1908). This distribution pattern minimizes self-shading and therefore maximizes light interception for photosynthesis (the accumulation response). In many plants, the photosynthetic machinery can handle only a certain amount of light. When the intensity is too high, the absorbed energy cannot be used for photochemistry and may cause severe damage to the chloroplasts. Under these conditions, the chloroplasts become relocated to the walls of the cells that are perpendicular to the direction of incident light. This arrangement minimizes light interception and therefore minimizes photodamage (the avoidance response). Both of these responses are mediated by the phototropins. Either phototropin can mediate the accumulation response (Sakai et al. 2001), but it is only phot2 that mediates the avoidance response (Kagawa et al. 2001, Jarillo et al. 2001).

Another response to blue light, mediated by the phototropins, is the opening of the stomata, small pores on the surface of leaves and stems that open in the light. Their opening permits uptake of CO_2 for photosynthesis, permits release of O_2, and increases water loss to drive transpiration and move water and mineral nutrients up from the roots. Kinoshita et al. (2001) demonstrated that in mutants of *Arabidopsis* that lack both phototropins, blue-light-activated stomatal opening is completely lacking.

Leaf expansion is also accelerated by blue light independent of photosynthesis (Van Volkenburgh and Cleland 1990, Van Volkenburgh et al. 1990). Mutants lacking both phototropins fail to show normal expansion (Sakamoto and Briggs 2002). Finally, the mature leaves of some species can adjust their orientation so that they lie in a plane at right angles to the incident sunlight to maximize photosynthesis (or in some instances parallel to the direction of strong light as a mechanism to avoid photodamage; Koller 2000), and this response probably is also activated by one or both phototropins.

Why Do Plants Have So Many Photoreceptors?

In many instances, two or more photoreceptors share partial functional redundancy, that is, they regulate the same developmental step or other response. Seed germination provides a good example. As mentioned above, not all seeds need light for germination. For those that do, however, it is phytochrome that mediates the activation of germination. Both phyA and phyB can mediate the response, but the phyA system is far more sensitive. The germination of *Arabidopsis* seeds usually is mediated by phyB, and several minutes of light are needed. However, if the seeds remain hydrated for many hours under the soil where the light is extremely dim, phyA is synthesized and light sensitivity goes up almost 1,000-fold (Shinomura et al. 1996). Thus the two photoreceptors are functionally redundant, but phyA is far more sensitive than phyB, and in this instance the two phytochromes do not function at the same time. PhyB functions early during seed hydration (imbibition), whereas phyA functions only very late during seed hydration.

An even more dramatic example of functional redundancy is the inhibition of hypocotyl elongation by light. This inhibition can be imposed by activating any one of the five phytochromes (Neff et al. 2000) or by activating either cry1 or cry2 (Lin et al. 1998) and at least one phototropin (Folta and Spalding 2001). In a dark-grown seedling, phyA may be present at levels 50 times higher than those of phyB. As mentioned previously, phyA is activated by far lower light levels than phyB. Because very low levels of light can penetrate all but the densest soils, these light levels may be sufficient to activate phyA, inhibit stem elongation, and begin diverting those resources that are still available into making leaves and the photosynthetic machinery essential for the plant's survival. Therefore, phyA provides an early warning for the plant that it is time to switch developmental priorities. Cry2 is more sensitive than cry1 in this response, although the difference is not nearly as dramatic as for the two phytochromes.

Seedlings germinated in complete darkness use scarcely any of their stored reserves to synthesize and assemble the many pigments, proteins, and lipids needed to carry out the complex photochemical and biochemical reactions of photosynthesis. Once they receive light, however, leaves begin to develop and expand, the full photosynthetic machinery is rapidly developed, and the plant begins to turn green and carry out photosynthesis. Whole cadres of genes are activated (e.g., those encoding proteins essential for photosynthesis), and other suites of genes whose encoded proteins are no longer needed are switched off. In this manner the seedlings become self-sufficient and no longer need to depend on stored reserves. All of these light-induced changes are collectively called photomorphogenesis.

Here again, there is extensive functional redundancy. Any one of the phytochromes and either cryptochrome can mediate this dark-to-light developmental transition, at least in part. Phot1 may also play a minor role in the process. That evolution has provided plants with this manifold redundancy underlines the importance of the dramatic developmental transition from the dark-growth pattern to the light-growth pattern. Thus, a mutation causing a loss of any one of these photoreceptors is not of fatal consequence because others are adequate to do the job.

It is noteworthy that not all of the abovementioned responses necessarily occur in all plant species. For example, with seed germination, it is generally only very small seeds that germinate in response to light. Typically, the germination of larger seeds is light independent. This adaptation makes sense in that small seeds have extremely limited reserves. It is important for the seedlings to reach the light and develop photosynthetic capacity extremely quickly. If small seeds germinated in darkness too far under the soil, few seedlings would make it to the surface before exhausting their reserves. Likewise, the regulation of flowering by daylength, a phytochrome-mediated process but also involving FKF1 at least in *Arabidopsis*, is not ubiquitous among higher plants. Even within a given species certain cultivars or varieties are sensitive to daylength (photoperiod) and others are not.

Why Is Daylength Measurement Important to Plants?

More than 80 years ago, Garner and Allard (1920) first reported the phenomenon of photoperiodism. A tobacco cultivar, Maryland Mammoth, failed to flower on long days but flowered readily on short days. Spinach and radish showed exactly the opposite behavior, flowering on long days but not on short days. They designated these species long-day plants and short-day plants. Three years later, Garner and Allard (1923) added two more response categories: intermediate plants that flowered only when the day was neither too long nor too short (e.g., a species of goldenrod) and day-neutral plants that simply flowered when they reached a certain size independent of daylength (a large number of species).

In 1938, Hamner and Bonner published an extremely important finding: a light break as short as one minute in the middle of a long night would prevent the cocklebur (*Xanthium pensylvanicum*) from flowering. Subsequent studies demonstrated that if the light pulse was red light, flowering was inhibited, but if the red light was followed by far-red, the effect of the red light pulse was canceled and the plants flowered (Borthwick et al. 1952a). Downs (1956) then demonstrated that a red light pulse in the middle of a long night would induce a long-day plant to flower, a response exactly opposite that of *Xanthium*. The effect was fully reversible by a far-red pulse. Thus both of these responses showed the red/far-red reversibility diagnostic for phytochrome-mediated responses. Sage (1992) provides an excellent review of all of the early work leading to our understanding of the role of daylength in flowering and the role of phytochrome as the cogent photoreceptor. As noted, Imaizumi et al. (2003) have demonstrated an important role for the blue-light receptor FKF1, in addition to the phytochromes, in the long-day flowering response of *Arabidopsis*.

Later workers found that daylight extension with artificial light would do the same thing: inhibit flowering of short-day plants and promote it in long-day plants. Florists regularly use either a night break or a daylength extended with artificial lighting to prevent short-day plants such as chrysanthemums from flowering. They can then remove groups of plants from these artificial long-day conditions at intervals and extend the period of time over which they have flowering plants for sale. Vince-Prue's 1994 review of photoperiodism and flowering is still timely.

By the mid-1950s it was recognized that daylength also plays a role in determining the dormancy state of many woody shrubs and trees (Downs and Borthwick 1956a, 1956b, Downs and Piringer 1958), a phenomenon likely mediated by the phytochromes, as with flowering. Many woody plants normally undergo a flush of vegetative growth induced by some environmental change, such as an increase in temperature or the onset of a rainy season. However, in due time the buds stop producing normal leaves and commence instead to produce bud scales. Ultimately, the buds cease producing any new organs and enter complete dormancy. Seedlings on long days continue growing far longer than seedlings on short days before the buds go into dormancy. Hence, artificially extending the daylength can significantly increase the length of time during which active growth and production of true leaves takes place. This technique is of great use to horticulturists because it enables them to increase the size of seedlings grown for any fixed period of time and to have larger plants for the market.

Later in the season in temperate regions, shortening photoperiods lead to the complex physiological and biochemical changes that cause leaf color change and finally leaf abscission (Olmsted 1951). Thus, daylength changes can play a role twice: regulating entrance of buds into dormancy and regulating leaf senescence and abscission. In both cases, phytochromes are likely to be the relevant photoreceptors, although a role for a homolog of FKF1 (or FKF itself) could theoretically be involved as well.

As with flowering, there are differences between species, with some species far more sensitive than others and some scarcely sensitive at all. The dearth of recent research in this area is strikingly underlined in the review by Vince-Prue (1994). Out of thirty-three pages, only a single page is devoted to seasonal responses.

Outdoor Artificial Lighting and Plant Physiology

Most of the work on plant photobiology has focused on the effects of light on activating the complex developmental switch from the growth pattern found in the dark to that found in the light. In darkness, the growth pattern maximizes stem growth and elongation to reach the light, investing almost no resources in leaf expansion or development of the photosynthetic machinery. In the light, the growth pattern maximizes leaf formation, leaf expansion, and the development of the photosynthetic apparatus and very much reduces stem elongation. Indeed, it is largely through studies of the changes induced when seedlings are taken from darkness and placed in the light that we have gained the present extensive knowledge of plant photoreceptors summarized in this chapter. Information on how artificial lighting might affect these processes in nature is almost completely lacking, but it is possible that it might hasten the switch of seedlings from the growth pattern in the dark to that in the light (with unknown consequences for the plant or its ecological relationships).

Undoubtedly, outdoor artificial lighting can affect plants beyond the seedling stage as well. Although plants whose flowering response is day neutral will not have their development affected by the extension of their photoperiod, they will be able to perform additional photosynthesis using streetlights as a light source. Because the relative intensity of street lighting is far below the intensity levels experienced during the day, the amount of additional photosynthesis may be trivial but could be marginally beneficial to the plant. In contrast, plants that are sensitive to photoperiod in their flowering response, their entrance into bud dormancy, or their initiation of leaf senescence in response to daylength changes could conceivably be negatively affected by outdoor artificial lighting. The extension of a short day even with fairly dim light can yield a long-day response in some instances.

In an effort to determine the effect of five types of light sources used for outdoor lighting (incandescent, high pressure sodium vapor, metal halide, fluorescent, and clear mercury vapor), Cathey and Campbell (1975a, 1975b) investigated the effect of providing a wide range of plant species with each of these light sources at a level of one footcandle (about 10 lux) during a 16-hour night. Included in the study were known short-day plants, known long-day plants, smaller horticulturally important plants such as chrysanthemums and carnations, and saplings of tree species such as paper birch, Norway maple, and American elm. Flowering was delayed in some short-day plants, vegetative growth was enhanced in several tree species, and flowering was promoted in some long-day plants. Some species, however, showed no measurable response at all. The strength of response to the various types of light sources was, in decreasing order, incandescent, high pressure sodium vapor, metal halide, cool white fluorescent, and clear mercury vapor. Cathey and Campbell (1975a) also performed some experiments that used higher light intensities, with quite similar results. Thus, plants differ widely in their sensitivity to short days extended with artificial light, and different light sources have different degrees of influence. Their second article (Cathey and Campbell 1975b) provides lists of those species that are most sensitive, those that are less sensitive, and those that are insensitive to daylength extended with artificial light. These experiments were performed under growth room conditions and only with saplings of tree species instead of mature trees (for obvious reasons) at a time when only a single photoreceptor, phytochrome A, was characterized. Cathey and Campbell's work, however, represents the best attempt to date to evaluate effects of outdoor lighting sources on plant development.

Still needed is a determination of the threshold below which sensitive species no longer respond to the various artificial light sources. Branches of sycamores growing on the University of California campus in Berkeley near streetlights often keep their leaves into the late fall and winter, while branches of trees somewhat farther away lose their leaves. Likewise, liquidambar trees growing on the Stanford University campus show a response to street lighting. One tree had a full complement of leaves (although partially senescent and brightly colored) in February 2002 on the side of the tree toward a streetlight. Only a few meters away, on the other side of the tree, all of the leaves had long since dropped. Thus, the effect of the streetlight was highly localized. The light intensity dropped below a threshold value within a couple of meters. Although these examples show effects of outdoor lighting on tree physiology, it is not possible to know what consequences this influence has for the plant. The light may extend the period during which photosynthesis can take place, if only trivially, or prevent dormancy and expose the tree to the somewhat harsher climate of the winter months while it is still physiologically active.

Cathey and Campbell (1975b) reported that plane trees exposed to high pressure sodium vapor lighting exhibited rapid and late season growth but then suffered severe winter dieback compared with trees screened from lighting. They also noted that depressed chlorophyll formation in leaves and the expansion of leaves associated with continuous lighting increases the sensitivity of trees to damage by pollution (Cathey and Campbell 1975b).

Information is lacking on what light intensities might be needed to prevent initiation of bud dormancy or leaf senescence for sycamore, liquidambar, or any other tree species that needs short days to enter dormancy. Indeed, different ecotypes of the same species may well vary greatly in their light sensitivity. The cocklebur (*Xanthium pensylvanicum*) is a classic example of extreme sensitivity to interruption of the long night with a very brief pulse of light (Hamner and Bonner 1938). It turned out, however, that the ecotype of *Xanthium pensylvanicum* that was used was uncharacteristically sensitive to light conditions. In many places *Xanthium pensylvanicum* plants flower right on schedule as the days shorten, despite growing along busy highways, subjected almost continuously through the night to the light from vehicle headlamps. On the other hand, security lighting outside certain prisons in Ohio prevents normal development of soybean plants as far as 100 feet (30 m) from the light sources (according to an e-mail exchange with prison officials). The cultivar of soybean (a plant needing short days to flower) that is planted in these fields appears to be unusually sensitive to night lighting.

Conclusion

As sessile organisms, higher plants rely heavily on environmental signals to guide their development. Among the most important environmental signals are those that come from the light environment. Thus, in the course of evolution, plants have acquired a wide range of photoreceptors that perceive and respond to light signals in the ultraviolet, blue, red, and near-infrared (far-red) regions of the electromagnetic spectrum. In the model plant *Arabidopsis thaliana*, 11 different photoreceptors have been characterized. Those absorbing and responding to UV-A and blue light include three cryptochromes, cry1, cry2, and cry3 (originally designated cryDASH), two phototropins, phot1 and phot2, and an additional photoreceptor, very recently characterized, designated FKF1. Those absorbing in the red and far-red regions of the spectrum are the five phytochromes, phyA through phyE. There is also evidence of one or more photoreceptors that sense and respond to UV-B, although these remain to be characterized. The photoreceptors allow the plant to measure and respond to four parameters of their light environment: light spectral quality, light intensity, light direction, and light duration. Sometimes these photoreceptors act independently, sometimes redundantly, sometimes cooperatively, sometimes antagonistically, sometimes at the same stage of development, and sometimes at different stages of development. Moreover, some of these responses are incredibly sensitive, responding to levels of light that the human eye can barely perceive, whereas others are activated only by high light intensities. Among the many processes affected by light are seed germination, stem elongation, leaf expansion, conversion from a vegetative state to a flowering state, flower development, fruit development, cessation of leaf production (bud dormancy), and leaf senescence and abscission.

Without doubt artificial lighting affects plants. Not so clear, however, is whether artificial lighting poses any short-term consequences to an individual plant or long-term consequences to any particular species in nature. Research has barely begun to elucidate the effects on plants themselves or on the ecological relation that characterize plants in nature. It is unknown what effects altered plant physiology might have on pollination mutualisms or herbivore defense or on plants as habitat for other species. The scanty research observations to date do provide a starting point, however, and the absence of evidence of effects, whatever their nature, should not construed as evidence of their absence. Research is badly needed in this neglected area of plant biology.

Literature Cited

Ahmad, M., and A. R. Cashmore. 1993. *HY4* gene of *A. thaliana* encodes a protein with characteristics of a blue-light photoreceptor. *Nature* 366:162–166.

Baskin, T. I., and M. Iino. 1987. An action spectrum in the blue and ultraviolet for phototropism in alfalfa. *Photochemistry and Photobiology* 46:127–136.

Batschauer, A. 2005. Plant cryptochromes: their genes, biochemistry, and physiological roles. Pages 211–246 in W. R. Briggs and J. L. Spudich (eds.), *Handbook of photosensory receptors*. Wiley-VCH Verlag, Weinheim, Germany.

Borthwick, H. A., S. B. Hendricks, and M. W. Parker. 1952a. The reaction controlling floral initiation. *Proceedings of the National Academy of Sciences of the United States of America* 38:929–934.

Borthwick, H. A., S. B. Hendricks, M. W. Parker, E. H. Toole, and V. K. Toole. 1952b. A reversible photoreaction controlling seed germination. *Proceedings of the National Academy of Sciences of the United States of America* 38:662–666.

Briggs, W. R. In press. Blue/UV-A receptors: historical overview. In E. Schäfer and F. Nagy (eds.), *Photomorphogenesis in plants and bacteria: function and signal transduction mechanisms*. Third edition. Springer, Dordrecht, The Netherlands.

Briggs, W. R., C. F. Beck, A. R. Cashmore, J. M. Christie, J. Hughes, J. A. Jarillo, T. Kagawa, H. Kanegae, E. Liscum, A. Nagatani, K. Okada, M. Salomon, W. Rüdiger, T. Sakai, M. Takano, M. Wada, and J. C. Watson. 2001a. The phototropin family of photoreceptors [Letter to the editor]. *Plant Cell* 13:993–997.

Briggs, W. R., and J. M. Christie. 2002. Phototropins 1 and 2: versatile plant blue-light receptors. *Trends in Plant Science* 7:204–210.

Briggs, W. R., J. M. Christie, and M. Salomon. 2001b. Phototropins: a new family of flavin-binding blue light receptors in plants. *Antioxidants & Redox Signaling* 3:775–788.

Briggs, W. R., and E. Huala. 1999. Blue-light photoreceptors in higher plants. *Annual Review of Cell Developmental Biology* 15:33–62.

Briggs, W. R., and M. Iino. 1983. Blue-light-absorbing photoreceptors in plants. *Philosophical Transactions of the Royal Society of London. Series B, Biological Sciences* 303:347–359.

Brudler, R., K. Hitomi, H. Daiyasu, H. Toh, K. Kucho, M. Ishiura, M. Kanehisa, V. A. Roberts, T. Todo, J. A. Tainer, and E. D. Getzoff. 2003. Identification of a new cryptochrome class: structure, function, and evolution. *Molecular Cell* 11:59–67.

Butler, W. L., K. H. Norris, H. W. Siegelman, and S. B. Hendricks. 1959. Detection, assay, and preliminary purification of the pigment controlling photoresponsive development of plants. *Proceedings of the National Academy of Sciences of the United States of America* 45:1703–1708.

Cashmore, A. R. 2005. Plant cryptochromes and signaling. Pages 247–258 in W. R. Briggs and J. L. Spudich (eds.), *Handbook of photosensory receptors*. Wiley-VCH Verlag, Weinheim, Germany.

Cathey, H. M., and L. E. Campbell. 1975a. Effectiveness of five vision-lighting sources on photo-regulation of 22 species of ornamental plants. *Journal of the American Society for Horticultural Science* 100:65–71.

Cathey, H. M., and L. E. Campbell. 1975b. Security lighting and its impact on the landscape. *Journal of Arboriculture* 1:181–187.

Christie, J. M., and W. R. Briggs. 2005. Blue light sensing and signaling by the phototropins. Pages 277–303 in W. R. Briggs and J. L. Spudich (eds.), *Handbook of photosensory receptors*. Wiley-VCH Verlag, Weinheim, Germany.

Christie, J. M., P. Reymond, G. K. Powell, P. Bernasconi, A. A. Raibekas, E. Liscum, and W. R. Briggs. 1998. *Arabidopsis* NPH1: a flavoprotein with the properties of a photoreceptor for phototropism. *Science* 282:1698–1701.

Christie, J. M., M. Salomon, K. Nozue, M. Wada, and W. R. Briggs. 1999. LOV (light, oxygen, or voltage) domains of the blue-light photoreceptor phototropin (nph1): binding sites for the chromophore flavin mononucleotide. *Proceedings of the National Academy of Sciences of the United States of America* 96:8779–8783.

Clack, T., S. Mathews, and R. A. Sharrock. 1994. The phytochrome apoprotein family in *Arabidopsis* is encoded by five genes: the sequences and expression of *PHYD* and *PHYE*. *Plant Molecular Biology* 25:413–427.

Downs, R. J. 1956. Photoreversibility of flower initiation. *Plant Physiology* 31: 279–284.

Downs, R. J., and H. A. Borthwick. 1956a. Effect of photoperiod upon the vegetative growth of *Weigela florida* var. *variegata*. *Proceedings of the American Society for Horticultural Science* 68:518–521.

Downs, R. J., and H. A. Borthwick. 1956b. Effects of photoperiod on growth of trees. *Botanical Gazette* 117:310–326.

Downs, R. J., and A. A. Piringer Jr. 1958. Effects of photoperiod and kind of supplemental light on vegetative growth of pines. *Forest Science* 4:185–195.

Flint, L. H. 1934. Light in relation to dormancy and germination in lettuce seed. *Science* 80:38–40.

Flint, L. H., and E. D. McAlister. 1935. Wave lengths of radiation in the visible spectrum inhibiting the germination of light-sensitive lettuce seed. *Smithsonian Miscellaneous Collections* 94(5):1–11.

Flint, L. H., and E. D. McAlister. 1937. Wave lengths of radiation in the visible spectrum promoting the germination of light-sensitive lettuce seed. *Smithsonian Miscellaneous Collections* 96(2):1–8.

Folta, K. M., and E. P. Spalding. 2001. Unexpected roles for cryptochrome 2 and phototropin revealed by high-resolution analysis of blue light-mediated hypocotyl growth inhibition. *Plant Journal* 26:471–478.

Garner, W. W., and H. A. Allard. 1920. Effect of the relative length of day and night and other factors of the environment on growth and reproduction in plants. *Journal of Agricultural Research* 18:553–606.

Garner, W. W., and H. A. Allard. 1923. Further studies in photoperiodism, the response of the plant to relative length of day and night. *Journal of Agricultural Research* 23:871–920.

Gressel, J. 1979. Blue light photoreception. *Photochemistry and Photobiology* 30:749–754.

Hamner, K. C., and J. Bonner. 1938. Photoperiodism in relation to hormones as factors in floral initiation and development. *Botanical Gazette* 100:388–431.

Hoffman, P. D., A. Batschauer, and J. B. Hays. 1996. *PHH1*, a novel gene from *Arabidopsis thaliana* that encodes a protein similar to plant blue-light photoreceptors and microbial photolyases. *Molecular and General Genetics* 253:259–265.

Huala, E., P. W. Oeller, E. Liscum, I.-S. Han, E. Larsen, and W. R. Briggs. 1997. *Arabidopsis* NPH1: a protein kinase with a putative redox-sensing domain. *Science* 278:2120–2123.

Huq, E., and P. H. Quail. 2005. Phytochrome [PQ1] signaling. Pages 151–170 in W. R. Briggs and J. L. Spudich (eds.), *Handbook of photosensory receptors*. Wiley-VCH Verlag, Weinheim, Germany.

Imaizumi, T., H. G. Tran, T. E. Swartz, W. R. Briggs, and S. A. Kay. 2003. FKF1 is essential for photoperiodic-specific light signalling in *Arabidopsis*. *Nature* 426:302–306.

Jarillo, J. A., M. Ahmad, and A. R. Cashmore. 1998. NPL1 (Accession No.

AF053941): a second member of the NPH serine/threonine kinase family of *Arabidopsis* (PGR 98-100). *Plant Physiology* 117:719.

Jarillo, J. A., H. Gabrys, J. Capel, J. M. Alonso, J. R. Ecker, and A. R. Cashmore. 2001. Phototropin-related NPL1 controls chloroplast relocation induced by blue light. *Nature* 410:952–954.

Kagawa, T., T. Sakai, N. Suetsugu, K. Oikawa, S. Ishiguro, T. Kato, S. Tabata, K. Okada, and M. Wada. 2001. *Arabidopsis* NPL1: a phototropin homolog controlling the chloroplast high-light avoidance response. *Science* 291:2138–2141.

Khurana, J. P., and K. L. Poff. 1989. Mutants of *Arabidopsis thaliana* with altered phototropism. *Planta* 178:400–406.

Kinoshita, T., M. Doi, N. Suetsugu, T. Kagawa, M. Wada, and K. Shimazaki. 2001. Phot1 and phot2 mediate blue light regulation of stomatal opening. *Nature* 414:656–660.

Kleine, T., P. Lockhart, and A. Batschauer. 2003. An *Arabidopsis* protein closely related to *Synechocystis* cryptochrome is targeted to organelles. *Plant Journal* 35:93–103.

Koller, D. 2000. Plants in search of sunlight. *Advances in Botanical Research* 33:35–131.

Koornneef, M., E. Rolff, and C. J. P. Spruit. 1980. Genetic control of light-inhibited hypocotyl elongation in *Arabidopsis thaliana* (L.) Heynh. *Zeitschrift für Pflanzenphysiologie* 100:147–160.

Lascève, G., J. Leymarie, M. A. Olney, E. Liscum, J. M. Christie, A. Vavasseur, and W. R. Briggs. 1999. *Arabidopsis* contains at least four independent blue-light-activated signal transduction pathways. *Plant Physiology* 120:605–614.

Lin, C., M. Ahmad, J. Chan, and A. R. Cashmore. 1996. CRY2: a second member of the *Arabidopsis* cryptochrome gene family (Accession No. U43397) (PGR 96-001). *Plant Physiology* 110:1047.

Lin, C., D. E. Robertson, M. Ahmad, A. A. Raibekas, M. S. Jorns, P. L. Dutton, and A. R. Cashmore. 1995. Association of flavin adenine dinucleotide with the *Arabidopsis* blue light receptor CRY1. *Science* 269:968–970.

Lin, C., H. Yang, H. Guo, T. Mockler, J. Chen, and A. R. Cashmore. 1998. Enhancement of blue-light sensitivity of *Arabidopsis* seedlings by a blue light receptor cryptochrome 2. *Proceedings of the National Academy of Sciences of the United States of America* 95:2686–2690.

Liscum, E., and W. R. Briggs. 1995. Mutations in the *NPH1* locus of *Arabidopsis* disrupt the perception of phototropic stimuli. *Plant Cell* 7:473–485.

Liscum, E., and W. R. Briggs. 1996. Mutations of *Arabidopsis* in potential transduction and response components of the phototropic signaling pathway. *Plant Physiology* 112:291–296.

Malhotra, K., S.-T. Kim, A. Batschauer, L. Dawut, and A. Sancar. 1995. Putative blue-light photoreceptors from *Arabidopsis thaliana* and *Sinapis alba* with a high degree of sequence homology to DNA photolyase contain the two photolyase cofactors but lack DNA repair activity. *Biochemistry* 34:6892–6899.

Neff, M. M., C. Fankhauser, and J. Chory. 2000. Light: an indicator of time and place. *Genes & Development* 14:257–271.

Olmsted, C. E. 1951. Experiments on photoperiodism, dormancy, and leaf age and abscission in sugar maple. *Botanical Gazette* 112:365–393.

Poggioli, S. 1817. Della influenza che ha il raggio magnetico sulla vegetazione delle piante [Of the influence of magnetic beams on the vegetation of plants]. *Opuscoli Scientifici* 1:9–23.

Reymond, P., T. W. Short, W. R. Briggs, and K. L. Poff. 1992. Light-induced phosphorylation of a membrane protein plays an early role in signal transduction for phototropism in *Arabidopsis thaliana*. *Proceedings of the National Academy of Sciences of the United States of America* 89:4718–4721.

Sachs, J. 1864. Wirkungen farbigen Lichts auf Pflanzen [Effects of colored light on plants]. *Botanische Zeitung* 47:353–358.

Sachs, J. 1887. *Lectures on the physiology of plants*. Clarendon Press, Oxford.

Sage, L. C. 1992. *Pigment of the imagination: a history of phytochrome research*. Academic Press, San Diego.

Sakai, T., T. Kagawa, M. Kasahara, T. E. Swartz, J. M. Christie, W. R. Briggs, M. Wada, and K. Okada. 2001. *Arabidopsis* nph1 and npl1: blue light receptors that mediate both phototropism and chloroplast relocation. *Proceedings of the National Academy of Sciences of the United States of America* 98:6969–6974.

Sakamoto, K., and W. R. Briggs. 2002. Cellular and subcellular localization of phototropin 1. *Plant Cell* 14:1723–1735.

Salomon, M., J. M. Christie, E. Knieb, U. Lempert, and W. R. Briggs. 2000. Photochemical and mutational analysis of the FMN-binding domains of the plant blue light receptor, phototropin. *Biochemistry* 39:9401–9410.

Senger, H., and W. R. Briggs. 1981. The blue light receptor(s): primary reactions and subsequent metabolic changes. Pages 1–38 in K. C. Smith (ed.), *Photochemical and photobiological reviews*. Volume 6. Plenum Press, New York.

Senn, G. 1908. *Die Gestalts- und Lageveränderung der Pflanzen-Chromatophoren* [The change of position and shape of plant chromatophores]. Wilhelm Engelmann, Leipzig.

Sharrock, R. A., and P. H. Quail. 1989. Novel phytochrome sequences in *Arabidopsis thaliana*: structure, evolution, and differential expression of a plant regulatory photoreceptor family. *Genes & Development* 3:1745–1757.

Shinomura, T., A. Nagatani, H. Hanzawa, M. Kubota, M. Watanabe, and M. Furuya. 1996. Action spectra for phytochrome A- and B-specific photoinduction of seed germination in *Arabidopsis thaliana*. *Proceedings of the National Academy of Sciences of the United States of America* 93:8129–8133.

Short, T. W., and W. R. Briggs. 1994. The transduction of blue light signals in higher plants. *Annual Review of Plant Physiology and Plant Molecular Biology* 45:143–171.

Smith, H. 1995. Physiological and ecological function within the phytochrome family. *Annual Review of Plant Physiology and Plant Molecular Biology* 46:289–315.

Smith, H. 2000. Phytochromes and light signal perception by plants—an emerging synthesis. *Nature* 407:585–591.

Tu, S.-L., and J. C. Lagarias. 2005. The phytochromes. Pages 121–149 in W. R.

Briggs and J. L. Spudich (eds.), *Handbook of photosensory receptors*. Wiley-VCH Verlag, Weinheim, Germany.

Van Volkenburgh, E., and R. E. Cleland. 1990. Light-stimulated cell expansion in bean (*Phaseolus vulgaris* L.) leaves. I. Growth can occur without photosynthesis. *Planta* 182:72–76.

Van Volkenburgh, E., R. E. Cleland, and M. Watanabe. 1990. Light-stimulated cell expansion in bean (*Phaseolus vulgaris* L.) leaves. II. Quantity and quality of light required. *Planta* 182:77–80.

Vince-Prue, D. 1994. The duration of light and photoperiodic responses. Pages 447–490 in R. E. Kendrick and G. H. M. Kronenberg (eds.), *Photomorphogenesis in plants*. Second edition. Kluwer Academic Publishers, Dordrecht, The Netherlands.

Chapter 17

Synthesis

Travis Longcore and Catherine Rich

As diurnal creatures, humans have long sought methods to illuminate periods of darkness. In preindustrial times, artificial light was generated by burning various materials, including wood, oil, and even dried fish. Although these methods of lighting certainly influenced animal behavior and local ecology, such effects were limited in scale. The relatively recent invention of electric lights, and their rapid proliferation, however, have resulted in a wholesale transformation of the nighttime environment over significant portions of the Earth's surface.

Each of the contributed chapters in this book describes the effects of artificial night lighting on a specific taxonomic group. Similar mechanisms are found in each of these chapters, and the collection of these reviews allows for an examination of processes. In this final chapter we consider these processes within the nested hierarchy of physiological ecology, behavioral and population ecology, community ecology, and ecosystem ecology. The boundaries within this hierarchy are not always distinct, but it provides an organizing framework to discuss effects from the individual to ecosystem scale.

Physiological Ecology

Alterations in natural patterns of light and dark can disrupt physiological processes. Most evidence of such disruption is found in studies of animals subjected to artificial lighting in laboratory conditions; only a few physiological studies (e.g., Rees 1982) involve exposure of animals to artificial night lighting in the wild. Physiological effects of lighting on vertebrates may result from disruption of signals associated with daylength, thereby changing the daily, monthly, or yearly timing of certain physiological processes or inhibiting regular physiological processes that normally occur in darkness (see Chapters 2, 6, 9, and 10, this volume).

Some physiological changes, including entering reproductive condition and preparing to migrate or hibernate, are associated with light cues. Egg laying and molt in birds are closely associated with photoperiod, as has been demonstrated extensively for poultry and also for wild birds in laboratory conditions (see Chapter 6, this volume). In a study of wild birds in the field, Rees (1982) showed that birds foraging under artificial lights laid down fat more quickly than normal and underwent physiological changes preparing them early for migration.

Exposure to illumination, even quite dim light, during normal scotophase can disrupt the production of hormones such as melatonin and prolactin. This phenomenon has been documented in humans, where depressed nighttime hormone production is implicated in accelerated tumor growth (Stevens and Rea 2001). Melatonin plays a significant role in mediating seasonal changes in physiology and behavior. Humans exposed to the constantly increased illumination of urban life exhibit altered seasonal variations in melatonin production (Wehr 1997, Stevens and Rea 2001). Chronic exposure to artificial illumination may also contribute to increased breast cancer incidence in shift workers as a result of this disruption of hormonal cycling (Hansen 2001). Similar studies have not been conducted on nonhuman animals, but corresponding changes in melatonin production have been documented for other vertebrates, including mammals (Foster and Provencio 1999), birds (Kliger et al. 2000), amphibians (see Chapters 9 and 10, this volume), and fishes (showing a lunar cycle in melatonin production; Rahman et al. 2004). Melatonin is also produced in invertebrates, including arthropods, crustaceans, mollusks, planarians, dinoflagellates, and anthozoans (see Mechawar and Anctil 1997). It is not yet clear, however, whether disruption of hormone production results in significant effects on individuals or populations in the wild. Such effects are possible; hormone production affects critical behaviors such as the timing of migration and reproduction. Small changes in the timing of these events resulting from a light signal that is essentially "wrong" may decrease fitness.

Behavioral and Population Ecology

Artificial night lighting produces demonstrable effects on the behavioral and population ecology of organisms in natural settings. As a whole, these effects derive from changes in orientation (including disorientation and misorientation) and attraction or repulsion caused by lights, which in turn may affect foraging, reproduction, migration, and communication.

Orientation and Attraction/Repulsion

Orientation and disorientation or misorientation are responses to ambient illumination (i.e., the amount of light incident on objects in an environment). In contrast, attraction and repulsion occur in reaction to the light sources themselves and therefore are responses to luminance or the brightness of the source of light (Health Council of the Netherlands 2000).

Increased illumination may extend diurnal or crepuscular behaviors into the night by improving an animal's ability to orient itself. For example, many normally diurnal birds (e.g., Hill 1992) and reptiles (Schwartz and Henderson 1991) forage under artificial lights. This has been termed the "night-light niche" for reptiles and seems beneficial for those species that can exploit it, but not for their prey (Schwartz and Henderson 1991; see Chapter 8, this volume).

In addition to foraging, orientation under artificial illumination may induce other behaviors, such as territorial singing in birds (Bergen and Abs 1997) or territorial displays in salamanders (see Chapter 10, this volume). For example, male northern mockingbirds (*Mimus polyglottos*) sing at night before mating, but only sing at night after mating in artificially lighted areas (Derrickson 1988) or during the full moon. The effect of these light-induced behaviors on fitness is unknown.

Constant artificial night lighting may also disorient organisms accustomed to navigating in a dark environment. The best-known example of this is the disorientation of hatchling sea turtles emerging from nests on sandy beaches (see Chapter 7, this volume). Under normal circumstances, hatchlings move away from low, dark silhouettes (historically dune vegetation), allowing them to crawl quickly to the ocean. With beachfront lighting, however, the silhouettes that would have cued movement are no longer perceived, resulting in disorientation (Salmon et al. 1995; see Chapter 7, this volume). Lighting also affects the egg-laying behavior of female sea turtles (see Salmon 2003, Witherington 1997; Chapter 7, this volume).

Changes in lighting level may disrupt orientation in nocturnal animals. The range of anatomical adaptations to allow night vision is broad (Park 1940), and rapid increases in light can temporarily blind animals. For frogs, rapid increase in illumination causes a reduction in visual capability from which the recovery time may be minutes to hours (Buchanan 1993; see Chapter 9, this volume). Once adapted to a light, frogs may be attracted to it as well (Figure 17.1; Jaeger and Hailman 1973; see Chapter 9, this volume).

Figure 17.1 Attraction of frogs to a candle set out on a small raft. Illustration by Charles Copeland of an experiment in northern Maine or Canada described by Long (1901). Twelve or fifteen bullfrogs (*Rana catesbeiana*) climbed onto the small raft before it flipped over.

In addition to disorientation, birds may be entrapped by lights (Chapters 4 and 5, this volume; Evans Ogden 1996). Once a bird is within a lighted zone at night, it will not leave the lighted area. Large numbers of nocturnally migrating birds are therefore imperiled when meteorological conditions bring them close to lights, such as during inclement weather. Within the sphere of lights, birds may collide with each other or with a structure, become exhausted, or be taken by predators. Furthermore, the birds that are waylaid by lights in urban areas at night often die in collisions with windows as they try to escape during the day. Artificial lighting has "attracted" birds to smokestacks, lighthouses (Squires and Hanson 1918), broadcast towers (Evans Ogden 1996), boats (Dick and Davidson 1978), greenhouses (Abt and Schultz 1995), oil platforms (Wiese et al. 2001), and other structures, resulting in direct mortality during migration.

Many groups of insects, of which moths are one well-known example (Frank 1988; see Chapter 13, this volume), are attracted to lights. Other taxa attracted to lights include lacewings, beetles, bugs, caddisflies, craneflies, midges, hoverflies, wasps, and bush crickets (Figure 17.2; see Chapter 12, this volume; Eisenbeis and Hassel 2000, Kolligs 2000). Attraction

Figure 17.2. Thousands of mayflies carpet the ground around a security light at Millecoquins Point in Naubinway on the Upper Peninsula of Michigan. Courtesy of P. J. DeVries.

depends on the spectrum of light—insect collectors use ultraviolet light because of its attractive properties—and the characteristics of other lights in the vicinity (see Chapters 12 and 13, this volume).

Nonflying arthropods vary in their reaction to lights. Some nocturnal spiders are negatively phototactic, whereas others will exploit light if available (Nakamura and Yamashita 1997). Some insects are always positively phototactic as an adaptive behavior and others always photonegative (Summers 1997). In arthropods these responses may also be influenced by the frequent correlations between light, humidity, and temperature.

Reproduction

Reproductive behaviors may be altered by artificial night lighting. For example, female *Physalaemus pustulosus* frogs are less selective about mate choice when lighting is increased, presumably preferring to mate quickly and avoid the increased predation risk of mating activity (Rand et al. 1997). Lighting also may inhibit amphibian movement to and from breeding areas by stimulating phototactic behavior. Buchanan (Chapter 9, this volume) reports that frogs in an experimental enclosure stopped mating activity during night football games; lights from a nearby stadium increased sky glow. Mating choruses resumed only when the enclosure was covered to shield the frogs from the light. De Molenaar et al. (2000; see Chapter 6, this volume) reported evidence that artificial lighting affects the choice of nest site in black-tailed godwit (*Limosa l. limosa*) in wet grassland habitats.

Highly synchronized reproductive events may be vulnerable to disruption by artificial lighting. Moser et al. (2004) described the predawn mating flights of the ant *Atta texana*. In the absence of artificial lighting, flights of different colonies are always synchronized approximately 15 minutes before dawn. The researchers observed that in two areas with visible nighttime lighting the flight was delayed until after dawn and suggested that this may decrease reproductive success. Many taxonomic groups exhibit reproductive behavior synchronized to lunar cycles, including fish species (Taylor 1984, Walker 1949), eels (Tsukamoto et al. 2003), marine polychaetes (Bentley et al. 2001), and mayflies (Hora 1927, Hartland-Rowe 1955). Because reproduction is often timed to the darkest part of the month at the new moon to reduce predation, artificial lighting in the form of direct glare or sky glow can either desynchronize mating or increase predation during the event. Such reduction of reproductive success threatens survival of species at risk. For example, artificial lighting from nearby urban development is thought to disrupt the mating and oviposition activity of the endangered Salt Creek tiger beetle (*Cicindela nevadica lincolniana*; Allgeier and Higley 2004). Laboratory research demonstrated that the tiger beetles are attracted to lights producing illumination dimmer than that experienced in their habitat in the wild, suggesting that light pollution may already be adversely affecting the species (Allgeier and Higley 2004).

Artificial light may work synergistically with other factors to affect reproduction. Adult mayflies are positively phototactic and swarm around lights adjacent to their aquatic habitats. Under normal conditions female mayflies are attracted to the polarized light reflected off water bodies, ensuring that their eggs are deposited in the water. But asphalt roads also reflect polarized light and under many circumstances may be even more attractive than real streams (Kriska et al. 1998). Future research should disentangle the effect of artificial lighting itself and the effect of asphalt reflecting polarized light from both natural and artificial sources. The co-occurrence of asphalt roads and parking lots with artificial lights is a lethal combination for mayflies.

Just as the behaviors of animals that prefer to engage in reproductive activity in the dark of night may be disrupted, bright lights may stimulate reproductive activity in diurnal breeders. California market squid (*Loligo opalescens*) were long thought to spawn predominantly at night because of the frequent observation of this behavior by observers on lighted fishing and research vessels (Forsythe et al. 2004). Documentation from an unlighted, remotely operated vehicle has now shown that squid spawning normally is a diurnal activity but is induced by bright lights at night (Forsythe et al. 2004). The shift of this behavior from day to night has the obvious consequence of high mortality from the fishery usually associated with the lights but may also affect sexual selection and other aspects of reproductive behavior.

Communication

Intraspecific and interspecific visual communication may be influenced by artificial night lighting. Some species use light to communicate, and their communication therefore is especially susceptible to disruption. Female glowworms attract males up to 45 m (148 ft) away with bioluminescent flashes; the presence of artificial lighting reduces the visibility of these communications. Lloyd (Chapter 14, this volume) describes how the complex visual communication system of fireflies could be impaired by stray light.

Artificial lighting could also alter communication patterns as a secondary effect. Coyotes (*Canis latrans*) group howl and group-yip howl more during the new moon when it is darkest. Communication is necessary to reduce trespassing from other packs or to assemble packs to hunt larger prey during dark conditions (Bender et al. 1996). Sky glow could increase ambient illumination to eliminate this pattern in affected areas.

Because of the central role of vision in orientation of most animals, it is not surprising that artificial lighting alters behavior. This may have direct negative consequences for some species, whereas for other species the influence may seem to be positive. Such "positive" effects, however, may have negative consequences in the context of community ecology.

Community Ecology

The behaviors exhibited by individual animals in response to ambient illumination (orientation, disorientation, misorientation) and to luminance (attraction, repulsion) influence community interactions, of which competition and predation are examples.

Competition

Artificial lighting could disrupt the interactions of groups of species that show resource partitioning across illumination gradients. For example, in natural communities some foraging times are partitioned between species with different lighting level preferences. The squirrel treefrog (*Hyla squirrela*) is able to orient and forage at lighting levels as low as 10^{-5} lux and under natural conditions typically stops foraging at illuminations above 10^{-3} lux (Buchanan 1998). The western toad (*Bufo boreas*) forages only at illuminations between 10^{-1} and 10^{-5} lux, whereas the tailed frog (*Ascaphus truei*) forages only during the darkest part of the night, below 10^{-5} lux (Hailman 1984). Although these three species are not necessarily sympatric and differ in other niche dimensions, their behavior illustrates a division of the light gradient by foragers.

Many bat species are attracted to the insects that congregate around light sources (see Chapter 3, this volume). Although it may seem that this is a "positive" effect, the increased food concentration benefits only those species that exploit light sources and therefore results in altered community structure (see Chapter 3, this volume). Faster-flying species of bats congregate at lights to feed on insects, but slower-flying species avoid lights (Blake et al. 1994, Rydell and Baagøe 1996; see Chapter 3, this vol-

Figure 17.3. Crowned hornbill (*Tockus alboterminatus*) hawking insects at a light at the Kibale Forest National Park, Uganda. Courtesy of P. J. DeVries.

ume). Artificial lighting in an area therefore would provide an advantage to faster-flying species.

Changes in competitive communities occur as diurnal species move into the "night-light niche" (Schwartz and Henderson 1991; see Chapter 8, this volume). This concept as originally described applies to reptiles but easily extends to other taxa, such as spiders (see Chapter 13, this volume) and birds (Figure 17.3; Goertz et al. 1980, Sick and Teixeira 1981, Hill 1992, Frey 1993).

Predation

Although it may seem a benefit for diurnal species to be able to forage longer under artificial lights, any gains from increased activity time can be offset by increased predation risk (Gotthard 2000). The balance between gains from extended foraging time and risk of increased predation is a central topic for research on small mammals, reptiles, and birds (Kotler 1984, Lima 1998; see Chapter 2, this volume). Small rodents forage less at high illumination levels (Lima 1998), a tendency also exhibited by some lagomorphs (Gilbert and Boutin 1991), marsupials (Julien-Laferrière 1997), snakes (Klauber 1939; see Chapter 8, this volume), bats (Rydell 1992; see Chapter 3, this volume), fishes (Gibson 1978; see Chapter 11, this volume), aquatic invertebrates (Moore et al. 2001; see Chapter 15, this volume), and certainly other taxa.

Unexpected changes in light conditions can disrupt predator–prey relationships. Gliwicz (1986, 1999) described high predation by fish on zooplankton during nights when the full moon rose hours after sunset. Zooplankton had migrated to the surface to forage under cover of darkness, only to be illuminated by the rising moon and subjected to intense predation. This "lunar light trap" (Gliwicz 1986) illustrates a natural occurrence, but unexpected illumination from human sources similarly would disrupt predator–prey interactions, often to the benefit of the predator.

Constant artificial night lighting disrupts predator–prey relationships, which is consistent with the documented importance of natural light regimes in mediating such interactions. Harbor seals congregated under artificial lights to eat juvenile salmonids migrating downstream; turning the lights off reduced predation levels (Yurk and Trites 2000). Nighttime illumination at urban crow roosts was higher than at control sites; crows presumably congregate at such sites to reduce predation from owls (Gorenzel and Salmon 1995). Desert rodents reduced foraging activity when exposed to the light of a single camp lantern (Kotler 1984). Mercury vapor lights in particular disrupt the interaction between bats and tympanate moths by interfering with moth detection of ultrasonic chirps used by bats in echolocation, leaving moths unable to take their normal evasive action (Svensson and Rydell 1998, Frank 1988; see Chapter 13, this volume).

We conclude from these examples and the foregoing chapters that community structure will be altered where artificial night lighting affects interspecific interactions. A "perpetual full moon" from artificial lights will favor light-tolerant species and exclude others. If the darkest natural conditions never occur, species that maximize foraging during the new moon eventually could be compromised, at risk of failing to meet monthly energy budgets. The resulting community structure will be simplified; these changes in turn could affect ecosystem characteristics.

Ecosystem Ecology

The cumulative effects of behavioral changes induced by artificial light on competition and predation have the potential to disrupt key ecosystem functions. The spillover effects from artificial night lighting on aquatic invertebrates illustrates this point. Diel vertical migration of zooplankton presumably results from a need to avoid predation under lighted conditions, so many zooplankton forage near water surfaces only under dark conditions (Gliwicz 1986, Lima 1998; see Chapter 15, this volume). Light dimmer than that of a half moon (less than 10^{-1} lux) is sufficient to influence the vertical distribution of some aquatic invertebrates, and indeed patterns of diel vertical migration change with the lunar cycle (Dodson 1990).

Moore et al. (2001; see Chapter 15, this volume) documented the effect of artificial light on the diel migration of the zooplankton *Daphnia* in the wild. Artificial illumination decreased the magnitude of diel migrations, in both the range of vertical movement and the number of individuals migrating. Moore et al. hypothesized that this disruption of diel vertical migration by artificial lighting may detrimentally affect ecosystem health. With fewer zooplankton migrating to the surface to graze, algae populations may increase. Such algal blooms would then have a series of adverse effects on water quality (Moore et al. 2001).

The reverberating effects of community changes caused by artificial lighting could influence other ecosystem functions. Outcomes are not yet predictable, but all indications are that light-influenced ecosystems will suffer from significant changes attributable to artificial light alone and to artificial light in combination with other disturbances. Even remote areas may experience increased illumination from sky glow, but the most noticeable effects will be found in areas where lights are close to natural habitats.

Conclusion

The introduction to this book opens with a question, "What if we woke up one morning only to realize that all of the conservation planning of the last thirty years told only half the story—the daytime story?" The many examples and mechanisms discussed in these chapters provide ample evidence that consideration of the nighttime should be an integral part of conservation planning. To do this requires incorporating the scientific knowledge now available and continued research into the influence of artificial night lighting on natural communities.

The effects of artificial night lighting on species, habitats, and ecosystems will be mitigated in the development process only if the legal framework to regulate environmental impacts more generally provides a mechanism to do so. Most environmental disclosure laws, such as the National Environmental Policy Act, encourage the use of checklists of potential adverse environmental effects as a screening tool to identify projects warranting further evaluation. Such checklists usually consider night lighting from an aesthetic standpoint only. For more thorough environmental analysis, it would be appropriate to consider artificial night lighting also in an ecological context. A project site itself may not have biological resources, but the lighting from the project might have significant adverse consequences for species adjacent to the site (direct glare) or some distance away (sky glow).

The mechanisms described in this book by which artificial night lighting influences species and habitats provide guidance for the types of impacts that should be addressed in legally mandated environmental review or during voluntary planning. For species that are susceptible to attraction or misorientation by lights, shielding or filtering lights may be appropriate, but mitigation must be guided by an understanding of the particular species that might be affected. For example, yellow light attracts the fewest insects (see Chapters 12 and 13, this volume) and juvenile turtles (see Chapter 7, this volume) but disorients some salamanders (see Chapter 10, this volume) and reduces foraging of small mammals (see Chapter 2, this volume; Bird et al. 2004). Conservation planners should pay special attention to aquatic habitats because of the central role of light in structuring aquatic communities and the demonstrated sensitivity of aquatic organisms to artificial light (see Chapters 5, 9–12, and 15, this volume). Open habitats, such as grasslands, scrublands, dunes, and deserts, are likewise vulnerable (see Chapters 2 and 6, this volume). The influence of artificial lighting on animal movement pathways, whether they be terrestrial (see Chapter 2, this volume), aquatic (see Chapter 11, this volume), or aerial (see Chapters 4, 5, 12, and 13, this volume), deserves special attention in conservation planning.

Much remains to be learned about the ecology of organisms under natural nighttime lighting conditions. There is some urgency because fewer and fewer reference sites are available. Cinzano et al. (2001) calculated that although 18.7% of Earth's surface is subjected to artificial night sky brightness 10% greater than natural night sky brightness, that number is 61.8% for the United States and 85.3% for the European Union. For many regionally limited natural communities it is already impossible to find sites with naturally dark conditions. Potential control sites for research on the effects of artificial night lighting are rapidly being enveloped in what Cinzano et al. (2001) aptly term "a luminous fog."

Artificial night lighting should be considered by all ecological field researchers because of its effects on animal behavior, which in turn may influence the number of individuals counted by any given method. For example, fish may move considerably within water bodies in response to ambient lighting levels. The effect of light on activity is so profound that hydroacoustic estimates of abundance of one species of small fish were eight times higher under a new moon than under a full moon (Gaudreau and Boisclair 2000). Acoustic monitoring of bird migration must also consider the influence of artificial lighting. Graber (1968) and Graber and Cochran (1960) noted that the calling rate of birds increases around lighted structures and cautioned against using calls as indicators of abundance without accounting for the increased calling rates in response to disorientation by lights. Researchers must be sure to account for differences in ambient nighttime illumination at different locations, especially during the new moon, when illumination from artificial lighting is more likely to be greater than moonlight.

Progress in further understanding the ecological consequences of artificial night lighting will be achieved through a combination of experimental and observational studies. This will require that ecologists collaborate with physical scientists and engineers to improve equipment to measure light characteristics at ecologically relevant levels under diverse field conditions. Illumination during the darkest overcast night is about 10^{-5} lux, and a starry sky before moonrise is about 10^{-4} lux. Instruments must detect this lower range of natural illumination or risk missing crucial differences between natural and artificial conditions. Such measurements should be included routinely as part of environmental monitoring protocols so that researchers have the data necessary to disentangle the confounding and cumulative effects of other facets of human disturbance with which artificial night lighting often is correlated, such as roads, urban development, noise, exotic species, animal harvest, and resource extraction. Progress will also depend on well-designed experiments on natural populations, preferably in the powerful before–after–control–impact pair design (see Chapters 2 and 6, this volume).

Widespread, permanent night lighting has been introduced in an extremely short period of human history, one that corresponds with the exploitation of fossil fuels and the rise of industrial society. In this short time, night itself has been transformed across much of the Earth, with consequent lethal and sublethal effects on species in many habitats and taxonomic groups. Essentially, artificial night lighting is homogenizing the range of physical conditions present in natural ecosystems. For an increasing proportion of the Earth's surface, the darkest conditions of night no longer occur.

As we consider the effects of human activity on the rest of the species inhabiting this planet, we are challenged to overcome the bias of our own senses and to interpret the world as perceived not by us, but by other organisms. It is all too easy to ignore those conditions that we cannot, or do not, see. For us, the difference between the darkest night and the light of a quarter moon is rarely observed and of little consequence. Polarized light and ultraviolet wavelengths are invisible altogether. Although we are accustomed to investigating unseen attributes in the microscopic realm, it takes more effort to remember that creatures who share our same perceptual scale can experience the world quite differently from the way we do. So let us be reminded, as we light the world to suit our needs and whims, that doing so may come at the expense of other living beings, some of whom detect subtle gradations of light to which we are blind, and for whom the night is home.

Acknowledgments

Portions of this chapter, including photographs by P. J. DeVries, appeared in *Frontiers in Ecology and the Environment* (Longcore and Rich 2004).

Literature Cited

Abt, K. F., and G. Schultz. 1995. Auswirkungen der Lichtemissionen einer Großgewächshausanlage auf den nächtlichen Vogelzug [Impact of light emissions from a large illuminated greenhouse on nocturnal bird migration]. *Corax* 16:17–29.

Allgeier, W. J., and L. G. Higley. 2004. Preference for light sources in two tiger beetle species, *Cicindela nevadica lincolniana* and *Cicindela togata globicollis*. Poster presented at Expanding the Ark: The Emerging Science and Practice of Invertebrate Conservation, March 25–26, American Museum of Natural History Center for Biodiversity and Conservation.

Bender, D. J., E. M. Bayne, and R. M. Brigham. 1996. Lunar condition influences coyote (*Canis latrans*) howling. *American Midland Naturalist* 136:413–417.

Bentley, M. G., P. J. W. Olive, and K. Last. 2001. Sexual satellites, moonlight and the nuptial dances of worms: the influence of the moon on the reproduction of marine animals. *Earth, Moon and Planets* 85/86:67–84.

Bergen, F., and M. Abs. 1997. Verhaltensökologische Studie zur Gesangsaktivität von Blaumeise (*Parus caeruleus*), Kohlmeise (*Parus major*) und Buchfink (*Fringilla coelebs*) in einer Großstadt [Etho-ecological study of the singing activity of the blue tit (*Parus caeruleus*), great tit (*Parus major*) and chaffinch (*Fringilla coelebs*) in a large city]. *Journal für Ornithologie* 138:451–467.

Bird, B. L., L. C. Branch, and D. L. Miller. 2004. Effects of coastal lighting on foraging behavior of beach mice. *Conservation Biology* 18:1435–1439.

Blake, D., A. M. Hutson, P. A. Racey, J. Rydell, and J. R. Speakman. 1994. Use of lamplit roads by foraging bats in southern England. *Journal of Zoology, London* 234:453–462.

Buchanan, B. W. 1993. Effects of enhanced lighting on the behaviour of nocturnal frogs. *Animal Behaviour* 45:893–899.

Buchanan, B. W. 1998. Low-illumination prey detection by squirrel treefrogs. *Journal of Herpetology* 32:270–274.

Cinzano, P., F. Falchi, and C. D. Elvidge. 2001. The first world atlas of the artificial night sky brightness. *Monthly Notices of the Royal Astronomical Society* 328:689–707.

Derrickson, K. C. 1988. Variation in repertoire presentation in northern mockingbirds. *Condor* 90:592–606.

Dick, M. H., and W. Davidson. 1978. Fishing vessel endangered by crested auklet landings. *Condor* 80:235–236.

Dodson, S. 1990. Predicting diel vertical migration of zooplankton. *Limnology and Oceanography* 35:1195–1200.

Eisenbeis, G., and F. Hassel. 2000. Zur Anziehung nachtaktiver Insekten durch Straßenlaternen: eine Studie kommunaler Beleuchtungseinrichtungen in der Agrarlandschaft Rheinhessens [Attraction of nocturnal insects to street lights: a study of municipal lighting systems in a rural area of Rheinhessen (Germany)]. *Natur und Landschaft* 75(4):145–156.

Evans Ogden, L. J. 1996. *Collision course: the hazards of lighted structures and windows to migrating birds*. World Wildlife Fund Canada and the Fatal Light Awareness Program, Toronto, Canada.

Forsythe, J., N. Kangas, and R. T. Hanlon. 2004. Does the California market squid (*Loligo opalescens*) spawn naturally during the day or at night? A note on the successful use of ROVs to obtain basic fisheries biology data. *Fishery Bulletin* 102:389–392.

Foster, R. G., and I. Provencio. 1999. The regulation of vertebrate biological clocks by light. Pages 223–243 in S. N. Archer, M. B. A. Djamgoz, E. R. Loew, J. C. Partridge, and S. Vallerga (eds.), *Adaptive mechanisms in the ecology of vision*. Kluwer Academic Publishers, Dordrecht, The Netherlands.

Frank, K. D. 1988. Impact of outdoor lighting on moths: an assessment. *Journal of the Lepidopterists' Society* 42:63–93.

Frey, J. K. 1993. Nocturnal foraging by scissor-tailed flycatchers under artificial light. *Western Birds* 24:200.

Gaudreau, N., and D. Boisclair. 2000. Influence of moon phase on acoustic estimates of the abundance of fish performing daily horizontal migration in a small oligotrophic lake. *Canadian Journal of Fisheries and Aquatic Sciences* 57:581–590.

Gibson, R. N. 1978. Lunar and tidal rhythms in fish. Pages 201–213 in J. E. Thorpe (ed.), *Rhythmic activity of fishes*. Academic Press, London.

Gilbert, B. S., and S. Boutin. 1991. Effect of moonlight on winter activity of snowshoe hares. *Arctic and Alpine Research* 23:61–65.

Gliwicz, Z. M. 1986. A lunar cycle in zooplankton. *Ecology* 67:883–897.

Gliwicz, Z. M. 1999. Predictability of seasonal and diel events in tropical and temperate lakes and reservoirs. Pages 99–124 in J. G. Tundisi and

M. Straškraba (eds.), *Theoretical reservoir ecology and its applications*. International Institute of Ecology, Brazilian Academy of Sciences and Backhuys Publishers, São Carlos.

Goertz, J. W., A. S. Morris, and S. M. Morris. 1980. Ruby-throated hummingbirds feed at night with the aid of artificial light. *Wilson Bulletin* 92:398–399.

Gorenzel, W. P., and T. P. Salmon. 1995. Characteristics of American crow urban roosts in California. *Journal of Wildlife Management* 59:638–645.

Gotthard, K. 2000. Increased risk of predation as a cost of high growth rate: an experimental test in a butterfly. *Journal of Animal Ecology* 69:896–902.

Graber, R. R. 1968. Nocturnal migration in Illinois: different points of view. *Wilson Bulletin* 80:36–71.

Graber, R. R., and W. W. Cochran. 1960. Evaluation of an aural record of nocturnal migration. *Wilson Bulletin* 72:253–273.

Hailman, J. P. 1984. Bimodal nocturnal activity of the western toad (*Bufo boreas*) in relation to ambient illumination. *Copeia* 1984:283–290.

Hansen, J. 2001. Light at night, shiftwork, and breast cancer risk. *Journal of the National Cancer Institute* 93:1513–1515.

Hartland-Rowe, R. 1955. Lunar rhythm in the emergence of an Ephemeropteran. *Nature* 176:657.

Health Council of the Netherlands. 2000. *Impact of outdoor lighting on man and nature*. Publication No. 2000/25E. Health Council of the Netherlands, The Hague.

Hill, D. 1992. *The impact of noise and artificial light on waterfowl behaviour: a review and synthesis of available literature*. British Trust for Ornithology Report No. 61, Norfolk, UK.

Hora, S. L. 1927. Lunar periodicity in the reproduction of insects. *Journal and Proceedings of the Asiatic Society of Bengal* (NS) 23:339–341.

Jaeger, R. G., and J. P. Hailman. 1973. Effects of intensity on the phototactic responses of adult anuran amphibians: a comparative survey. *Zeitschrift für Tierpsychologie* 33:352–407.

Julien-Laferrière, D. 1997. The influence of moonlight on activity of woolly opossums (*Caluromys philander*). *Journal of Mammalogy* 78:251–255.

Klauber, L. M. 1939. Studies of reptile life in the arid southwest. Part I. Night collecting on the desert with ecological statistics. *Bulletins of the Zoological Society of San Diego* 14:7–64.

Kliger, C. A., A. E. Gehad, R. M. Hulet, W. B. Roush, H. S. Lillehoj, and M. M. Mashaly. 2000. Effects of photoperiod and melatonin on lymphocyte activities in male broiler chickens. *Poultry Science* 78:18–25.

Kolligs, D. 2000. Ökologische Auswirkungen künstlicher Lichtquellen auf nachtaktive Insekten, insbesondere Schmetterlinge (Lepidoptera) [Ecological effects of artificial light sources on nocturnally active insects, in particular on moths (Lepidoptera)]. *Faunistisch–Ökologische Mitteilungen* Supplement 28:1–136.

Kotler, B. P. 1984. Risk of predation and the structure of desert rodent communities. *Ecology* 65:689–701.

Kriska, G., G. Horváth, and S. Andrikovics. 1998. Why do mayflies lay their eggs en masse on dry asphalt roads? Water-imitating polarized light reflected from asphalt attracts Ephemeroptera. *Journal of Experimental Biology* 201:2273–2286.

Lima, S. L. 1998. Stress and decision making under the risk of predation: recent developments from behavioral, reproductive, and ecological perspectives. *Advances in the Study of Behavior* 27:215–290.

Long, W. J. 1901. *Wilderness ways*. Ginn & Company, Boston.

Longcore, T., and C. Rich. 2004. Ecological light pollution. *Frontiers in Ecology and the Environment* 2:191–198.

Mechawar, N., and M. Anctil. 1997. Melatonin in a primitive metazoan: seasonal changes of levels and immunohistochemical visualization in neurons. *Journal of Comparative Neurology* 387:243–254.

Molenaar, J. G. de, D. A. Jonkers, and M. E. Sanders. 2000. *Road illumination and nature. III. Local influence of road lights on a black-tailed godwit (*Limosa l. limosa*) population*. DWW Ontsnipperingsreeks deel 38A, Delft.

Moore, M. V., S. M. Pierce, H. M. Walsh, S. K. Kvalvik, and J. D. Lim. 2001. Urban light pollution alters the diel vertical migration of *Daphnia*. *Verhandlungen der Internationalen Vereinigung für Theoretische und Angewandte Limnologie* 27:779–782.

Moser, J. C., J. D. Reeve, J. M. S. Bento, T. M. C. Della Lucia, R. S. Cameron, and N. M. Heck. 2004. Eye size and behaviour of day- and night-flying leafcutting ant alates. *Journal of Zoology, London* 264:69–75.

Nakamura, T., and S. Yamashita. 1997. Phototactic behavior of nocturnal and diurnal spiders: negative and positive phototaxes. *Zoological Science* 14:199–203.

Park, O. 1940. Nocturnalism: the development of a problem. *Ecological Monographs* 10:485–536.

Rahman, M. S., B.-H. Kim, A. Takemura, C.-B. Park, and Y.-D. Lee. 2004. Effects of moonlight exposure on plasma melatonin rhythms in the seagrass rabbitfish, *Siganus canaliculatus*. *Journal of Biological Rhythms* 19:325–334.

Rand, A. S., M. E. Bridarolli, L. Dries, and M. J. Ryan. 1997. Light levels influence female choice in Túngara frogs: predation risk assessment? *Copeia* 1997:447–450.

Rees, E. C. 1982. The effect of photoperiod on the timing of spring migration in the Bewick's swan. *Wildfowl* 33:119–132.

Rydell, J. 1992. Exploitation of insects around streetlamps by bats in Sweden. *Functional Ecology* 6:744–750.

Rydell, J., and H. J. Baagøe. 1996. Gatlampor ökar fladdermössens predation på fjärilar [Streetlamps increase bat predation on moths]. *Entomologisk Tidskrift* 117:129–135.

Salmon, M. 2003. Artificial night lighting and turtles. *Biologist* 50:163–168.

Salmon, M., M. G. Tolbert, D. P. Painter, M. Goff, and R. Reiners. 1995. Behavior of loggerhead sea turtles on an urban beach. II. Hatchling orientation. *Journal of Herpetology* 29:568–576.

Schwartz, A., and R. W. Henderson. 1991. *Amphibians and reptiles of the West Indies: descriptions, distributions, and natural history*. University of Florida Press, Gainesville.

Sick, H., and D. M. Teixeira. 1981. Nocturnal activities of Brazilian hummingbirds and flycatchers at artificial illumination. *Auk* 98:191–192.

Squires, W. A., and H. E. Hanson. 1918. The destruction of birds at the lighthouses on the coast of California. *Condor* 20:6–10.

Stevens, R. G., and M. S. Rea. 2001. Light in the built environment: potential role of circadian disruption in endocrine disruption and breast cancer. *Cancer Causes and Control* 12:279–287.

Summers, C. G. 1997. Phototactic behavior of *Bemisia argentifolii* (Homoptera: Aleyrodidae) crawlers. *Annals of the Entomological Society of America* 90:372–379.

Svensson, A. M., and J. Rydell. 1998. Mercury vapour lamps interfere with the bat defence of tympanate moths (*Operophtera* spp.; Geometridae). *Animal Behaviour* 55:223–226.

Taylor, M. H. 1984. Lunar synchronization of fish reproduction. *Transactions of the American Fisheries Society* 113:484–493.

Tsukamoto, K., T. Otake, N. Mochioka, T.-W. Lee, H. Fricke, T. Inagaki, J. Aoyama, S. Ishikawa, S. Kimura, M. J. Miller, H. Hasumoto, M. Oya, and Y. Suzuki. 2003. Seamounts, new moon and eel spawning: the search for the spawning site of the Japanese eel. *Environmental Biology of Fishes* 66:221–229.

Walker, B. W. 1949. *Periodicity of spawning by the grunion,* Leuresthes tenuis, *an atherine fish*. Ph.D. thesis, University of California, Los Angeles.

Wehr, T. A. 1997. Melatonin and seasonal rhythms. *Journal of Biological Rhythms* 12:518–527.

Wiese, F. K., W. A. Montevecchi, G. K. Davoren, F. Huettmann, A. W. Diamond, and J. Linke. 2001. Seabirds at risk around offshore oil platforms in the north-west Atlantic. *Marine Pollution Bulletin* 42:1285–1290.

Witherington, B. E. 1997. The problem of photopollution for sea turtles and other nocturnal animals. Pages 303–328 in J. R. Clemmons and R. Buchholz (eds.), *Behavioral approaches to conservation in the wild*. Cambridge University Press, Cambridge.

Yurk, H., and A. W. Trites. 2000. Experimental attempts to reduce predation by harbor seals on out-migrating juvenile salmonids. *Transactions of the American Fisheries Society* 129:1360–1366.

About the Contributors

PAUL BEIER is a professor of conservation biology and wildlife ecology in the School of Forestry at Northern Arizona University. His current work focuses on ecoregional conservation planning, including least cost corridor analysis, and collaborative, science-based "missing linkages" efforts in Arizona and southern California.

CARROLL G. BELSER has studied the biology of wading birds, red-cockaded woodpeckers, and eastern bluebirds. She is now a research associate in the Clemson University Radar Ornithology Laboratory.

WINSLOW R. BRIGGS is professor emeritus, Department of Biological Sciences, Stanford University, and director emeritus, Department of Plant Biology, Carnegie Institution of Washington. He has been a member of the National Academy of Sciences since 1974.

BRYANT W. BUCHANAN is an associate professor of biology at Utica College in upstate New York, where he conducts research on light pollution and amphibian sensory biology and evolution and teaches courses in general biology, evolution, genetics, and vertebrate biology.

MELANI S. CHEERS, a developmental biology researcher at Carnegie Mellon University, graduated magna cum laude from Mount Holyoke College with degrees in anthropology and biology. She gathered light pollution data on Boston-area lakes under a National Science Foundation undergraduate fellowship at Wellesley College in summer 1999.

PROFESSOR GERHARD EISENBEIS of the Johannes Gutenberg University, Mainz is an entomologist with research interests in morphology, ecophysiology, and ecology of soil arthropods. His most popular publications are the *Atlas on the Biology of Soil Arthropods* (Springer 1987) and the *Biological Atlas of Aquatic Insects* (Apollo Books 2002). His recent research is focused on the behavior of nocturnal insects around illumination sources.

ROBERT N. FISHER is a research scientist with the U.S. Geological Survey's Western Ecological Research Center in San Diego, California. His research includes a variety of fields (biodiversity, evolution, and natural history with a focus on reptiles and amphibians) that are relevant to understanding human impacts on the environment. Dr. Fisher is also interested in development, testing, and implementation of inventory and monitoring protocols and integrated data management.

KENNETH D. FRANK is a physician at the Philadelphia Veterans Affairs Medical Center and the University of Pennsylvania School of Medicine. The effect of artificial lighting on moths has interested him since childhood.

SIDNEY A. GAUTHREAUX JR. has studied behavioral and physiological aspects of bird migration since the late 1950s. He is a professor of biological sciences at Clemson University and director of the Clemson University Radar Ornithology Laboratory.

DICK A. JONKERS specializes in impact studies, such as those concerning traffic noise and airports, and advises in conflicts between society and nature.

SUSAN J. KOHLER earned a Ph.D. from the University of California at Berkeley. She worked at Wellesley College as an advanced instrumentation specialist, where she helped to develop instruments for measuring artificial light at night. She is now a research scientist at General Electric, where she is investigating applications of magnetic resonance spectroscopy.

JAMES E. LLOYD is professor emeritus at the Department of Entomology and Nematology at the University of Florida, Gainesville. He conducts research on the biosystematics (historical ecology) of Lampyridae, with special focus on ecological aspects and influences of sexual signals and the significance of sexual signal information for understanding species and the species problem, in this taxonomic group and as a theoretical consideration.

TRAVIS LONGCORE is science director of The Urban Wildlands Group and research assistant professor of geography at the University of Southern California Center for Sustainable Cities. He is regularly a lecturer at UCLA for the Department of Geography, the Department of Ecology and Evolutionary Biology, and the Institute of the Environment. His research interests are in the conservation of biodiversity and ecological processes in urban and urbanizing areas.

JOHANNES G. DE MOLENAAR completed his Ph.D. in 1972 on the vegetation of southeast Greenland. Since 1970 he has specialized in impact studies, ranging from nature management and agrienvironment schemes to military activities and physical planning.

WILLIAM A. MONTEVECCHI is a professor of psychology, biology, and ocean sciences at Memorial University of Newfoundland, and an associate of the Museum of Comparative Zoology of Harvard University. He conducts research on marine birds as indicators of changing environmental conditions and has authored about 200 scientific papers and the book *Newfoundland Birds: Exploitation, Study, and Conservation* (Nuttall Ornithological Club, Cambridge, Massachusetts).

MARIANNE V. MOORE, an associate professor in the Department of Biological Sciences at Wellesley College, is a limnologist interested in how artificial night lighting affects the behavior, spatial distribution, and species interactions of lake and estuarine organisms. She and her colleagues recently developed techniques for quantifying the spectral quality and intensity of artificial night lighting above and below water.

BARBARA NIGHTINGALE holds a master's degree in marine affairs from the University of Washington and has written extensively on the effects of overwater structures on salmon and other fishes in marine habitats of the Northwest. She is an environmental planner with the Aquatic Resources Division of the Washington State Department of Natural Resources.

GAD PERRY is a professor of conservation biology at Texas Tech University. His work focuses on reptiles and amphibians, primarily in the tropics.

CATHERINE RICH holds an A.B. with honors from the University of California at Berkeley, a J.D. from the UCLA School of Law, and an M.A. in geography from UCLA. She is executive officer of The Urban Wildlands Group, a conservation nonprofit that she co-founded with Travis Longcore. She is interested in understanding how animals and plants experience the assaults of the modern world and is dedicated to communicating this knowledge to a wide audience.

JENS RYDELL completed a Ph.D. in animal ecology at Lund University (Sweden) in 1990. His subsequent research has concentrated on the interactions between bats and insects and, more generally, on the evolution of predation and predator defense.

MICHAEL SALMON is a research professor in the Department of Biological Sciences, Florida Atlantic University. He maintains an active research program in animal behavior, with special reference to the behavior of marine turtles and crustaceans.

MARIA E. SANDERS completed her Ph.D. in 1999 on integrating knowledge of vegetation, nature management, and environment, applying geographic information systems and advanced statistical methods to support nature conservation and management, and she continues to conduct research in this field.

CHARLES A. SIMENSTAD is a research associate professor and coordinator of the Wetland Ecosystem Team at the School of Aquatic and Fishery Sciences, University of Washington. His research focuses on the functional role of estuarine and coastal habitats to support juvenile Pacific salmon and other fish and wildlife and the application of this knowledge to the restoration of tidal habitats at landscape scales.

SHARON E. WISE is an associate professor of biology at Utica College in New York. Her research examines the behavioral ecology of amphibians, especially salamanders, including factors that influence aggression and foraging activities. Her recent work includes the effect of light pollution on foraging behavior of *Plethodon cinereus* and development in *Xenopus laevis*.

Index

f refers to pages with figures
t refers to pages with tables